T0155887

Communications in Computer and Information Science 1467

More information about this series at http://www.springer.com/series/7899

Minrui Fei · Luonan Chen ·
Shiwei Ma · Xin Li (Eds.)

Intelligent Life System Modelling, Image Processing and Analysis

7th International Conference on Life System Modeling
and Simulation, LSMS 2021
and 7th International Conference on Intelligent Computing
for Sustainable Energy and Environment, ICSEE 2021
Hangzhou, China, October 30 – November 1, 2021
Proceedings, Part I

Editors
Minrui Fei
Shanghai University
Shanghai, China

Luonan Chen
University of Tokyo
Tokyo, Japan

Shiwei Ma
Shanghai University
Shanghai, China

Xin Li
Shanghai University
Shanghai, China

ISSN 1865-0929 ISSN 1865-0937 (electronic)
Communications in Computer and Information Science
ISBN 978-981-16-7206-4 ISBN 978-981-16-7207-1 (eBook)
https://doi.org/10.1007/978-981-16-7207-1

This Springer imprint is published by the registered company Springer Nature Singapore Pte Ltd.
The registered company address is: 152 Beach Road, #21-01/04 Gateway East, Singapore 189721, Singapore

Preface

This book series constitutes the proceedings of the 2021 International Conference on Life System Modeling and Simulation (LSMS 2021) and the 2021 International Conference on Intelligent Computing for Sustainable Energy and Environment (ICSEE 2021), which were held during October 30 – November 1, 2021, in Hangzhou, China. The LSMS and ICSEE international conference series aim to bring together international researchers and practitioners in the fields of advanced methods for life system modeling and simulation, advanced intelligent computing theory and methodologies, and engineering applications for achieving net zero across all sectors to meet the global climate change challenge. These events are built upon the success of previous LSMS conferences held in Shanghai, Wuxi, and Nanjing in 2004, 2007, 2010, 2014, and 2017, and ICSEE conferences held in Wuxi, Shanghai, Nanjing, and Chongqing in 2010, 2014, 2017, and 2018, respectively, and are based on large-scale UK-China collaboration projects on sustainable energy. Due to the COVID-19 pandemic situation, the themed workshops as part of these two conferences were organized online in 2020.

At LSMS 2021 and ICSEE 2021, technical exchanges within the research community took the form of keynote speeches and panel discussions, as well as oral and poster presentations. The LSMS 2021 and ICSEE 2021 conferences received over 430 submissions from authors in 11 countries and regions. All papers went through a rigorous peer review procedure and each paper received at least three review reports. Based on the review reports, the Program Committee finally selected 159 high-quality papers for presentation at LSMS 2021 and ICSEE 2021. These papers cover 18 topics and are included in three volumes of CCIS proceedings published by Springer. This volume of CCIS includes 29 papers covering four relevant topics.

The organizers of LSMS 2021 and ICSEE 2021 would like to acknowledge the enormous contribution of the Program Committee and the referees for their efforts in reviewing and soliciting the papers, and the Publication Committee for their editorial work. We would also like to thank the editorial team from Springer for their support and guidance. Particular thanks go to all the authors, without their high-quality submissions and presentations the conferences would not have been successful.

Finally, we would like to express our gratitude to our sponsors and organizers, listed on the following pages.

October 2021

Minrui Fei
Kang Li
Qinglong Han

Organization

Honorary Chairs

Wang, XiaoFan	Shanghai University, China
Umezu, Mitsuo	Waseda University, Japan

General Chairs

Fei, Minrui	Shanghai University, China
Li, Kang	University of Leeds, UK
Han, Qing-Long	Swinburne University of Technology, Australia

International Program Committee

Chairs

Ma, Shiwei	China Simulation Federation, China
Coombs, Tim	University of Cambridge, UK
Peng, Chen	Shanghai University, China
Chen, Luonan	University of Tokyo, Japan
Sun, Jian	China Jiliang University, China
McLoone, Sean	Queen's University Belfast, UK
Tian, Yuchu	Queensland University of Technology, Australia
He, Jinghan	Beijing Jiaotong University
Zhang, Baolin	Qingdao University of Science and Technology, China

Local Chairs

Aleksandar Rakić	University of Belgrade, Serbia
Athanasopoulos, Nikolaos	Queen's University Belfast, UK
Cheng, Long	Institute of Automation, Chinese Academy of Sciences, China
Dehghan, Shahab	Imperial College London, UK
Ding, Jingliang	Northeastern University, China
Ding, Ke	Jiangxi University of Finance and Economics, China
Duan, Lunbo	Southeast University, China
Fang, Qing	Yamagata University, Japan
Feng, Wei	Shenzhen Institute of Advanced Technology, Chinese Academy of Sciences, China
Fridman, Emilia	Tel Aviv University, Israel
Gao, Shangce	University of Toyama, Japan
Ge, Xiao-Hua	Swinburne University of Technology, Australia
Gupta M. M.	University of Saskatchewan, Canada

Gu, Xingsheng	East China University of Science and Technology, China
Han, Daojun	Henan University, China
Han, Shiyuan	University of Jinan, China
Hunger, Axel	University of Duisburg-Essen, Germany
Hong, Xia	University of Reading, UK
Hou, Weiyan	Zhengzhou University, China
Jia, Xinchun	Shanxi University, China
Jiang, Zhouting	China Jiliang University, China
Jiang, Wei	Southeast University, China
Lam, Hak-Keung	King's College London, UK
Li, Juan	Qingdao Agricultural University, China
Li, Ning	Shanghai Jiao Tong University, China
Li, Wei	Central South University, China
Li, Yong	Hunan University, China
Liu, Wanquan	Curtin University, Australia
Liu, Yanli	Tianjin University, China
Ma, Fumin	Nanjing University of Finance & Economics, China
Ma, Lei	Southwest University, China
Maione, Guido	Technical University of Bari, Italy
Na, Jing	Kunming University of Science and Technology, China
Naeem, Wasif	Queen's University Belfast, UK
Park, Jessie	Yeungnam University, South Korea
Qin, Yong	Beijing Jiaotong University, China
Su, Zhou	Shanghai University, China
Tang, Xiaopeng	Hong Kong University of Science and Technology, Hong Kong, China
Tang, Wenhu	South China University of Technology, China
Wang, Shuangxing	Beijing Jiaotong University, China
Xu, Peter	University of Auckland, New Zealand
Yan, Tianhong	China Jiliang University, China
Yang, Dongsheng	Northeast University, China
Yang, Fuwen	Griffith University, Australia
Yang, Taicheng	University of Sussex, UK
Yu, Wen	National Polytechnic Institute, Mexico
Zeng, Xiaojun	University of Manchester, UK
Zhang, Wenjun	University of Saskatchewan, Canada
Zhang, Jianhua	North China Electric Power University, China
Zhang, Kun	Nantong University, China
Zhang, Tengfei	Nanjing University of Posts and Telecommunications, China
Zhao, Wenxiao	Chinese Academy of Science, China
Zhu, Shuqian	Shandong University, China

Members

Huang, Deqing	Southwest Jiaotong University, China
Jahromi, Amir Abiri	University of Leeds, UK
Jiang, Lin	University of Liverpool, UK
Jiang, Ming	Anhui Polytechnic University, China
Kong, Jiangxu	China Jiliang University, China
Li, MingLi	China Jiliang University, China
Li, Chuanfeng	Luoyang Institute of Science and Technology, China
Li, Chuanjiang	Harbin Institute of Technology, China
Li, Donghai	Tsinghua University, China
Li, Tongtao	Henan University of Technology, China
Li, Xiang	University of Leeds, UK
Li, Xiaoou	CINVESTAV-IPN, Mexico
Li, Xin	Shanghai University, China
Li, Zukui	University of Alberta, Canada
Liu Jinfeng	University of Alberta, Canada
Liu, Kailong	University of Warwick, UK
Liu, Mandan	East China University of Science and Technology, China
Liu, Tingzhang	Shanghai University, China
Liu, Xueyi	China Jiliang University, China
Liu, Yang	Harbin Institute of Technology, China
Long, Teng	University of Cambridge, UK
Luo, Minxia	China Jiliang University, China
Ma, Hongjun	Northeastern University, China
Ma, Yue	Beijing Institute of Technology, China
Menhas, Muhammad Ilyas	Mirpur University of Science and Technology, Pakistan
Naeem, Wasif	Queen's University Belfast, UK
Nie, Shengdong	University of Shanghai for Science and Technology, China
Niu, Qun	Shanghai University, China
Pan, Hui	Shanghai University of Electric Power, China
Qian, Hong	Shanghai University of Electric Power, China
Ren, Xiaoqiang	Shanghai University, China
Rong, Qiguo	Peking University, China
Song, Shiji	Tsinghua University, China
Song, Yang	Shanghai University, China
Sun, Qin	Shanghai University, China
Sun, Xin	Shanghai University, China
Sun, Zhiqiang	East China University of Science and Technology, China
Teng, Fei	Imperial College London, UK
Teng, Huaqiang	Shanghai Instrument Research Institute, China
Tian, Zhongbei	University of Birmingham, UK
Tu, Xiaowei	Shanghai University, China
Wang, Binrui	China Jiliang University, China
Wang, Qin	China Jiliang University, China

Zhou, Bowen	Northeast University, China
Zhou, Huiyu	University of Leicester, UK
Zhou, Peng	Shanghai University, China
Zhou, Wenju	Ludong University, China
Zhou, Zhenghua	China Jiliang University, China
Zhu, Jianhong	Nantong University, China

Organization Committee

Chairs

Qian, Lijuan	China Jiliang University, China
Li, Ni	Beihang University, China
Li, Xin	Shanghai University, China
Sadegh, Azizi	University of Leeds, UK
Zhang, Xian-Ming	Swinburne University of Technology, Australia
Trautmann, Toralf	Centre for Applied Research and Technology, Germany

Members

Chen, Zhi	China Jiliang University, China
Du, Dajun	Shanghai University, China
Song, Yang	Shanghai University, China
Sun, Xin	Shanghai University, China
Sun, Qing	Shanghai University, China
Wang, Yulong	Shanghai University, China
Zheng, Min	Shanghai University, China
Zhou, Peng	Shanghai University, China
Zhang, Kun	Shanghai University, China

Special Session Chairs

Wang, Ling	Shanghai University, China
Meng, Fanlin	University of Essex, UK
Chen, Wanmi	Shanghai University, China
Li, Ruijiao	Fudan University, China
Yang, Zhile	SIAT, Chinese Academy of Sciences, China

Publication Chairs

Niu, Qun	Shanghai University, China
Zhou, Huiyu	University of Leicester, UK

Publicity Chair

Yang, Erfu	University of Strathclyde, UK

Registration Chairs

Song, Yang	Shanghai University, China
Liu, Kailong	University of Warwick, UK

Secretary-generals

Sun, Xin	Shanghai University, China
Gan, Shaojun	Beijing University of Technology, China

Sponsors

China Simulation Federation (CSF)
China Instrument and Control Society (CIS)
IEEE Systems, Man and Cybernetics Society Technical Committee on Systems Biology
IEEE CC Ireland Chapter

Cooperating Organizations

Shanghai University, China
University of Leeds, UK
China Jiliang University, China
Swinburne University of Technology, Australia
Life System Modeling and Simulation Technical Committee of the CSF, China
Embedded Instrument and System Technical Committee of the China Instrument and Control Society, China

Co-sponsors

Shanghai Association for System Simulation, China
Shanghai Instrument and Control Society, China
Zhejiang Association of Automation (ZJAA), China
Shanghai Association of Automation, China

Supporting Organizations

Queen's University Belfast, UK
Nanjing University of Posts and Telecommunications, China
University of Jinan, China
University of Essex, UK
Queensland University of Technology, Australia
Central South University, China
Tsinghua University, China
Peking University, China
University of Hull, UK

Beijing Jiaotong University, China
Nantong University, China
Shenzhen Institute of Advanced Technology, Chinese Academy of Sciences, China
Shanghai Key Laboratory of Power Station Automation Technology, China
Complex Networked System Intelligent Measurement and Control Base, Ministry of Education, China
UK China University Consortium on Engineering Education and Research
Anhui Key Laboratory of Electric Drive and Control, China

Contents – Part I

Computational Method in Taxonomy Study and Neural Dynamics

Intelligent Medical Apparatus, Clinical Applications and Intelligent Design of Biochips

Contents – Part II

Intelligent Methods in Developing Electric Vehicles, Energy Conversion Devices and Equipment

Intelligent Control Methods in Energy Infrastructure Development and Distributed Power Generation Systems

Contents – Part III

Advanced Neural Network Theory and Algorithms

Advanced Computational Methods and Applications

Fuzzy, Neural, and Fuzzy-Neuro Hybrids

Intelligent Modelling, Monitoring, and Control

Intelligent Manufacturing, Autonomous Systems, Intelligent Robotic Systems

Medical Imaging and Analysis Using Intelligent Computing

Research on Reconstruction of CT Images Based on AA-R2Unet in Inferior Mesenteric Artery Classification

Peixia Xu[1], Meirong Wang[2,3], Yu Han[1], Jinghao Chen[2,3], YuanFan Zhu[5], Kun Zhang[1,3,4](✉), and Bosheng He[2,3](✉)

[1] School of Electrical Engineering, Nantong University, Nantong, China
[2] Department of Radiology, The Second Affiliated Hospital of Nantong University, Nantong, China
[3] Nantong Key Laboratory of Intelligent Medicine Innovation and Transformation, Nantong, China
[4] Nantong Key Laboratory of Intelligent Control and Intelligent Computing, Nantong, China
[5] School of Management Science and Engineering, ChongqingTechnology and Business University, Chongqing, China

Abstract. In the process of colorectal cancer surgery, dispute on whether to retain the left colonic artery continues to this day. The classification of the inferior mesenteric artery has instructional significance for clinic which has been proved. Therefore, realizing relative position among vessels is an essential step for the diagnosis and therapy of colorectal cancer disease. The method of 3D reconstruction based on CT sequence images in hospital is time-consuming and laborious. In this paper, we propose a kind of AA-R2Unet based on Unet to realize automatic segmentation of the inferior mesenteric artery. We use attention augment rather than the convolution to achieve the global feature, stacking predictions with 2D features to achieve the function of 3D visualization. The improved Focal Loss function we proposed focuses more on our target vessel. AA-R2Uet achieved a per case precision of 93.34%, recall of 90.28%, and 90.06% of F1 score.

Keywords: CT sequence images · AA-R2Unet · Focal loss · 3D visualization · IMA classi-fication

1 Introduction

In recent years, with the increasing of colorectal cancer, lymphadenectomy at the root of the inferior mesenteric artery (IMA) has become routine [1]. In the operation of middle and lower rectal cancer, it is a common practice that we choose to preserve the left colonic artery (LCA). However, whether LCA should be retained is controversial currently for the consideration of the actual distribution of patients [2]. In 1997, Japanese scholar Yada et al. First classified IMA into four types according to the starting point of root among LCA, sigmoid artery (SA), and superior rectal artery (SRA), these four classifications are shown in Fig. 1. Researches have shown that the variable anatomy of

M. Fei et al. (Eds.): LSMS 2021/ICSEE 2021, CCIS 1467, pp. 3–12, 2021.
https://doi.org/10.1007/978-981-16-7207-1_1

the inferior mesenteric artery and its branches is significant in radiological interpretation and colorectal surgery from both an oncological perspective when considering resection as well as in colonic ischemia. Nowadays even with preoperative images, it can still be a challenge to identify these structures due to intraoperative individual conditions. To determine the specific location of the left colonic artery before surgery, three-dimensional reconstruction of the inferior mesenteric artery has become the current hot spot. Locating the root of LCA is the key to preserve LCA. If not handled properly, the intestinal tract connected with it will have the problem of insufficient blood supply, leading to intestinal necrosis [3].

Previously developed 3D-reconstruction technology of medical images based on CT images rely mostly on multiplanar reconstruction (MPR), maximum intensity projection (MIP), volume walkthrough (VRT), virtual endoscopy (VE), etc. Although these methods can also reconstruct IMA and achieve the function of classification, unfortunately, it wastes a lot of time and effort as well as the requirement of reconstruction ability which increases the burden of doctors in some degree. Figure 4 shows IMA processed by MPR. In some scientific literature, some methods have been used for the three-dimensional reconstruction of blood vessels. For instance, a centerline-based method for automatic segmentation of cerebral blood vessels [4]. Although this method retains the element of vessel diameter, there are tolerances in the determination of the centerline which leads to the deviation between the model established by this method and the actual model (Fig. 2).

With the development of convolutional neural networks, Unet which was proposed in 2015 performs well in medical image processing and is widely used in medical image segmentation. To reconstruct the 3D model which approaches the real model, we proposed AA-R2Unet, the overall flowchart of our work can be seen in Fig. 3. The main contribution of our proposed approach includes:

(1) We propose a deep learning segmentation neural network named AA-R2Unet based on Unet. In the last two layers, we use self-attention mechanism to replace the original convolution path to achieve the global information. This mechanism converts the original single mode into several independent parallel modes, making it available for model to learning features in subspaces.
(2) The improved Focal Loss function we proposed solves the problem of unbalanced pixel between target and background which is more targeted in IMA segmentation. In the condition of ensuring accuracy, we segment the original CT cross-sectional image sequence, segment our target IMA, conserve the original information of CT image as possible.
(3) we use 3D visualization which transforms two-dimensional segmentation image into three-dimension to assist the judgment of IMA classification and realize the function of automatic segmentation.

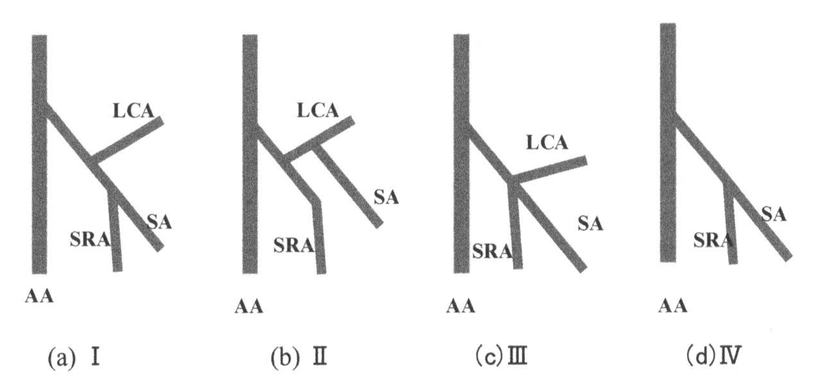

Fig. 1. Four classifications for IMA.

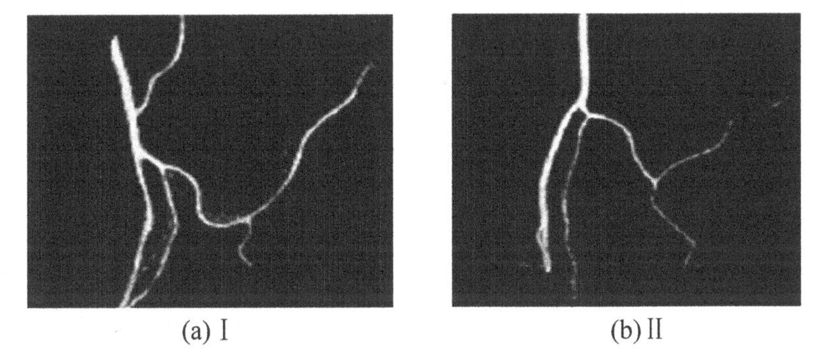

Fig. 2. IMA processed by MPR.

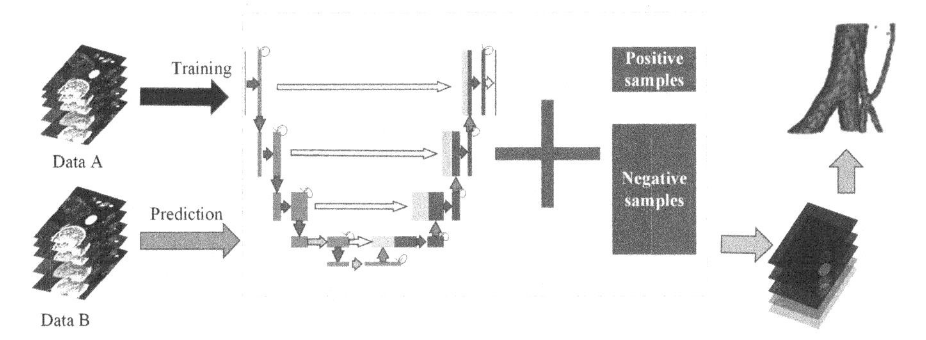

Fig. 3. Overall system flow chart.

2 Model for IMA Segmentation

2.1 Model Set-Up

With the deepening of the neural network, it is found that the networks with deep layers usually performs badly. Contrary to expectations, the accuracy of classification indicates a trend of decreasing instead of keeping flat what we called degradation problems [5].

Recurrent Residual (R2U) Unet [6] is proposed to solve the problem of vanishing gradient in Unet and performs well in medical image segmentation. However, the convolution operation of neural networks has another defect of missing the global information. To overcome this problem, we propose AA-R2Unet based on R2Unet, the framework of our model is shown as follows in Fig. 4. To obtain the global information, we use attention augment to replace the traditional convolution in the last two layers of the contraction path and transfer it to the expanding path to improve the ability of feature learning. Based on this, our model can be improved more efficiently and deeply. Considering that the last layer contains the most feature information and a large number matrix needed to be created for storing the attention weight, which consumes memory space, we choose not to replace all the convolution layers.

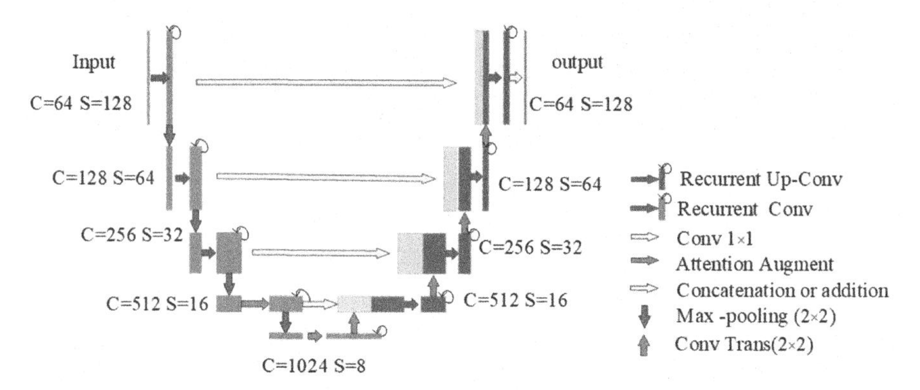

Fig. 4. The framework of AA-R2Unet.

2.2 Attention Augment

Supposing that there is an input sequence $X = [X_1, X_2, ..., X_N]$, a query vector query (Q) is defined to query some information from X, we need to extract some information from N number of elements in X more or less, so it involves a retrieval, that is key (K) and value (V). The framework of attention argument [7] can be seen in Fig. 5. A feature map is defined as (W, H, C_f) where W represents the width, H means height, C_f signifies the depth of the feature, new generated matrix QVK of shape (W, H, $2 \times d_k + d_v$) will then appear by 1×1 convolution. Dividing matrix Q, K, V which have been separated from the channel of depth into N matrix for the subsequent training, d_k, d_k, d_v represent the channel depth of Q, K, V respectively. This multi-headed attention converts the original single mode into several independent parallel modes, making it available for model to learning features in subspaces. To calculate the weight matrix, we convert the matrix Q, K, V into tensor with one dimension namely Flat Q $(W \times H, d_k)$, Flat K $(W \times H, d_k)$, Flat V $(W \times H, d_v)$. Multiplication between Flat Q and Flat K is retained in attention argumentation, in addition it involves a relative position which can calculate the relative position of each point in the feature map through matrix Q. S_w, S_h refer to the relative position of matrix W (WH, WH, 1) and matrix H (WH, WH, 1) which originate from model training. Adding these three matrix we have given and then dividing the factor $\sqrt{d_k}$, we will achieve the matrix of feature weight after the processing of softmax. The attention feature matrix we got in the end experience multiply with matrix V and 1×1 convolution after reshaping in original size.

The formula for calculating attention feature matrix is given in Eq (1).

$$AF = Conv\,(\,Soft\max\,(\frac{QK^T + S_H + S_W}{\sqrt{d_k}})\,V\,). \tag{1}$$

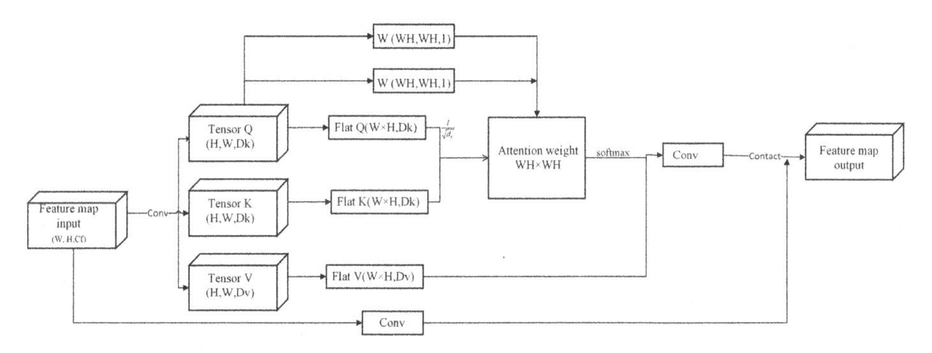

Fig. 5. The framework of attention augment.

2.3 Parameter Adjustment in Focal Loss

Cross entropy [8] used to be utilized as a loss function in Unet. To a binary classification problem, cross-entropy is given by Eq (2).

$$L = -y\log y' - (1 - y') = \begin{cases} -\log y' \\ -\log(1 - y') \end{cases}. \tag{2}$$

Focal Loss is designed for the imbalance problem in classification. To a binary classification problem, Focal Loss is given in Eq (3).

$$L_{focal-loss} = \begin{cases} -\alpha(1 - y')^{\gamma}\log y' \\ -(1 - \alpha)y'^{\gamma}\log(1 - y') \end{cases}. \tag{3}$$

In the base of cross-entropy, Focal Loss adds one factor. Where gamma >0 reduces the loss of samples that are easy to classify, paying more attention to samples incorrectly classified. Another newly added factor-alpha is used to balance the proportion between positive samples and negative samples.

Fixed parameter makes it hard for loss function to perform well in Convergence. In this paper, the focal loss is complete by a self-adaptive parameter. After massive experiences, we come to the conclusion that when gamma =2, Focal Loss performs well. Training data are trained in batches, we calculate the massive of positive and negative samples as alpha in each batch. The formula for calculating is given in Eq (4).

$$L_{focal-loss} = \begin{cases} -\alpha(1-y')^2 \log y' \\ -(1-\alpha)y'^2 \log(1-y') \end{cases}, \alpha = \begin{cases} \alpha_P = \frac{|P|+|N|}{|P|} \; y = 1 \\ \alpha_N = \frac{|P|+|N|}{|N|} \; y = 0 \end{cases}. \quad (4)$$

Num 0 and 1 is the result of binary, 1 is correspond to the pixel 255, P represents the sum of 1 in mask while N stands for that of 0.

3 Results and Performance Comparison

3.1 Dataset

The data set we obtained in this experiment comes from Nantong first people's hospital. In this paper, 24 patients' sequence were used as training dataset for training, the other 8 groups were used for prediction.

Image Preprocessing: The original image we got from the hospital was in DICOM format. Using ITK-SNAP to mask the label for our CT sequences in DICOM. Separating the RGB channel of the label and the original images, taking the channel B, G of the original images and channel R of label images, the new fused images of three channels are our training data, the concrete flow chart is shown in the Fig. 6.

Data Augmentation: For deep learning needs enormous amount of data. For training while the cases we got from hospital are less, one way to solve this problem is data augment including rotation, clipping, mirror and other operations.

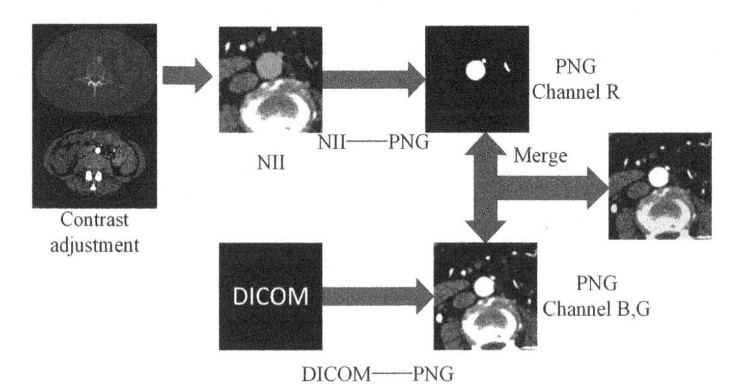

Fig. 6. Flow chart of preprocessing.

Contrast adjustment: When using ITK-SNAP, we should focus on changing the window width and window level to adjust the contrast of images for the clearly displayed of target vessels. In this experiment, we can clearly observe the target blood vessels by adjusting the window width and window level to 150 and 600 respectively.

3.2 Experiments

The platform used in this experiment comes from the deep learning computing platform with two RTX-3090ti 24GB graphics cards. The operating system and version are ubuntu20.04, the machine learning environment configuration is torch 1.7.0cuda11.1 and the program compiling environment is python3.6.12.

Based on the original Unet model, we trained for about 5 h to observe the segmentation effect, unfortunately the result turns out to be bad. We have taken two measures to solve this problem. One is to introduce Focal Loss, the other is to expand the target area relatively, cutting images of 512 × 512 into 128 × 128. Figure 7 shows the different between prediction of 512 × 512 and 128 × 128. This is a typical imbalance between positive and negative samples. After cutting into 128 × 128 images, this proportion is 3% to 4% and this is 15 times higher than 512 × 512 which can be seen in Table 2.

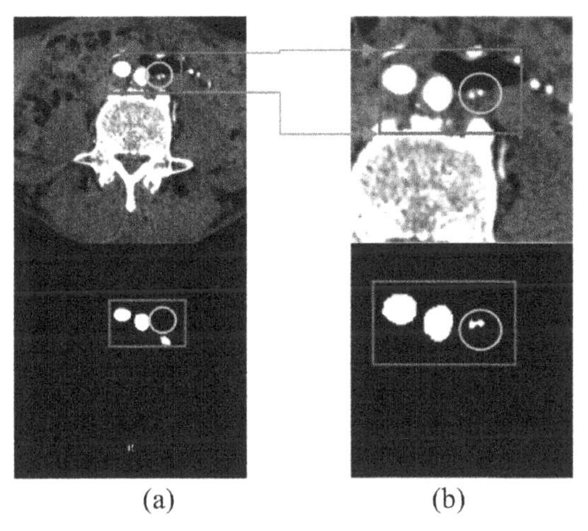

Fig. 7. Predition results of different sizes. (a) 512 × 512. (b) 128 × 128. Boxes with red represent the same region, line in blue is used to align regions.

In order to verify the improvement of this proposed model, we compared models between AA-R2Unet with Focal Loss, AA-R2Unet and Unet. Results are shown in Table 1.

Penalty for not segmenting the abdominal aorta is much larger than that of IMA. For example, if one layer segments abdominal aorta without segmenting IMA, the score hardly change to accuracy. In the following part, we will analyze the performance of different network through the generated 3D model. Figure 8 shows the model generated

Table 1. Comparisons between three models.

Patient num	Network	Accuracy	Precision	Recall	F1score
3930261	Unet	99.4%	91.43%	91.35%	90.83%
	AA-R2Unet	99.40%	87.50%	95.83%	91.28%
	AA-R2Unet + Focal Loss	99.57%	92.23%	94.72%	93.39%
3901808	Unet	99.19%	96.58%	75.25%	83.93%
	AA-R2Unet	99.39%	93.43%	85.54%	89.23%
	AA-R2Unet + Focal Loss	99.26%	95.98%	78.34%	85.91%

by patients 3930261 and 3901808 of the above three models. We can find that the original Unet perform the worst. Although Unet performs well in the traditional segmentation, it does not perform very well in the micro segmentation of CT sequence.

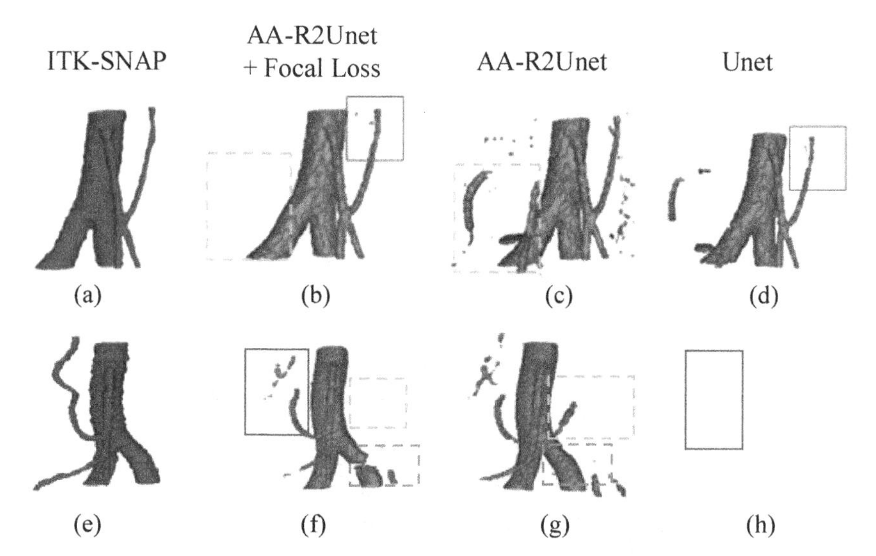

Fig. 8. Model of two patients. Pictures in the first row are models of patient 3930261 and the other is 3902808.

Compared traditional Unet with ours, we find that the traditional Unet fails to segment LCA which is a significant factor for classification, it maybe implicit in patient 3930261. However, if you observe red boxes you will find that the branch of LCA has the trend to break which is quite obvious in patient 3901808 where LCA almost completely disappeared.

Compared picture (c, g) with (b, f), as what has shown in orange boxes, it is intuitive that more false parts and noise exist in model AA-R2Unet without Focal Loss. Compared (f) with (g), model AA-R2Unet without Focal Loss performs better on the segmentation of abdominal aorta than that we proposed while it remained unchanged on

IMA segmentation. But it is much better than original Unet since the target information is available. The model we proposed can improve the ability of segmentation and reduce the possibility of superfluous clutter at the same time where I though is valuable.

Based on the problem of data samples, only three types of IMA cases are included in our eight predictive datasets. Comparing the model marked by doctors with the model generated by our prediction, as is shown in the Fig. 9, we can see that our proposed method is indeed helpful for auxiliary classification.

4 Conclusion and Future Work

In this paper, AA-R2Unet is proposed to segment IMA. We add attention argument mode to achieve the global feature as well as the improved Focal Loss which focus more on the region of vessel IMA.

There is still some insufficiency in our work, problems such as fracture of branches and incomplete separation still exist which can be seen in Fig. 10 combined (e, f) in Fig. 9, it seems that our model does not perform well on curved vessels. One main reason may be the loss of context information, we'll consider to introduce LSTM [10] in our following work or try to track these undetected vessels.

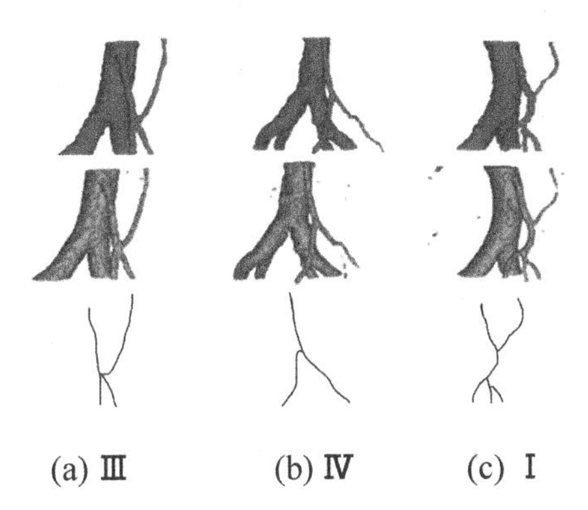

Fig. 9. Three classifications in our prediction dataset.

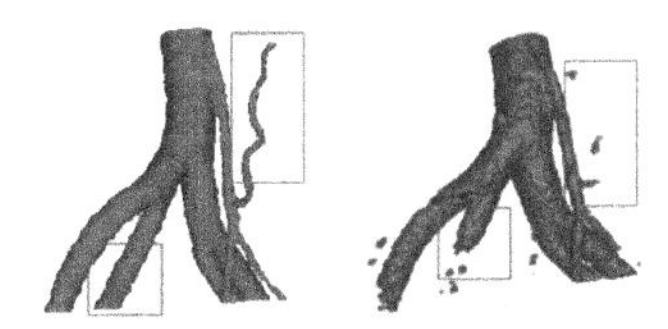

Fig. 10. Bad example.

Acknowledgments. This work was financially supported by Nantong Health Commission in 2020, QA2020002, and Key clinical project of Nantong University, 2019LZ002.

References

1. Inoue, H., Sasaki, K., Nozawa, H., et al.: Therapeutic significance of D3 issection for low rectal cancer: a comparison of dissections between the lateral pelvic lymph nodes and the lymph nodes along the root of the inferior mesenteric artery in a multicenter retrospective cohort study. Int. J. Colorectal Dis. 1–8 (2020)
2. Scott-Conner, C.E.H.: Low anterior resection for rectal cancer. In: Chassin's Operative Strategy in Colon and Rectal Surgery, pp. 86–88. Springer, New York (2002)

3. Abe, T., Ujiie, A., Taguchi, Y., et al.: Anomalous inferior mesenteric artery supplying the ascending, transverse, descending, and sigmoid colons. J. Anat. Sci. Int. **93**(1), 144–148 (2018)
4. Novikov, A.A., Major, D., Wimmer, M., et al.: Automated anatomy-based tracking of systemic arteries in arbitrary field-of-view CTA scans. J. IEEE Trans. Med. Imaging **36**(6), 1359–1371 (2017)
5. He, K., Zhang, X., Ren, S., et al.: Deep residual learning for image recognition. CVPR 770–778 (2016)
6. Alom, M.Z., Yakopcic, C., Taha, T.M., et al.: Nuclei segmentation with recurrent residual convolutional neural networks based U-Net (R2U-Net). In: NAECON 2018-IEEE National Aerospace and Electronics Conference, pp. 228–233. IEEE (2018)
7. Fan, D.P., Zhou, T., Ji, G.P., et al.: Inf-net: automatic covid-19 lung infection segmentation from CT images. J. IEEE Trans. Med. Imaging **39**(8), 2626–2637 (2020)
8. Luna, M., Park, S.H.: 3D patchwise U-Net with transition layers for MR brain segmentation. MICCAI 394–403 (2018)
9. Zhang, Y., Li, H., Du, J., et al.: 3D multi-attention guided multi-task learning network for automatic gastric tumor segmentation and lymph node classification. J. IEEE Trans. Med. Imaging **40**(6), 1618–1631 (2021)
10. Dantas, H., Warren, D.J., Wendelken, S.M., et al.: Deep learning movement intent decoders trained with dataset aggregation for prosthetic limb control. J. IEEE Trans. Biomed. Eng. **66**(11), 3192–3203 (2019)

Research on CT Image Grading of Superior Mesenteric Artery Based on AA Res-Unet

Yu Han[1], Jinghao Chen[2,3], Peixia Xu[1], Meirong Wang[2,3], YuanFan Zhu[5], Kun Zhang[1,3,4](✉), and Bosheng He[2,3](✉)

[1] School of Electrical Engineering, Nantong University, Nantong, China
[2] Department of Radiology, The Second Affiliated Hospital of Nantong University, Nantong, China
[3] Nantong Key Laboratory of Intelligent Medicine Innovation and Transformation, Nantong, China
[4] Nantong Key Laboratory of Intelligent Control and Intelligent Computing, Nantong, China
[5] School of Management Science and Engineering, ChongqingTechnology and Business University, Chongqing, China

Abstract. In recent years, the incidence of Crohn's disease (CD) is on the rise, which calls for more accurate and less invasive diagnostic tools. Crohn's disease usually need to be confirmed by colonoscopy, and this paper puts forward a network model AA Res-Unet based on the fundamental Unet, can realize automatic of superior mesenteric artery CT image segmentation, and further to realize the automatic grading function, help doctors by observing the superior mesenteric artery features, early diagnosis of disease. Image segmentation achieved a per case precision of 98.76%, recall of 98.00%, and 98.34% of F1 score.

Keywords: AA Res-Unet · Vessel segmentation · Crohn · Vessel classification

1 Introduction

Crohn's disease (CD) is an inflammatory bowel disease with unknown pathogenesis that chronically and repeatedly involves any part of the digestive tract [1]. In recent years, with the progress of medical technology, the popularization of multi-slice computed tomography enterography (MSCTE) and colonoscopy, and the progress of clinical treatment, the mortality rate of CD has been reduced. Imaging manifestations of CD are often non-specific, including intestinal wall thickening, intestinal dilatation, mesenteric edema, mesenteric effusion, and mesenteric vascular hyperplasia [2, 4]. Because the imaging manifestations of CD greatly overlap with other abdominal lesions, it is a long-term problem to timely and accurately diagnose and evaluate the lesions to provide clinical diagnostic basis and improve the treatment and prognosis of patients with CD [3].

Compared with traditional CT, dual-source dual-energy CT (DSDECT) can obtain material attenuation data at high and low energy X-ray lines through two sets of independent tube-detector systems [5, 6], so as to make qualitative and quantitative diagnosis of lesions [7].

Y. Han and J. Chen—Contributed equally to this work and should be considered co-first authors.

M. Fei et al. (Eds.): LSMS 2021/ICSEE 2021, CCIS 1467, pp. 13–23, 2021.
https://doi.org/10.1007/978-981-16-7207-1_2

In addition, DECT uses Mono+ technology to generate virtual monospectrum images, so as to display more clearly small intestinal straight vessels in the energy level range of 40 keV ~ 190 keV [8]. Therefore, DSDECT can not only obtain the absolute iodine value (iodine map) of intestinal wall enhancement through the separation of substances purified by energy spectrum, but also quantitatively evaluate the degree of inflammation in Crohn's intestinal wall [9]. It can also improve the image contrast of micro-small intestinal supplying blood vessels through photonic single energy spectrum technology, and clearly show the boundary between the inflammatory bowel and normal intestinal tissue [9]. At the same time, DECT can clearly show proliferative small vessels below grade 4 of superior mesenteric artery on low KeV images and iodine images, which can improve the diagnostic efficiency [10]. With the development of convolutional neural networks, Unet [11] which was proposed in 2015 performs well in medical image processing and is widely used in medical image segmentation.

By improving the image contrast of the blood supplying vessels in the fine small intestine, the boundary between the inflammatory intestinal canal and normal intestinal tissue caused by Crohn's disease was clearly shown, and the proliferation of small vessels below the fourth level of superior mesenteric artery was clearly shown, while the proliferation of lateral mesenteric artery and small recta vessels (branches of superior mesenteric artery) in Crohn's disease was more common [12, 13]. Therefore, the subjective and objective assessment of angiogenesis is particularly important for Crohn's disease to improve the diagnostic efficacy of intestinal wall ischemia [14]. Semi-automated quantitative superior mesenteric artery grading numbers will enable quantification of subjective indicators such as activity [15].

The whole process of the experiment is shown in Fig. 1. The preprocessed image is used as the input image of the new network AA Res-Unet (the specific model is introduced in the next section). After fusion and enlargement, the new model data is trained to achieve automatic segmentation of superior mesenteric artery images. Then, image thinning algorithm and SVM algorithm were used to improve the effect of vascular classification and realize the purpose of pre-diagnosis [16].

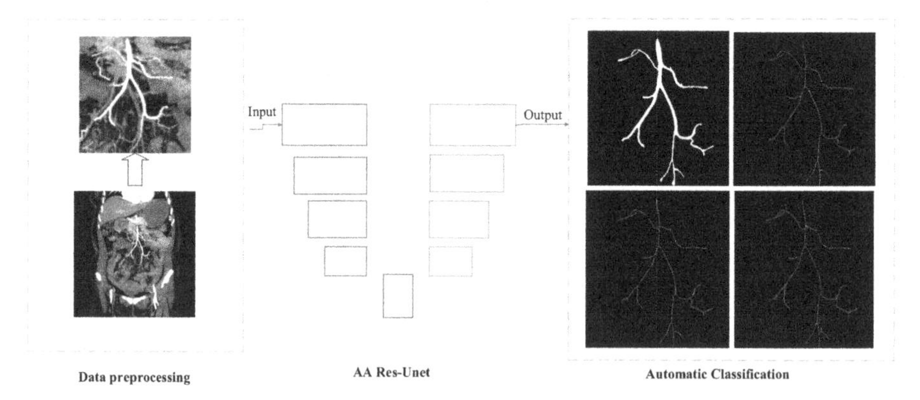

Fig. 1. Overall system flow chart.

2 Methods

2.1 Attention Augment

Attention mechanism first appeared in Natural Language Processing (NLP), which is mainly used to give different weights (i.e., different degrees of attention) to words in combination with contextual information [17]. In 2019, Google Brain applied the self-attention mechanism in natural language processing to the convolutional neural network [18].

The specific structure of Attention Augment is shown in Fig. 2. The essence of Attention Augment is to obtain a series of key-value pair mappings through query. The calculation formula of attention eigenmatrix is as follows named Equation 1:

$$O = Conv\,(\,Softmax\,(\frac{QK^T + S_H + S_W}{\sqrt{d_k}})\,V\,). \tag{1}$$

Attention Augment refers to the self-attention mechanism of natural language processing. Attention feature images are obtained by calculating the Attention weight of input features, and the calculated Attention feature images are joined together with the original convolution results to enhance the learning ability of convolution operation on global information. To improve the recognition accuracy of the model.

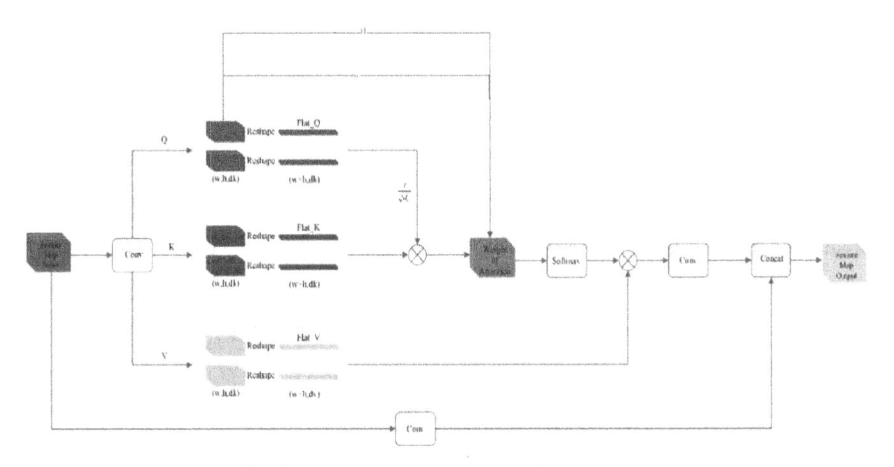

Fig. 2. The framework of attention augment.

2.2 AA Res-Unet

However, the experiment found that the deep network has the problem of model degradation, that is, it is not the deeper the model is, the better the prediction results are, because the deeper the network is, the more difficult it is to train.

Residual structure is specific as shown in Fig. 3. Identity mapping represents identity mapping, that is, the features of shallow network are copied to form new features with residuals [19]. The identity mapping is also called skip connection or shortcut connection.

The Res-Unet network used in this paper introduces the residual connection on the basis of the Unet structure. It is also well suited to medical problems [20].The specific structure of AA Res-Unet model proposed in this paper is shown in Fig. 3.

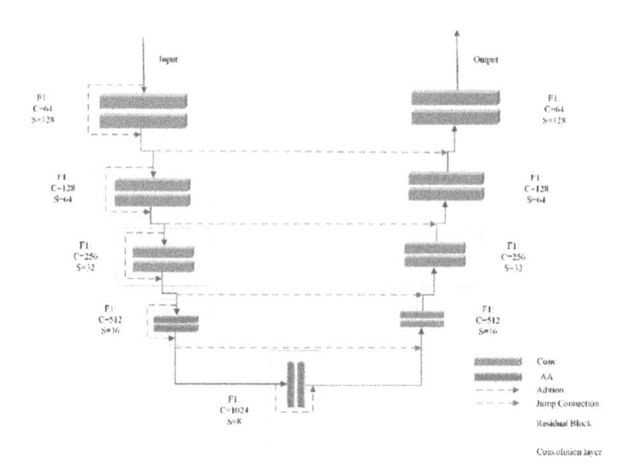

Fig. 3. AA Res-Unet architecture.

2.3 Loss Function

The Focal loss is proposed in the field of target detection, in order to solve the problem of serious imbalance between positive and negative samples. α is introduced to solve the problem of uneven proportion of positive and negative samples. And adding and will make the loss of the sample with large prediction probability smaller, while the sample with smaller prediction probability will become larger, thus enhancing the attention of the positive sample. We use and which are verified the best results during the experiment. Focal loss is expressed in Equation 2.

$$\mathrm{L}_3 = \begin{array}{c} -\frac{1}{N}\sum_{i=1}^{N}(\alpha y_i(1-p_i)^{\gamma}\log p_i \\ +(1-\alpha)(1-y_i)p_i^{\gamma}\log(1-p_i)) \end{array}. \tag{2}$$

Focal loss is more concerned with difficult to classify samples, and less concerned with easy to classify samples. But it also has the following shortcomings: Focal loss is easy to overfit, the value of loss is very large, and the adjustment is not convenient.

2.4 Image Classification

Skeleton is an important shape descriptor for object representation and recognition since it can preserve both topological and geometrical information. Zhang's parallel fast thinning algorithm is a classic image thinning algorithm, which can effectively carry out preliminary skeleton operation on the image.

Breakpoint connection operation is needed after skeletal treatment. The specific algorithm is as follows: first, traverse the whole graph to find possible endpoints and nodes. The specific algorithm for finding these points is described in detail later in this paper. To judge the distance between the end-point and node in the image, if the distance is smaller than the threshold value T, it is a breakpoint and needs to draw points to connect. T is the width of the blood vessel at the node. Generally, the width of one or two pixels is set. The strict calculation formula is shown in Equation 3:

$$T = W_{i1} + \max W_{jp},\ p \in \{1, 2, 3\}. \tag{3}$$

W_{i1} is the width of the divided blood vessel, and W_{jp} is the width of the three main blood vessels at the bifurcation point. After the breakpoint connection treatment, the abdominal arteriogram was repaired, which could lay the foundation for the follow-ing procedures and improve the final grading accuracy.

The specific method can refer to some methods to judge the connectivity of the center point in the image thinning algorithm. For this reason, we use SVM to help judge whether the detected point is an intersection point, in which the data of each point is the pixel value of the 9 points of the nine grids where the target point is located, and the corresponding result is 1 or 0, so as to realize the dichotomy of all nodes.

There are roughly 18 obvious and simple situations, as shown in the Fig. 4. where the corresponding result of these nodes is 1.

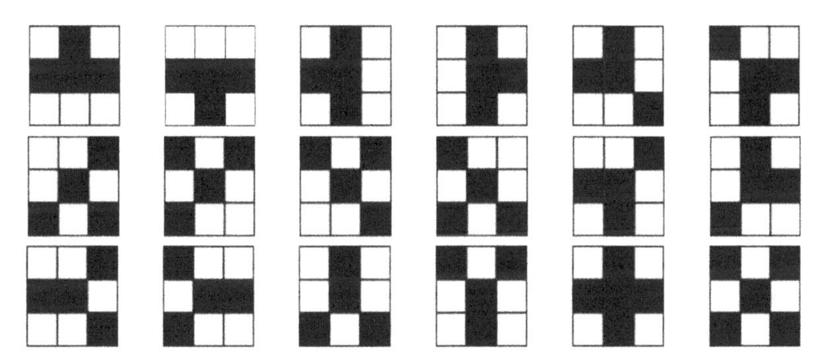

Fig. 4. The feature diagram in simple case.

It is worth noting that in subsequent experiments, it was found that these 18 cases were only the simplest case and not complete, but all cases could be simply added from these 18 simple cases. Therefore, we added some node data whose result should be 1 according to the SVM dichotomy, as shown in Fig. 5.

According to the information provided by the hospital experts, the location of the main vessel was confirmed, the value of the vessel trunk was marked as 0, The counting starts from 0 to 1, but the bifurcation points encountered in blood vessels of the same root do not add 1, and the number is marked on the angiography. In this way, the final output im-age can be easily identified as several levels of blood vessels. For example, above 0 is the first level blood vessel, between 0 and 1 is the second level blood vessel, between 1 and 2 is the third level blood vessel, and so on.

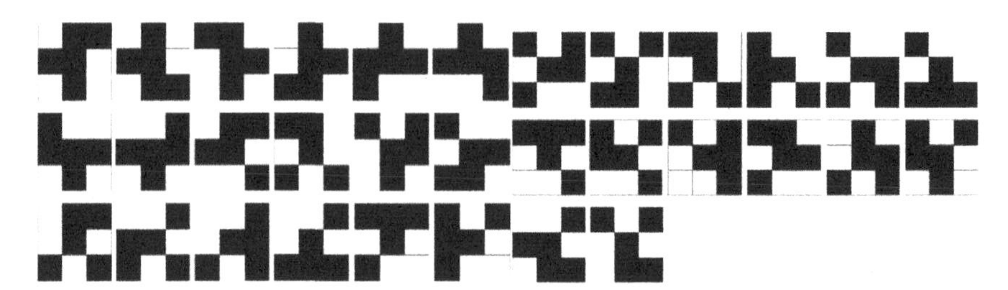

Fig. 5. The improved feature diagram.

3 Experiments And Results

3.1 Dataset

Data sources. Dual energy CT images of small intestine provided by the First People's Hospital of Nantong City were used in this paper. The data set consists of a CT original image and the corresponding binary standard image manually segmented by the expert. For the hospital, the third-generation dual-source CT (Somatom Force, Siemens Healthcare, Forchheim, Germany) was used to conduct dual-energy CTE images. Maximum intensityprojection (MIP) was performed on the superior mesenteric artery image, and the MIP image was reconstructed by Dual Energy Mono+ from 40 keV to 90 keV, increasing with each 10 keV. In a total of 6 groups of virtual monoenergetic imaging (VMI), we received a total of 179 of the above data, corresponding to 179 different patients.

Data selection. Medical science has provided us with images of six different energy levels from 40 kV to 90 kV shown in Fig. 6, We are paying attention to the prediction result of energy level. Since the new single energy software of medical image reconstruction by MIP image has a great influence on the image quality at different levels, the effect of vascular segmentation is directly affected. In Fig. 8, the final prediction results of medical image output at different levels of 40 kV and 90 kV are respectively predicted. 60 kV is the most appropriate energy level, the corresponding segmentation results are clear and complete, and the noise is almost invisible.

The initial images obtained by us are of different sizes.We compared the prediction clipped to 512 × 512 with the prediction clipped to 128 × 128. As shown in Fig. 7 below, the figure represents the comparison effect of prediction graphs with different input training picture sizes, where (a) is the original image, (b) is the prediction graph with input training picture size of 512 × 512, and (c) is the prediction graph with input training picture size of 128 × 128. It is not difficult to see that the effect of (c) is significantly better than that of (b). This may be because when the network depth of the model is fixed at 5, the input image of 512 × 512 fails to obtain enough feature information in the subsampling process, while the input image of 128 × 128 is decomposed to a small enough size to obtain more sufficient feature information. It should be noted that the comparison images are kept the same size in this article for better comparison. In fact, the original image and the input image are not the same size.

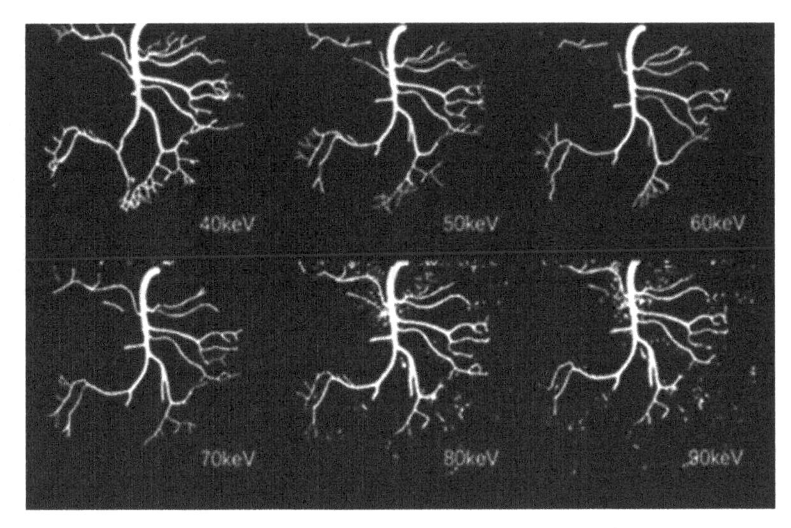

Fig. 6. Image comparison of superior mesenteric artery at different keV levels.

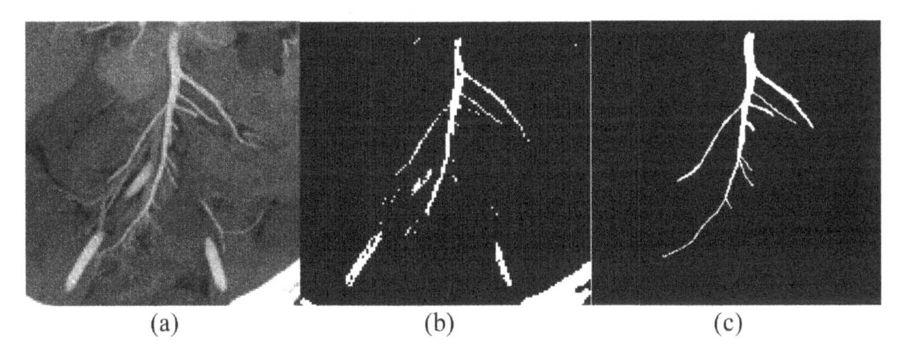

Fig. 7. The experimental comparison about different input.

The network depth of the model and the loss function used are also within the experimental range. Same as the figure above, in Fig. 7 (a) is the original image, (b) is the predicted result when the network depth is 5, and (c) is the predicted result when the network depth is 7. It can be seen that with the help of attention mechanism, the network depth has little influence on the image prediction result.

Data preprocessing. Acording to the three steps of the CT image preprocessing algorithm: histogram equalization, gamma transformation and normalization processing, the image is simply processed without affecting the image quality to improve the clarity and contrast of the image. After augmentation, we get a total of 20000 images for training, epoch number is set up as 50.

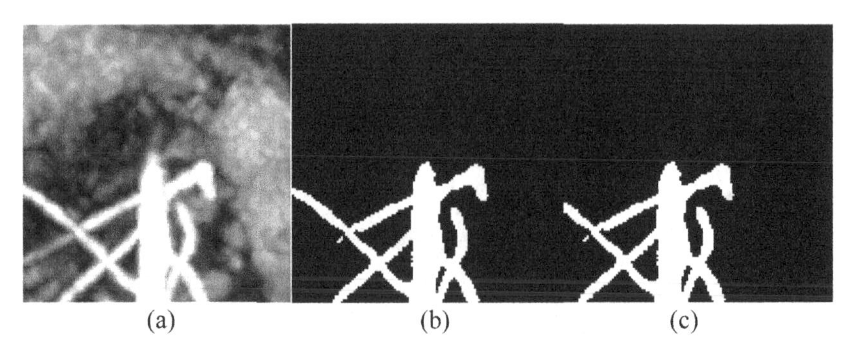

(a) (b) (c)

Fig. 8. Comparison of prediction graphs with different model layers.

3.2 Experimental Result

Vascular Segmentation Experiment. The experimental instrument in this paper is a remote server equipped with two 24GB video memory 3090 graphics cards each. The network model adopted is AA Res-Unet model, and the structure of the model is shown in Fig. 9. Attention mechanism and residual learning are added to the basic Unet model. The comparison between the improved model and the basic model is shown in Table 1 and Table 2.

The loss function has been improved from Softmax to Focal_Loss. The comparison effect is shown in the Fig. 9, of it, (a) is the original graph, (b) is the predicted result graph of the original loss function Softmax, and (c) is the predicted result graph of the loss function Focal_loss.

Table 1. Comparison of network model evaluation parameters.

Network	Accuracy	Precision	Recall	F1score
Unet	93.47%	94.05%	93.09%	90.23%
+AA Res-Unet	96.67%	98.68%	97.40%	98.07%
+Focal_loss	97.16%	98.76%	98.00%	98.34%

Table 2. Comparison of network model other evaluation parameters.

Network	IoU	mIoU	FWIoU
Unet	92.20%	82.77%	91.15%
+AA Res-Unet	96.22%	87.19%	93.88%
+Focal_loss	96.73%	89.46%	94.66%

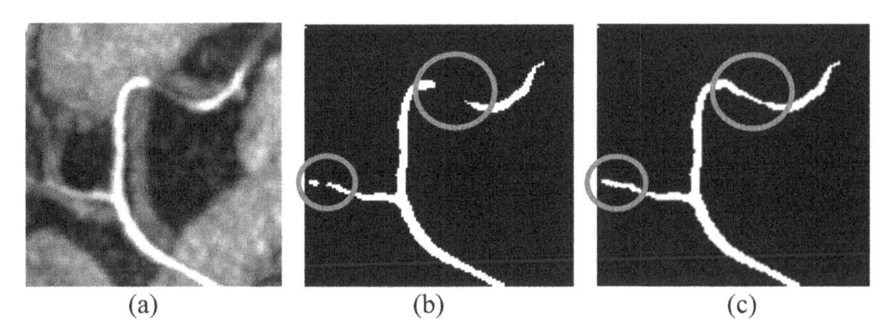

Fig. 9. Contrastive experiments of different loss functions.

Vascular Grading Experiment. The flow of vascular grading results is shown in the Fig. 10. (a) The segmented vessels; (b) The vessels after skeletonization; (c) The nodes marked by the optimized bifurcation point retrieval method; (d) The classified vessels marked outward starting from the main vessel being 0.

Fig. 10. Schematic diagram of vascular grading.

4 Conclusion and Future Work

In this paper, AA Res-Unet is proposed to segment SMA. We add attention argument mode to achieve the global feature as well as the improved focal loss which focus more on vessel SMA. Comparison results of proposed architecture compared to the traditional Unet and network with original focal loss shows that the proposed AA Res-Unet gives better result due to the advantages of enhancing the ability of segmentation and decreasing the existence of useless area. In addition, the improved vascular bifurcation point detection algorithm using SVM can effectively avoid some unreasonable situations and improve the detection accuracy.

At the same time, there are some improvements in the experiment. Image preprocessing: blood vessel images are two-dimensional images, which will lead to disjoint blood vessels in the actual three-dimensional situation. Intersections will appear in the two-dimensional perspective, resulting in errors. Intersection Search Algorithm: by relying on SVM to find all possibilities, a deep learning training model can be gradually improved to determine the intersection and bifurcation points.

Acknowledgments. This work was financially supported by Nantong Health Commission in 2020, QA2020002, and Key clinical project of Nantong University, 2019LZ002.

References

1. Roda, G., Chien, N.S., Kotze, P.G., et al.: Crohn's disease. Nat. Rev. Dis. Primers **6**, 22 (2020)
2. Ramos, G.P., Papadakis, K.A.: Mechanisms of disease: inflammatory bowel disease. Mayo Clin. Proc. Elsevier **94**(1), 155–165 (2019)
3. Torres, J., Bonovas, S., Doherty, G., et al.: ECCO guidelines on therapeutics in Crohn's disease: medical treatment. J. Crohn's Colitis **14**(1), 4–22 (2020)
4. Sulz, M.C., Burri, E., Michetti, P., et al.: Treatment algorithms for Crohn's disease. Digestion **101**(1), 42–56 (2020)
5. Javadi, S., et al.: Quantitative attenuation accuracy of virtual non-enhanced imaging compared to that of true non-enhanced imaging on dual-source dual-energy CT. Abdom. Radiol. **45**(4), 1100–1109 (2020). https://doi.org/10.1007/s00261-020-02415-8
6. Nestler, T., et al.: Diagnostic accuracy of third-generation dual-source dual-energy CT: a prospective trial and protocol for clinical implementation. World J. Urol. **37**(4), 735–741 (2018). https://doi.org/10.1007/s00345-018-2430-4
7. Schicchi, N., et al.: Third-generation dual-source dual-energy CT in pediatric congenital heart disease patients: state-of-the-art. Radiol. Med. (Torino) **124**(12), 1238–1252 (2019). https://doi.org/10.1007/s11547-019-01097-7
8. McGrath, T.A., et al.: Diagnostic accuracy of dual-energy computed tomography (DECT) to differentiate uric acid from non-uric acid calculi: systematic review and meta-analysis. Eur. Radiol. **30**(5), 2791–2801 (2019). https://doi.org/10.1007/s00330-019-06559-0
9. May, M.S., Wiesmueller, M., Heiss, R., et al.: Comparison of dual-and single-source dual-energy CT in head and neck imaging. Eur. Radiol. **29**(8), 4207–4214 (2019)
10. Shaqdan, K.W., Parakh, A., Kambadakone, A.R., Sahani, D.V.: Role of dual energy CT to improve diagnosis of non-traumatic abdominal vascular emergencies. Abdom. Radiol. **44**(2), 406–421 (2018). https://doi.org/10.1007/s00261-018-1741-7
11. Ronneberger, O., Fischer, P., Brox, T.: U-net Convolutionalneural networks for biomedical image segmentation. In: Springer International Publishing, pp. 234−241 (2015)
12. Mary, S.P., Thanikaiselvan, V.: Unified adaptive framework for contrast enhancement of blood vessels. Int. J. Electr. Comput. Eng. **10**(1), 2088−8708 (2020)
13. Ganss, A., Rampado, S., Savarino, E., et al.: Superior mesenteric artery syndrome: a prospective study in a single institution. J. Gastrointest. Surg. **23**(5), 997–1005 (2019)
14. Norsa, L., Bonaffini, P.A., Indriolo, A., et al.: Poor outcome of intestinal ischemic manifestations of COVID-19. Gastroenterology **159**(4), 1595–1597 (2020)
15. Pai, R.K., Jairath, V.: What is the role of histopathology in the evaluation of disease activity in Crohn's disease? Best Pract. Res. Clin. Gastroenterol. **38**, 101601 (2019)
16. Shankar, K., Lakshmanaprabu, S.K., Gupta, D., Maseleno, A., de Albuquerque, V.H.C.: Optimal feature-based multi-kernel SVM approach for thyroid disease classification. J. Supercomput. **76**(2), 1128–1143 (2018). https://doi.org/10.1007/s11227-018-2469-4

17. Bello, I., Zoph, B., Vaswani, A., et al.: Attention augmented convolutional networks. In: Proceedings of the IEEE/CVF International Conference on Computer Vision, pp. 3286–3295 (2019)
18. Nixon, J., Dusenberry, M.W., Zhang, L., et al.: Measuring calibration in deep learning. CVPR Workshops **2**(7) (2019)
19. Mousavi, S.M., Zhu, W., Sheng, Y., et al.: CRED: a deep residual network of convolutional and recurrent units for earthquake signal detection. Sci. Rep. **9**(1), 1–14 (2019)
20. Liang, G., Zheng, L.: A transfer learning method with deep residual network for pediatric pneumonia diagnosis. Comput. Methods Programs Biomed. **187**, 104964 (2020)

Endoscopic Single Image Super-Resolution Based on Transformer and Convolutional Neural Network

Xiaogang Gu[1], Feixiang Zhou[2(✉)], Rongfei Chen[1], Xinzhen Ren[1], and Wenju Zhou[1]

[1] School of Mechatronic Engineering and Automation, Shanghai University, Shanghai 200444, China

[2] School of Informatics, University of Leicester, Leicester L1 7RH, UK

fz64@leicester.ac.uk

Abstract. As we all know, the image quality is related to the size of the image sensor. However, the texture of blood vessels displayed poorly which caused by the small size of image sensor in a limited internal space of disposable medical endoscope. Therefore, different strategies, which are principally using statistical analysis to infer the value of pixels in high-resolution (HR) images, have been developed to enhance the quality of endoscopic images. It is inevitable to bring some negative phenomena such as blurry, blocking and ill-posed. In this paper, we propose to combine the Transformer model with the convolutional neural network in Super-Resolution (SR) reconstruction of blood vessel texture. The algorithm is better than the existing SR algorithms based on the CNN model quantitatively and qualitatively. Compared with the existing methods, the proposed method has significantly improved PSNR and SSIM in the public dataset, which proves the effectiveness of the Transformer network in SR reconstruction.

Keywords: Super-Resolution · Transformer · Convolutional neural network · Endoscope image

1 Introduction

With the advancement of modern medicine and technology, minimally invasive surgery is widely used. With the help of the endoscope placed in the patient's body, the surgeon can obtain more detailed images, and then can more accurately examine the condition of the internal organs. However, the unsatisfactory imaging of the equipment, we cannot obtain high-resolution and clear images of the texture of blood vessels. Dealing with this phenomenology, the Super-Resolution reconstruction methods have been proposed for a single image. Single-Image Super-Resolution (SISR) [1] reconstruction is a method that restores the single low-resolution (LR) image to the high-resolution (HR) image.

SISR reconstruction algorithms can be divided into two categories: traditional interpolation-based methods and learning-based methods. The traditional interpolation-based methods use the value of surrounding several pixels around a certain pixel to infer this pixel value in HR image, which calculated by the established mathematical

M. Fei et al. (Eds.): LSMS 2021/ICSEE 2021, CCIS 1467, pp. 24–32, 2021.
https://doi.org/10.1007/978-981-16-7207-1_3

relation. The learning-based SISR methods have better feature expressing capability and inference calculating speed, compared with traditional interpolation-based methods. However, most of the SR reconstruction methods are only based on convolutional neural networks (CNNs). Recently, with a large number of applications of Transformer in the field of vision which uses self-attention mechanism [2] to extract texture features, researchers have found that Transformer requires fewer prior hypotheses than CNN and performs better on large datasets. In order to improve the imaging quality of medical endoscopes, we propose a cross-scale Super-Resolution reconstruction algorithm based on the Transformer model for endoscopic images.

In short, the main contributions of this paper are as follows:

- Firstly, we propose an end-to-end super-resolution model, which combines the Transformer model and the deep CNN model to extract the texture features of the LR image.
- Secondly, the proposed model is applied to the blood vessel texture image enhancement of disposable medical endoscopes, so that HR images can be obtained.
- Finally, we propose a novel texture synthesis module to fuse the texture information of the original image and the migration image, so that the reconstructed images can retain more details.

2 Related Work

In this Section, some previous works related to SISR and Transformer involved in the paper are introduced below.

2.1 Single-Image Super-Resolution

The Single-Image Super-Resolution (SISIR) algorithm based on interpolation utilizes an interpolation kernel to approximate the loss of high-frequency image texture for restoring the reconstruction of HR images. Traditional interpolation algorithms mainly include nearest-neighbor interpolation, bilinear interpolation, and bicubic interpolation. However, the interpolation algorithms have a poor processing result in processing pixel mutations, such as edges and textures, and often appears jagged and blocking effects. The SISR algorithms based on learning model learns the mapping relationship between LR and HR images through the training set of images to complement the high-frequency information lost in LR, thereby reconstructing HR images. Dong et al. and Shi et al. used CNN to extract textures on low-resolution images to reconstruct HR images. Kim et al. introduced a recurrent neural network to complement the high-frequency information lost after upsampling by deepening the network. He et al. proposed the residual network, Resnet [3]. The staggered structure is to add a skipping connection to the network to skip the connection of some layers. Tai et al. introduced this method [4], using LR images and HR images to have cross-scale. It learns the difference between low-resolution and high-resolution, and adds the residual and low-resolution matrix to obtain the HR image.

2.2 Transformer for Vision

The Transformer models is an architecture used for sequence prediction [5]. It has been widely used in Natural-Language-Processing domain, such as machine translation [6], opinion extraction [7], text classification [8], question answering [9], speech recognition [10], OCR [11], etc. The self-attention mechanism in the Transformer model can explicitly simulate all pairwise interactions between elements in the sequence, making these architectures particularly suitable for specific constraints on ensemble predictions, such as removing duplicate predictions. Traditional CNN and RNN are abandoned in Transformer, and the entire network structure is entirely relyed on the attention mechanism.

3 Proposed Method

This paper proposes a SISR reconstruction algorithm for endoscopic images combined with Transformer and CNN model to solve the low resolution and blurry phenomenon in endoscopic images. The advanced part of the model focused on the use of the Transformer model to encode features of LR images, and the use of CNN to extract features of LR images. The architecture of the model is shown in Fig. 1. The architecture shown in the figure has three parts, feature extractor, texture migrator, and texture synthesizer. The CNN model is to extract preliminary details and Transformer model applied to obtain intrinsic texture in the texture extractor. The texture migration module measures the similarity of different detailed structures, and selects the local texture with the highest similarity for synthesis by comparing the similarity between the local textures. The texture synthesis module uses three different texture extraction operations to obtain images with three scales of 1×, 2×, and 4×, and uses the cross-scale self-similarity of the images to obtain high-frequency information of the image. The learnable texture extractor designed in this paper can be obtained by Transformer to obtain three different information of Q, K, and V, and then it will be fused with the texture information extracted by CNN (Fig. 2).

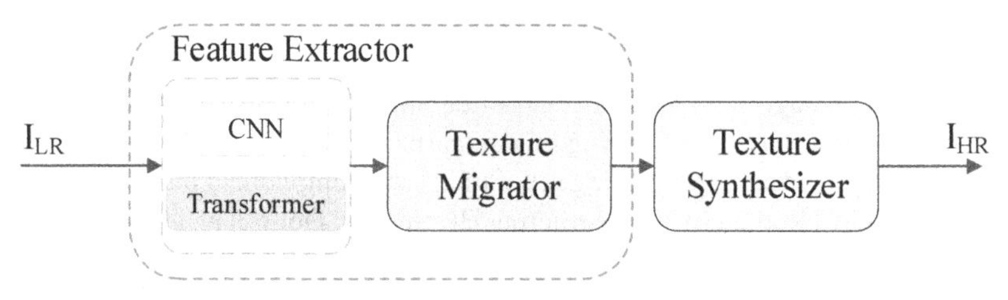

Fig. 1. Proposed network structure.

3.1 Texture Extraction and Migration

Firstly, we use a convolutional neural network to extract preliminary features on the original LR imagecited.

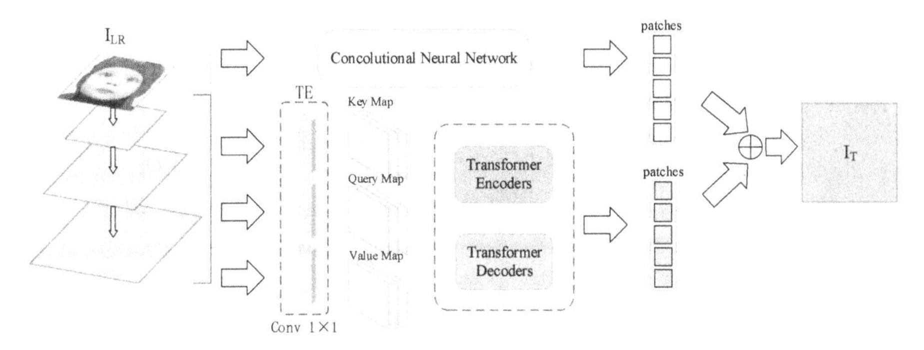

Fig. 2. Texture extraction network.

Secondly, the inputs of transformer-based texture extractor, the LR image is multi-sampled to obtain three different images. The texture extractor obtains three semantics of query map Q, key map K, and value map V from these three images. The calculation process is shown in formula 1.

$$\begin{aligned} Q &= TE(LR \uparrow) \\ K &= TE(LR \downarrow). \\ V &= TE(LR) \end{aligned} \tag{1}$$

Where TE is the operation performed by the feature extractor, and Q, K, and V are the basic components of the Transformer module, which represent the query, key, and value respectively, and are used for subsequent similarity calculations.

Finally, compared the similarity of different patches from CNN-based extractor and Transformer-based extractor, using the matrix inner product to calculate the similarity of the feature maps of Q and K. The calculation steps are shown in formula 2.

$$s_{i,j} = \left\langle \frac{q_i}{\|q_i\|}, \frac{k_i}{\|k_i\|} \right\rangle. \tag{2}$$

CNN and Transformer feature texture combination module, the formula is shown in formula 3.

$$\mathrm{I_T} = \mathrm{F} + \mathrm{Conv}(\mathrm{Concat}(\mathrm{F},\mathrm{T})) \odot \mathrm{S}. \tag{3}$$

Where F means preliminary characteristics of a LR image, T means the feature catched by the Transformer model OUTPUT, $\mathrm{I_T}$ is the texture patches fusion which includes F and T.

3.2 Texture Synthesis

Texture synthesis exchanges information between features of different scales through texture migration to synthesize a basic feature image $\mathrm{I_T}$, The image $\mathrm{I_T}$ undergoes further feature extraction through the residual network structure, and finally use the sub-pixel

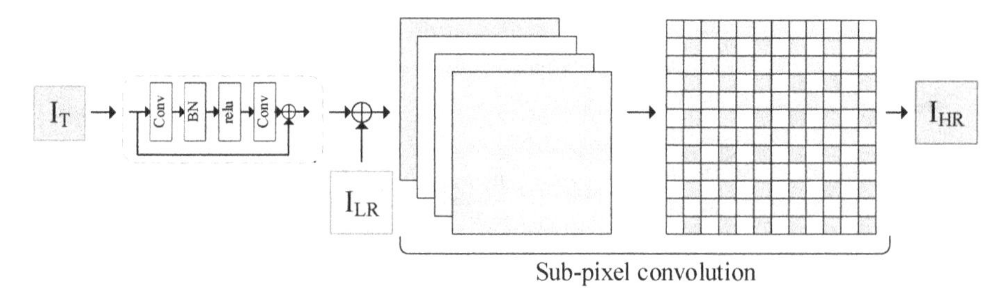

Fig. 3. Texture synthesis network.

convolution module to reconstruct the high-resolution image (Fig. 3). The calculation steps are shown in formula 4.

$$I_{HR}=[Res(I_{LR}\|I_T)+I_{LR}]\uparrow_{2N\times}. \quad (4)$$

Where Res(·) denotes Residual network operation,$\uparrow_{2N\times}$ denotes Sub-pixel convolution layer operations [12], 2N× denotes zoom factor.

In such a design, the texture features after the exchange are compared with the original low-pixel image to obtain a more powerful feature representation.

In order to obtain a better image reconstruction result during the training process, we train the network by using the L_1-norm as reconstruction loss.

$$L=\frac{1}{hwc}I_{HR}-I_{LR1}. \quad (5)$$

Where h and w denote the height and width of the evaluated images, c denotes the channel of the evaluated images.

4 Experiments

4.1 Datasets

The method proposed in the article was trained on the DIVerse 2K resolution high quality (DIV2K) images dataset. The DIV2K dataset contains a total of 1000 2K resolution RGB images, of which 800 are the training dataset, 100 are the validation dataset, and 100 are the test dataset. For testing the performance of the model, we utilize Set5 [13], Set14 [14], B100 [15], Urban100 [16], Manga109 [17] as the benchmark datasets.

4.2 Implement Detail

During the experimental training, we crop LR images which augmented by pyramid sampling algorithm with patch size 40 × 40. The initial learning rate of the proposed model is 1e-2, and finally becomes 1e-4. The β1, β2 and ϵ parameters of the Adam optimizer seted to 0.9, 0.999 and 1e-8, respectively. We deploy the model using PyTorch, and train it on the GPU which is Nvidia RTX2060.

4.3 Image Evaluation Standard

At present, the clarity evaluation of blood vessel texture is mainly measured by subjective feelings and objective standards. PSNR is used to judge the restoration degree of the processing result, the unit is dB, as shown in formulas 6 and 7. SSIM is used to compare the brightness, contrast and structure information of the result image with the original image, dimensionless, as shown in formula 8.

$$MSE = \frac{1}{M \times N} \sum_{j=1}^{M} \sum_{i=1}^{N} [x(i,j) - y(i,j)]^2 . \tag{6}$$

$$PSNR = 10 \cdot \lg\left(\frac{255^2}{MSE}\right). \tag{7}$$

$$SIM(x, y) = \frac{(2\mu_x\mu_y + c_1)(2\sigma_{xy} + c_2)}{\left(\mu_x^2 + \mu_y^2 + c_1\right)\left(\sigma_x^2 + \sigma_y^2 + c_2\right)}. \tag{8}$$

Among them, x represents the original image, y represents the processed image, MSE is the mean square error, M and N represent the size of the image, μ_x is the average value of x, μ_y is the average value of y, σ_x^2 is the variance of x, σ_y^2 is the variance of y, and σ_{xy} is the covariance of x and y. $c_1 = (255 * k_1)^2$, $c_2 = (255 * k_2)^2$ is a constant used to maintain stability, k_1 is 0.01, k_2 is 0.03.

4.4 Results of Blood Vessel Texture Reconstruction

Experiments have verified the PSNR and SSIM of different Super-Resolution algorithms, such as bicubic interpolation, SRCNN [18], VDSR [19], MDSR [20], RDN [21], RCAN [22].

Quantitative Evaluations. In Tables 1 and Table 2, we summarize the quantitative evaluation under the ×4 scaling factor. Compared with other super-resolution algorithms, our proposed method achieves the best performance in the benchmark datasets. In particular, under the large data set Manga109, PSNR and SSIM increased by 0.71 dB and 0.012, respectively. Therefore, these excellent performances prove the effectiveness of the Transformer model under a large amount of data.

Qualitative Evaluations. The visualized image result of the quality evaluation after the Super-Resolution reconstruction of the Set5 images is shown in Fig. 4. The proposed model has a better restoration effect in high-frequency details, such as eyelashes, wing sheaths, and needle caps, etc. In Fig. 5, we apply the proposed model to the reconstruction of blood vessel texture. It can be clearly seen that the blood vessel texture can be presented more clearly after using proposed method.

Table 1. The results of PSNR value measured on benchmark datasets compared with different Super-Resolution methods. The best results are in bold.

Algorithm	Set5	Set14	B100	Urban100	Manga109
Bicubic	28.42	26.00	25.96	23.14	24.89
SRCNN	30.48	27.49	26.9	24.52	27.58
VDSR	31.35	28.01	27.29	25.18	28.83
MDSR	31.34	26.95	26.56	25.51	28.93
RDN	31.24	27.58	26.356	25.38	29.24
RCAN	31.23	27.47	26.37	25.42	29.38
Proposed	**31.94**	**28.11**	**27.54**	**25.87**	**30.09**

Table 2. The results of SSIM value measured on benchmark datasets compared with different Super-Resolution methods. The best results are in bold.

Algorithm	Set5	Set14	B100	Urban100	Manga109
Bicubic	0.8104	0.7027	0.6675	0.6577	0.7866
SRCNN	0.8628	0.7503	0.7101	0.7221	0.8555
EDSR	0.8838	0.7674	0.7251	0.7524	0.8875
VDSR	0.8719	0.7601	0.7194	0.7831	0.8914
CSNLN	0.8738	0.7611	0.7195	0.7684	0.8941
USRNet	0.8901	0.7789	0.7203	0.7682	0.8952
Proposed	**0.8935**	**0.7842**	**0.7464**	**0.7844**	**0.9077**

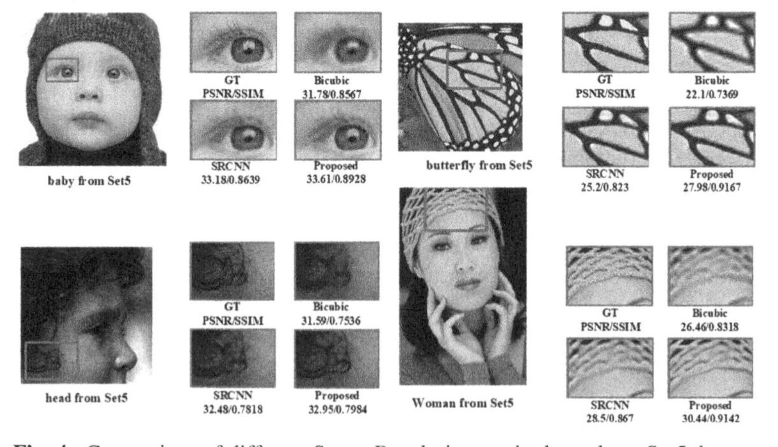

Fig. 4. Comparison of different Super-Resolution methods apply to Set5 dataset.

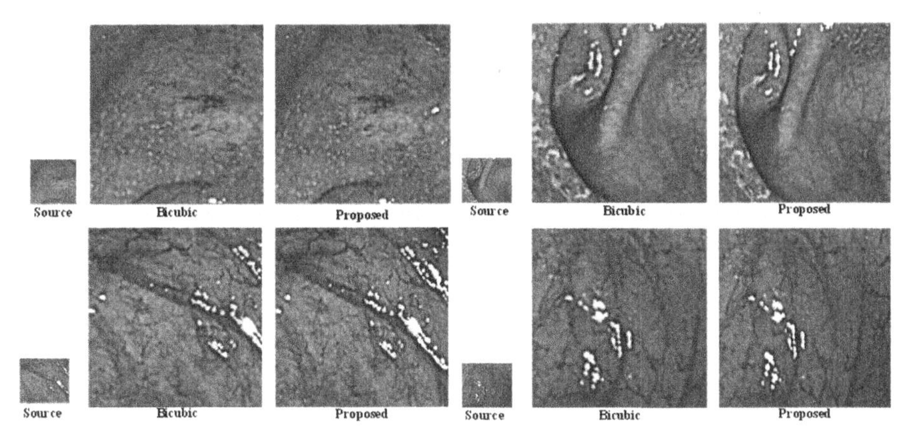

Fig. 5. Comparison of different networks apply to vessel texture reconstruction.

5 Conclusion

In this paper, we propose a Super-Resolution algorithm based on CNN-Transformer to solve the problem of insufficient resolution of endoscopic images. A multi-scale feature extraction and fusion algorithm designed to extract more detailed texture at different scales. Through the fusion of multi-scale feature, accurate local texture expression is achieved, and the LR image features are utilized to the greatest extent. The experiment shows that compared with the existing CNN network, added the Transformer network, local multi-scale features and hierarchical features to obtain HR images can be token full advantage. The testing result of the proposed method expound on benchmark datasets shows that the Transformer model has token more advantages in processing large amounts of data. Utilizing this method, the reconstructed blood vessel texture is more realistic that can be applied well in the surgical scene.

References

1. Irani, M., Peleg, S.: Improving resolution by image registration. CVGIP: Graph. Models Image Process. **53**(3) (1991)
2. Bahdanau, D., Cho, K., Bengio, Y.: Neural Machine Translation by Jointly Learning to Align and Translate (2014)
3. He, K., Zhang, X., Ren, S., Sun, J.: Deep residual learning for image recognition. In: 2016 IEEE Conference on Computer Vision and Pattern Recognition (CVPR) (2016)
4. Ying, T., Jian, Y., Liu, X.: Image Super-Resolution via deep recursive residual network. In: IEEE Conference on Computer Vision and Pattern Recognition. IEEE Press (2017)
5. Vaswani, A., et al.: Attention is all you need. J. Adv. Neural Inf. Process. Syst. **30**, 5998–6008 (2017)
6. Ott, M., Edunov, S., Grangier, D., Auli, M.: Scaling neural machine translation. In: WMT (2018)
7. Angelidis, S., Amplayo, R.K., Suhara, Y., Wang, X., Lapata, M.: Extractive Opinion Summarization in Quantized Transformer Spaces (2020)

8. Terechshenko, Z., et al.: A comparison of methods in political science text classification: transfer learning language models for politics. Soc. Sci. Electron. Publishing (2021)
9. Ngai, H., Park, Y., Chen, J., Parsapoor, M.: Transformer-Based Models for Question Answering on Covid19 (2021)
10. Lu, L., Liu, C., Li, J., Gong, Y.: Exploring Transformers for Large-Scale Speech Recognition (2020)
11. Yang, L., Wang, P., Li, H., Li, Z., Zhang, Y.: A Holistic Representation Guided Attention Network for Scene Text Recognition (2019)
12. Shi, W., Caballero, J., Huszár, F., Totz, J., Wang, Z.: Real-time single image and video Super-Resolution using an efficient sub-pixel convolutional neural network. In: 2016 IEEE Conference on Computer Vision and Pattern Recognition (CVPR). IEEE Press (2016)
13. Bevilacqua, M., Roumy, A., Guillemot, C., Morel, A.: Low-complexity single image super-resolution based on nonnegative neighbor embedding. BMVC (2012)
14. Zeyde, R., Elad, M., Protter, M.: On single image scale-up using sparse-representations. In: International Conference on Curves and Surfaces, pp. 711–730. Springer (2010)
15. Martin, D., Fowlkes, C., Tal, D., Malik, J.: A database of human segmented natural images and its application to evaluating segmentation algorithms and measuring ecological statistics. In: IEEE International Conference on Computer Vision, vol. 2, pp. 416–423. IEEE (2001)
16. Huang, J.B., Singh, A., Ahuja, N.: Single Image Super-Resolution from Transformed Self-Exemplars, pp. 5197–5206. IEEE Press (2015)
17. Matsui, Y., et al.: Sketch-based manga retrieval using manga109 dataset. Multimedia Tools Appl. **76**(20), 21811–21838 (2016). https://doi.org/10.1007/s11042-016-4020-z
18. Dong, C., Loy, C.C., He, K., Tang, X.: Learning a deep convolutional network for image Super-Resolution. In: Fleet, D., Pajdla, T., Schiele, B., Tuytelaars, T. (eds.) ECCV 2014. LNCS, vol. 8692, pp. 184–199. Springer, Cham (2014). https://doi.org/10.1007/978-3-319-10593-2_13
19. Kim, J., Lee, J.K., Lee, K.M.: Accurate image super-resolution using very deep convolutional networks. In: IEEE Conference on Computer Vision and Pattern Recognition (CVPR) Workshops, pp. 1646–1654. IEEE Press (2016)
20. Lim, B., Son, S., Kim, H., Nah, S., Lee, K.M.: Enhanced deep residual networks for single image Super-Resolution. In: 2017 IEEE Conference on Computer Vision and Pattern Recognition Workshops (CVPRW), The IEEE Conference on Computer Vision and Pattern Recognition (CVPR) Workshops. IEEE (2017)
21. Zhang, Y., Tian, Y., Kong, Y., Zhong, B., Fu, Y.: Residual dense network for image Super-Resolution. In: 2018 IEEE/CVF Conference on Computer Vision and Pattern Recognition. IEEE (2018)
22. Zhang, Y., Li, K., Li, K., Wang, L., Zhong, B., Fu, Y.: Image Super-Resolution Using Very Deep Residual Channel Attention Networks, pp. 286–301. Springer, New York (2018)

Facial Video-Based Remote Heart Rate Measurement via Spatial-Temporal Convolutional Network

Chunlin Xu(✉) and Xin Li

School of Mechatronic Engineering and Automation, Shanghai University, Shanghai 200072, China

Abstract. In this paper, in order to remotely measure the human heart rate (HR) with consumer-level cameras, we propose an end-to-end framework for reducing the influence of non-physiological signals (such as head movement and illumination variation). First, face tracking and skin segmentation are carried out on the input facial video, and then the designed model based on small-scale spatial-temporal convolutional network is used to classify the sub-regions of interest. Finally, the predicted results of each sub-block are aggregated to achieve the regression of the HR. Experiment on the LGI-PPGI dataset in four scenarios with increasing difficulty shows that the proposed approach achieves better performance compared with the benchmark.

Keywords: Heart rate · Facial video · Spatial-temporal convolutional network

1 Introduction

Human heart rate (HR) is of great significance for the monitoring of the individual psychological state and the prevention and treatment of chronic diseases. Traditional heart rate measurement methods, such as electrocardiogram (ECG) and photoplethysmography (PPG), are usually based on contact equipment which is not suitable for people with sensitive skin and long-term monitoring. In recent years, the study of low-cost and non-contact HR estimation utilizing imaging photoplethysmography (iPPG) has attracted many scholars. IPPG is also known as remote photoplethysmo-graphy (rPPG). It is realized by using a common digital camera to measure the tiny color changes of the skin surface caused by heartbeat pulses. Specifically, the circulation of blood through the cardiovascular system caused by the human heartbeat changes the blood oxygen content in the subcutaneous capillaries, resulting in variation in the skin absorption and reflection of the illumination spectrum. These subtle changes in skin reflection captured by the cameras are so related to the pattern of blood oxygen pulses that can be used as an important feature for identifying HR.

After years of development, various rPPG methods [1–4] based on different mechanisms and assumptions have been proposed. Although some methods have been proven to perform well under well-controlled conditions, obtaining rPPG pulses in the more

M. Fei et al. (Eds.): LSMS 2021/ICSEE 2021, CCIS 1467, pp. 33–42, 2021.
https://doi.org/10.1007/978-981-16-7207-1_4

complex realistic scenarios with motion artifacts and noise is still a challenge. The motion artifacts are mainly caused by two reasons: one is rigid motion, such as shaking the head; the other is non-rigid motion, including facial expressions and slight muscle jitters. In addition, the changes of scene illumination and occlusion will also introduce noise to the heart rate measurement. There are methods [5, 6] that hope to eliminate motion artifacts. With the rapid development of deep neural network, learning more robust physiological signal representations through data-driven methods has become a research hotspot.

In order to overcome the above limitations, we propose an end-to-end small-scale framework based on spatial-temporal convolutional network to measure the HR from the face video. A more complete physiological pulse representation of facial sequence can be extracted by fusing the skin structure features and the pixel variation in time domain. The proposed model is used to discriminate the feature maps from facial sub-blocks, and consequently the HR estimated value can be obtained by aggregating all the results. Finally, we perform our approach in scenarios with increasing difficulty on public LGI-PPGI datasets [7]. In summary, the main contributions of this work are as follows: (1) We propose a robust framework for estimating HR from face video; (2) The HR is estimated by aggregating the pulse categories of sub-regions, which addresses the shortcoming of high computational complexity of rPPG-related neural networks; (3) We compare the benchmark on the same dataset and prove the performance of this method.

2 Related Work

In recent years, researchers have made a lot of progress in non-contact rPPG extraction. In this section, we briefly review the existing work on remote HR measurement based on video analysis.

In 2008, Verkruysse et al. [1] first proved the feasibility of measuring PPG signals by common digital cameras. They verified that the color changes of the RGB channels in skin pixels contain physiological pulse information, and the green component features the strongest counterpart. Since then, to obtain more accurate pulse signals, more and more rPPG methods have been proposed. In 2010, Poh et al. [2] introduced a blind source separation algorithm to extract the rPPG pulse signal from face video. They calculated the average of RGB values over the face frame pixels, and separated pulse components from the observed signal by a joint approximate diagonalization of eigen-matrix (JADE)-based independent component analysis (ICA) algorithm. Then fast Fourier transform (FFT) and band-pass filter were utilized to screen out the largest spectral peak in normal cardiac frequency range (0.75 to 4 Hz, i.e. 42 to 240 bpm) as the HR estimation of the video. In 2017, Wang et al. [4] proposed an rPPG algorithm called POS, defining a vertical POS plane in the temporally normalized RGB space to purify the HR signals.

Although performing well in the experimental environment, these methods traditionally rely on assumptions such as linear combinations of source signals or specific skin reflection models, which may be invalid under conditions such as large facial movements or illumination variation. Therefore, there is still room for improve- ment in more robust HR measurement.

In addition to traditional methods, deep learning technologies have been gradually applied to remotely measure rPPG signals in recent years. In 2018, Chen et al. [8] used a convolutional attention neural network to establish a mapping between video frame differences and physiological impulse signal derivatives. Niu et al. [9] constructed the spatial-temporal facial feature representation by aggregating multiple sub-block ROI information, and then utilized residual neural network to estimate the rPPG signal. However, these methods require a strict pre-processing, leaving the global clues beyond the predefined ROI.

At present, many researchers have proposed deep learning frameworks for end-to-end HR measurement. Yu et al. [10] designed video enhancement networks (PhysNet and rPPGNet) for rPPG signal recovery, which were trained through supervised negative Pearson coefficient loss in the time domain. Spetlik et al. [11] designed a feature extractor and a HR estimator based on CNN to calculate the HR from facial sequence in two steps. Although large-scale end-to-end deep learning has made great progress in rPPG estimation, it is necessary to consider model complexity and computational cost. The purpose of this article is to introduce a small-scale framework for remote HR measurement using spatial-temporal convolution.

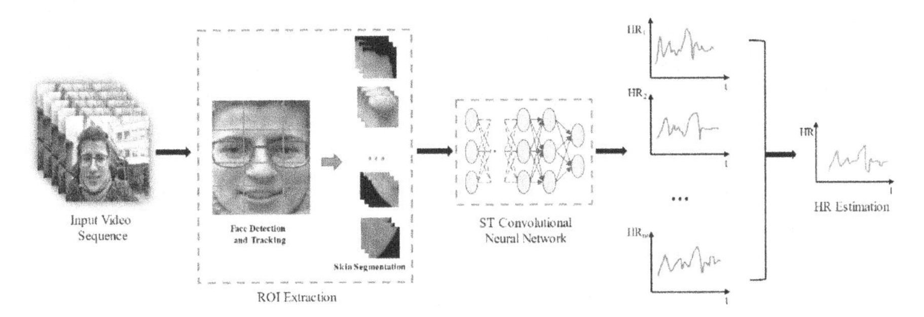

Fig. 1. The framework of the proposed HR estimation method.

3 Spatial-Temporal Network Based HR Measurement

Previous HR estimations are either based on artificially constructed features or through building large-scale deep networks, which typically rely on specific assumptions or require huge computing power. In this work, we propose a small-scale HR measurement approach based on spatial-temporal convolutional model without manual feature construction.

In this section, an overview of the proposed framework is presented through three procedures, as illustrated in Fig. 1. Face tracking is first performed on the input video stream to obtain face regions from which the ROI is extracted by skin segmentation. Then, the ROI is divided into small-scale samples as the input to the network model for learning the mapping from skin representations to pulse. The final HR value is regressed by aggregating all the results of sub-samples.

3.1 Face Tracking and Skin Segmentation

In this work, the open face detector MTCNN [12] is introduced to locate the face in the video for continuous ROI extraction. For purposes of faster detection, the Kalman filter approach is used to predict and track the face positions of the following frames.

The global face is selected for further processing. Since resizing operation may introduce noise to the HR signal, we directly address the original facial landmarks. Skin segmentation is utilized to remove the non-skin pixels. Given that the change of skin color with time caused by cardiac activity is the only information that can be directly applied in HR estimation, traditional rPPG methods treat the distribution of face pixel values in time domain as the representation of HR signal. To counter the impact of illumination artifacts and sensor noise, the spatial average value for RGB channels is the most common use, which has been illustrated its effectiveness, and related study [1] has shown that the green component contains the strongest rPPG signal.

For each input face sequence, we retain the global face pixels of green channel and part them into sub-blocks of size 28 × 28 for generating the spatial representation of the HR signal on skin. A fixed sliding window is used here to split video into separate sequences with the continuity of the sub-samples in the time domain preserved. Each sequence $v^i = \{v_1, v_2, ..., v_k\} \in \mathbb{R}^{1\times28\times28\times k}$ contains k frames, which are input to the spatial-temporal network learning the process of HR signal on different skin structures migration and change over time.

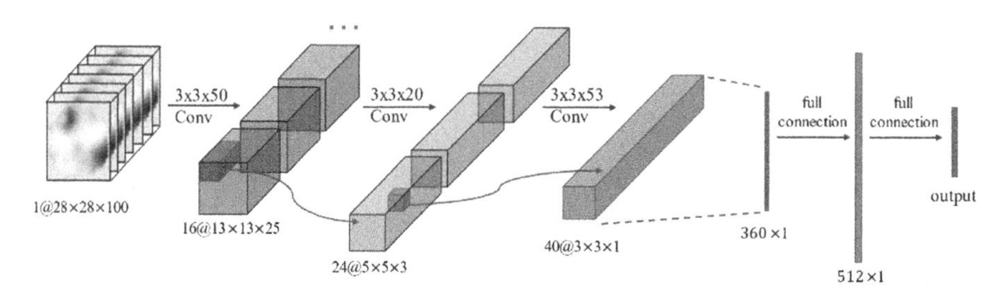

Fig. 2. Structure of our spatial-temporal convolutional neural network.

3.2 Spatial-Temporal Network

HR is calculated according to the feature sequences obtained in the previous section. The proposed network architecture is shown in Fig. 2. The foundation of it is 3D-CNN [13]. In the spatial domain, VGG-style CNN [14] is used to capture the pixel connection and texture structure in a frame. Except the last, each convolutional layer is followed by a pooling layer and an activation layer. The average pooling layer is selected, because down sampling multiple features for HR measurement conduces a higher signal-to-noise ratio. The rectified linear unit (ReLU) is used as hidden layer activation function to make the network nonlinear. The dropout operation is inserted after activation layer to avoid the overfitting, disabling 50% of the neurons. The obtained 360-dimensional feature vector is finally applied through the last fully connected layer to calculate the mapping scores of the pulse signal to each HR intervals. Such scores can be treated as the criterion of the HR estimation value of the input face sequence, and to a certain extent, as the weights of the pulsation component.

Since the HR category of each input video is the only label for training, and the output of our model is the score of each HR interval, for optimizing the difference between the predicted HR and the ground-truth, the KL divergence is selected as the loss function, that is, the cross-entropy loss function when the label is one-hot encoded. The equation is defined as:

$$loss(\hat{y}, \theta) = -\sum_{y_i \in y} y_i log \frac{\exp(\hat{y}_i)}{\sum_{\hat{y}_j \in \hat{y}} \exp(\hat{y}_j)}.(1)$$

where $loss(\hat{y}, \theta)$ donates the loss of prediction $\hat{y}$ of the model θ, y donates the ground-truth of the input sequence.

3.3 HR Estimation

The HR category of each face sub-region sequence is obtained from the model output. We assume that the expectation of the predicted HR distribution of the trained model is consistent with the counterpart of ground-truth. But the variance has nonzero value. In order to reduce the error, we aggregate the HR predictions of all face blocks as the final HR estimation:

$$H_p(i) = \frac{\sum_{k=1}^{m} \max\left(H_p^k(i), 0\right)}{m}.(2)$$

where $H_p(i)$ donates the HR in the i-th second, $H_p^k(i)$ donates the prediction of the k-th face block, m donates the total number of input face blocks.

4 Experiments

In this section, a public LGI-PPGI dataset with four different scenarios is introduced to measure the HR under realistic conditions. All the video steams used are processed by uniform standard to extract the HR labels. The trained model is compared with the baseline methods by utilizing five metrics.

4.1 Dataset

In this work, we evaluate our approach on open LGI-PPGI dataset with increasingly challenging sessions featuring participants of both genders, different skin tones and some have thick facial hair or glasses.

LGI-PPGI is a public database for remote physiological signal analysis. There is an amount of 100 uncompressed video recordings with approximately 200 min total duration, including 25 participants each of which is required to perform resting (no facial motion and static illumination), head as well as facial motions (only static illumination), talking in the city street (head and facial motions as well as natural varying illumination), and exercise in a gym (no further instructions given to the participant). The average frame rate is 25 FPS for the camera and 60 FPS for the pulse oximeter.

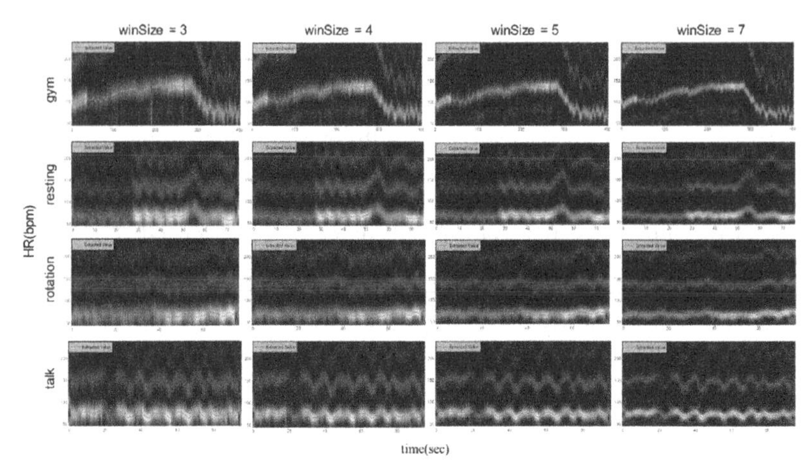

Fig. 3. Comparison of a subject's BVP spectrograms in different scenarios computed by STFT with different winSize. The rows (from down to top) reflect a resting, a face rotation, a bicycle ergometer exercise and a street talking scenario. The columns (from left to right) correspond to the increasing winSize.

4.2 The Ground Truth Signal

In this work, the ground-truth BVP signals of participants are available while the HRs of the videos are directly estimated by proposed method. Therefore, it is necessary to calculate the HR from BVP so as to train our model and compare it with the labels. The provided BVP signal is in the form of pulse sequences in the time domain. The approach for extracting the HR from such a signal is demonstrated in [15]. We perform the short time Fourier transform for the power spectrum density calculation of the BVP signal, from which the highest frequency component is tracked as label.

An important aspect rising from SFFT is the setting of time window size (winSize), which moves along the BVP signal to generate a succession of ground-truth HR sequences. The wider we assign the winSize, the higher the frequency resolution, but the lower the time resolution and the greater the computational complexity of model due to consistency of video and ground truth analysis.

The results of the HRs produced from the BVP signal are shown in Fig. 3. Different *winSizes* are performed on the same BVP sequence under four scenarios in LGI-PPGI dataset. As can be noted, a narrow time window leads to a large HR variation. We suppose that a lower frequency resolution causes noise being extracted as HR. When the winSize is equal to 4, a stable HR spectrum is obtained in all the scenarios, and for higher values, no significant changes are found.

According to the above analysis, we set the window size for video and ground truth analysis to 4 s in all experiments to make a trade-off between the accuracy and computational overhead.

4.3 Model Learning

HR is estimated from face sequence by the designed spatial-temporal convolutional network, which is trained using standard backpropagation algorithm. All weights of the model are initialized by the initialization method proposed by Glorot et al. [16]. Each face frame is zero-centered and is normalized by dividing it by 255. During training, the weights of model is updated according to the batch size of 256. Adam optimizer with an initial learning rate of 0.001 is used and the rate decay is performed over the increasing of iteration times by placing checkpoint. All the experiments are carried out on Ubuntu 20.04. The proposed network is developed from the deep learning framework of PyTorch. The processing of face frames is completed by using skvideo.

4.4 Evaluation

In our experiments, five kinds of metrics are introduced to evaluate the effectiveness of the proposed approach. The mean of measure error (ME), the average of the absolute error (MAE) and the root mean squared error ($RMSE$) are all based on the error $E_{HR}(i) = H_{est}(i) - H_{gt}(i)$, where H_{est} donates the HR estimated, H_{gt} donates the ground-truth from BVP signal. The fourth one is the mean of absolute percentage error denoted as $MERP$. The last one is the Pearson's correlation coefficient r used to indicate the linear correlation between the estimation and the ground truth.

Experiments are conducted on the LGI-PPGI dataset using the evaluation protocol above. Traditional methods including ICA [2], CHROM [3], SSR [17] and POS [4] is performed for comparison. The results of these methods are implemented through the open source toolkits [18]. All results for the four increasingly difficult scenarios are presented in Table 1, 2, 3 and 4.

From the results, we can see our approach achieves promising performance for all the evaluation indicators. As shown in Table 1 and 2, for the resting and rotation scenarios, MAE, RMSE, and MER from the proposed method outperform the HR estimation based on hand-crafted features. In addition, we obtain a set of high Pearson's correlation coefficients (0.94 and 0.82), which indicates a stronger linear correlation between the predicted HR and the ground truth.

It can be seen from Table 3 that in the realistic street talking scenario with the influence of large variations in head movement and illumination, the proposed method achieves better results with an MAE of 8.72 bpm and an $RMSE$ of 9.72 bpm. The above considerations indicate the success of the proposed small-scale network model and the effectiveness of our framework that performs well in challenges of LGI-PPGI dataset.

Bland-Altman plot (Fig. 4) is used for further evaluating the estimation consistencies between ground-truth and the predicted HR by individual approaches. The results show that the proposed spatial-temporal model achieves a better consistency and smaller standard error in a less constrained environment. In addition, for the session where the subject is keeping exercise, most of the methods fail while our method achieves a better consistency (as shown in Table 4). We suspect that the movement of the fingers during running affects the acquisition of real BVP signals. Therefore, non-contact measurement of human HR in realistic scenarios is still an important challenge. There is still room for taking care of the error in HR measurement of sport context.

Table 1. The results in resting scenario.

Method	ME (bpm)	MAE (bpm)	RMSE (bpm)	MER (bpm)	PCC
ICA	1.07	2.98	5.03	4.39%	0.44
CHROM	0.45	2.03	3.50	3.00%	0.68
SSR	0.47	1.81	3.68	2.64%	0.64
POS	**−0.21**	2.28	5.01	3.33%	0.63
Proposed	0.72	**1.50**	**1.91**	**2.28%**	**0.94**

Table 2. The results in facial rotation scenario.

Method	ME (bpm)	MAE (bpm)	RMSE (bpm)	MER (bpm)	PCC
ICA	1.82	4.94	7.61	7.23%	0.31
CHROM	3.86	8.82	15.30	13.02%	0.41
SSR	5.45	7.20	11.03	10.68%	0.45
POS	5.69	7.45	11.15	11.03%	0.49
Proposed	**0.66**	**1.31**	**1.69**	**1.93%**	**0.82**

Table 3. The results in street talking scenario.

Method	ME (bpm)	MAE (bpm)	RMSE (bpm)	MER (bpm)	PCC
ICA	**−0.70**	16.33	26.60	22.07%	0.08
CHROM	3.86	13.90	21.16	19.26%	−0.12
SSR	2.92	19.85	32.94	27.05%	0.03
POS	2.75	16.51	29.11	22.53%	0.20
Proposed	8.67	**8.72**	**9.72**	**12.25%**	**0.29**

Table 4. The results in exercise scenario.

Method	ME (bpm)	MAE (bpm)	RMSE (bpm)	MER (bpm)	PCC
ICA	45.22	47.31	49.28	38.48%	0.26
CHROM	17.15	40.43	48.52	33.22%	0.31
SSR	**−8.11**	28.65	45.19	26.54%	0.12
POS	13.03	41.42	51.18	34.15%	0.13
Proposed	−8.76	**21.25**	**25.47**	**18.9%**	**0.73**

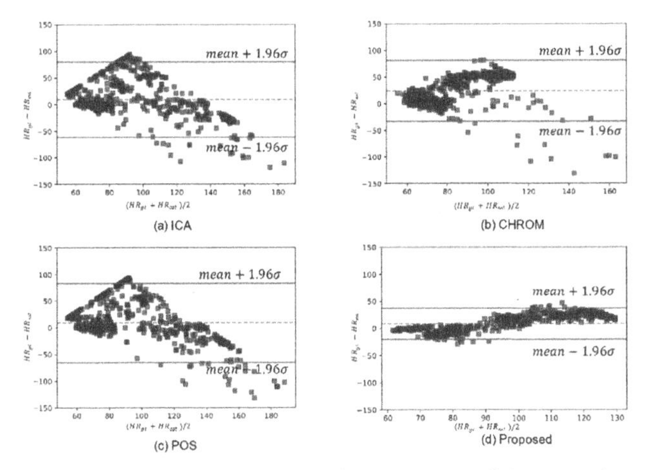

Fig. 4. The Bland-Altman plots demonstrating the agreement of the ground truth HR (*HRgt*) and the estimated HR (*HRest*) for (a) ICA, (b) CHROM, (c) POS and (d) proposed method on LGI-PPGI database. The red lines represent the mean and 95% limits of agreement.

5 Conclusion

In this paper, we propose a new end-to-end framework for human HR measurement, reducing the impact of non-physiological signals (such as head movement, lighting conditions). Our method is compared with the benchmark in four scenarios with increasing difficulty (resting, rotation angle, exercise on a treadmill, and dialogue on a real street) on LGI-PPGI dataset. HR is obtained by aggregating the predictions of sub-ROI sequences to decrease the computational cost. By learning the mapping of spatial and temporal facial representations to physiological signal, the proposed small-scale network achieves higher accuracy compared to other methods based on hand-crafted features under the condition of restricted head movement and illumination variation. In more complex scenarios, our approach shows a better consistency but large error. It is still a difficult challenge to estimate the accurate HR remotely in the scenario of strenuous exercise. In our future work, we will explore the unsupervised and semi-supervised learning technologies for HR measurement and extend the related work to the measurement of other physiological signals.

References

1. Verkruysse, W., Svaasand, L.O., Nelson, J.S.: Remote plethysmographic imaging using ambient light. Opt. Express **16**, 21434–21445 (2008)
2. Poh, M.Z., McDuff, D.J., Picard, R.W.: Non-contact, automated cardiac pulse measurements using video imaging and blind source separation. Opt. Express **18**, 10762–10774 (2010)
3. De, H.G., Jeanne, V.: Robust pulse rate from chrominance-based rPPG. IEEE Trans Biomed. Eng. **60**, 2878–2886 (2013)
4. Wang, W., Brinker, A.C., Stuijk, S., Haan, G.: Algorithmic principles of remote PPG. IEEE Trans. Biomed. Eng. **64**, 1479–1491 (2017)

5. Li, X., Chen, J., Zhao, G., Pietikainen, M.: Remote heart rate measurement from face videos under realistic situations. In: 27th IEEE Conference on Computer Vision and Pattern Recognition, pp. 4264–271. IEEE Press (2014)
6. Wang, W., Brinker, A.C., Stuijk, S., Haan, G.: Robust heart rate from fitness videos. Physiol. Meas. **38**, 1023–1044 (2017)
7. Pilz, C.S., Zaunseder, S., Krajewski, J., Blazek, V.: Local group invariance for heart rate estimation from face videos in the wild. In: 31st Meeting of the IEEE/CVF Conference on Computer Vision and Pattern Recognition Workshops, pp. 1335–1343. IEEE Press (2018)
8. Chen, W., McDuff, D.: DeepPhys: video-based physiological measurement using convolutional attention networks. In: Ferrari, V., Hebert, M., Sminchisescu, C., Weiss, Y. (eds.) ECCV 2018. LNCS, vol. 11206, pp. 356–373. Springer, Cham (2018). https://doi.org/10.1007/978-3-030-01216-8_22
9. Niu, X., Shan, S., Han, H., Chen, X.: RhythmNet: end-to-end heart rate estimation from face via spatial-temporal representation. IEEE Trans. Image Process. **29**, 2409–2423 (2020)
10. Yu, Z., Peng, W., Li, X., Hong, X., Zhao, G.: Remote heart rate measurement from highly compressed facial videos: an end-to-end deep learning solution with video enhancement. In: 17th IEEE/CVF International Conference on Computer Vision, pp. 151–160. IEEE Press (2019)
11. Spetlik, R., Franc, V., Cech, J., Matas, J.: Visual heart rate estimation with convolutional neural network. In: 29th British Machine Vision Conference (2018)
12. Cai, Z., Liu, Q., Wang, S., Yang, B.: Joint head pose estimation with multi-task cascaded convolutional networks for face alignment. In: 24th International Conference on Pattern Recognition, pp. 495–500. IEEE Press (2018)
13. Ji, S., Xu, W., Yang, M., Yu, K.: 3D convolutional neural networks for human action recognition. IEEE Trans. Pattern Anal. Mach. Intell. **35**, 221–231 (2013)
14. Simonyan, K., Zisserman, A.: Very deep convolutional networks for large-scale image recognition. In: 3rd International Conference on Learning Representations (2015)
15. Rajendra Acharya, U., Paul Joseph, K., Kannathal, N., Lim, C.M., Suri, J.S.: Heart rate variability: a review. Med. Bio. Eng. Comput. **44**, 1031–1051 (2006)
16. Glorot, X., Bordes, A., Bengio, Y.: Deep sparse rectifier neural networks. In: 14th International Conference on Artificial Intelligence and Statistics, pp. 315--323. IEEE Press (2011)
17. Wang, W., Stuijk, S., De Haan, G.: A novel algorithm for remote photoplethysmography: spatial subspace rotation. IEEE Trans. Biomed. Eng. **63**, 1974–1984 (2016)
18. Boccignone, G., Conte, D., Cuculo, V., D'Amelio, A., Grossi, G., Lanzarotti, R.: An open framework for remote-PPG methods and their assessment. IEEE Access. **8**, 216083–216103 (2020)

Retinal Vessel Segmentation Network Based on Patch-GAN

Chenghui Li, Jun Yao(✉), and Tianci Jiang

School of Mechatronic Engineering and Automation, Shanghai University, Shanghai 200444, China
grandone0529@shu.edu.cn

Abstract. In order to segment the blood vessels from the retina more accurately, a new retinal vessel segmentation network called Patch-GAN is proposed in this paper. The proposed Patch-GAN retinal vessel segmentation network combines the recently popular GAN network architecture to achieve precise segmentation of retinal vessels through adversarial training. The generator adopts U-Net structure, which consists of a symmetric Encoder-Decoder with skip connections. Short connection blocks are added to the encoder and decoder to prevent gradient dispersion. The discriminator uses the classical convolutional neural network and adds dense connection blocks to the discriminator network, which improves the discriminator's recognition ability and can train the adversarial model better. The proposed method has been tested on two public databases, STARE and DRIVE. The final test results show that the accuracy and F_1 score are 0.9658 and 0.8104 in the DRIVE dataset, and 0.9745 and 0.8356 in the STARE dataset, respectively. The visualization of experimental results shows that the method in this paper can detect smoother blood vessels and blood vessel margins.

Keywords: Retinal vessel segmentation · Patch-GAN · U-Net

1 Introduction

For the past few years, medical image processing technology has been widely used in many fields. Many researches have been conducted on medical images, and methods based on computer vision can automatically process medical images, which can greatly reduce the pressure on medical staff. In particular, because retinal blood vessel obstruction and diabetic retinopathy are the main causes of blindness [1], it is particularly important to detect these retinal vascular diseases in the early stages of onset to prevent blindness. Some studies have shown that the diameter and branch angle of the retinal vessel structure are highly related to coronary artery disease and diabetes [2]. We can check the fundus to understand the tissue condition of the blood vessels of the whole body, such as the sclerosis, bleeding, exudation, edema and hemangioma of the fundus blood vessels can reflect the nature and degree of certain diseases of the whole body. Currently, the method used by experts to examine retinal diseases is to manually separate blood vessels from images of the patient's retina. However, the process of manually

M. Fei et al. (Eds.): LSMS 2021/ICSEE 2021, CCIS 1467, pp. 43–53, 2021.
https://doi.org/10.1007/978-981-16-7207-1_5

segmenting retinal blood vessels takes a long time, and is easily affected by the doctor's subjective influence. When doctors are tired, they often make wrong judgments. Therefore, it is crucial to avoid the direct intervention of doctors to automatically perform retinal vessel segmentation [3].

Fundus images usually include the vascular tree, fovea, retina and optic disc [4]. The analysis of fundus images is very important for detecting early ocular diseases. Usually, doctors segment blood vessels from color fundus images, observe related fundus diseases and make corresponding diagnoses. However, since the color of the retinal blood vessels is very similar to the background color of the retina, it is difficult for us to achieve accurate segmentation. In order to use machine learning algorithms to achieve accurate automatic segmentation, many studies have been conducted, and finally it is found that using deep learning methods to segment retinal images achieves better performance. Deep learning is a neural network with a deeper perceptron layer [5]. The use of deep learning methods has excellent performance in image segmentation. For example, the full convolutional network (FCN) [6], which has outstanding performance in semantic segmentation, can solve the pixel-level classification problem. Another classic segmentation network is U-Net [7]. U-Net uses a symmetrical encoding and decoding structure and skip connections to achieve accurate image segmentation. The encoder is responsible for extracting the characteristic information of the input image, and the decoder is responsible for reconstructing the image and restoring the details of the image.

For retinal vessel segmentation, many methods have been proposed. Soares et al. [8] proposed a network to automatically segment the retinal vessels. This network used two-dimensional Gabor wavelet and pixel's intensity to obtain feature vectors. According to the feature vector of the pixel, the pixels in the fundus image were classified as vascular or non-vascular pixels to generate a segmentation result. Maninis et al. [9] provided a network to analyze both retinal blood vessels and optic discs segmentation. Compared to the second human annotator, which served as control, the network showed results that were more consistent with the gold standard. Orlando et al. [10] proposed a model to realize the processing of slender vessel structures. They used support vector machines (SVM) to learn the parameters in the network. Son et al. [11] proposed a method to generate the retinal vascular probability map using a generative adversarial network. Their method was based on the Keras framework. In [12] a method of combining CNN with fully CRFs was proposed, using an improved cross-entropy loss function to develop multi-scale CNN. In [13] a multiscale network with short connections was proposed. The segmentation performance of the network could be well improved by learning multi-scale features.

Although the retinal blood vessel segmentation method based on convolutional neural network has achieved better results than traditional methods to a certain extent, there are still some problems. For example, because the fundus image is complex, the blood vessel area is easily affected by other diseased areas, which leads to insufficient feature extraction ability of the encoder, unable to extract effective features, in addition, there are few samples used to train neural networks, which can easily lead to overfitting problems. In view of the above problems, a new type of blood vessel segmentation network called Patch-GAN is developed based on SUD-GAN [14]. The method has the following characteristics.

(1) Inspired by the residual network, each sub-module of U-Net is replaced in the form of residual connection. Short connections are added to the convolution block, which helps back propagation during training and can greatly improve the segmentation effect of the generator.
(2) Dense connection blocks are added into the middle of discriminator, which can improve the discriminating ability of the discriminator, and at the same time can better realize the selection of features and guide the generation of the generator.
(3) During training, the choice of loss function is very important, it can measure the quality of model prediction. Therefore, the WGAN-GP [15] is selected as the loss function in this paper, which can better train the model. The input image is divided into 128 $\times$ 128 small patches by random cropping, and these small blocks are trained. The training based on random clipping can create more training images, and it can effectively avoid the occurrence of overfitting. Finally, the experimental results indicate that the segmentation results of Patch-GAN are better than most existing techniques.

The main work of this paper is as follows. The second part discusses the method proposed in this paper. The third part compares Patch-GAN with previous studies. Finally, a conclusion is given.

2 The Proposed Patch-GAN

2.1 Network Architecture

The overall structure of Patch-GAN is shown in Fig. 1. The Patch-GAN is inspired by the idea of generative adversarial network (GAN) [16]. GAN is proposed by Goodfellow et al., which is the most popular research field of artificial intelligence in the past few years. It includes two parts. One of which is a generator, which mainly accomplishes the subject part of the task, and the other is a discriminator, which is used to evaluate the quality of the generator. The generator tries to trick the discriminator to get a higher score, while the discriminator tries its best to separate the generated images from the ground truth. The two networks train against each other and become stronger at the same time. In this paper, the generator segmented blood vessels from a given retinal image, while the discriminator needs to determine whether the input image comes from a probability map or the ground truth. In the training stage, the lower sampling layer in the generator is used to extract fundus features first, and then the segmented vessel is obtained through up-sampling. The ground truth and the samples segmented by the generator are input into the discriminator of the network together, the discriminator extracts the features of the input image through the deep convolution network to realize the discrimination of true and false samples.

In the generator, the goal is to segment into image that are as similar to ground truth as possible, and a symmetrical encoding and decoding structure is applied in the network to segment the input small block. The encoder corresponds to the image down-sampling process, and the decoder corresponds to the feature map up-sampling process, and there are jump connections between the corresponding encoder and the decoder. These jump connections can help the up-sampling layer to restore the detailed information of the image. The encoder extracts the feature information of the image through a typical convolutional neural network structure. In this paper, a total of 4 down-sampling is performed in the encoding part, and the down-sampling part uses 4 convolution blocks to extract the abstract features of the retinal fundus image. Each convolution block has two 3×3 convolutional layers, two instance normalization (IN) layers, and two Leaky-ReLU activation functions. The use of the Leaky-ReLU layer can prevent the gradient from disappearing. An improved short connection structure shown in Fig. 2 is added between the IN layers in each convolution block of the encoder, which can effectively improve the gradient disappearance or gradient explosion caused by the multilayer network. After each pooling operation, the feature map size decreases and the number of channels doubles. The decoder performs up-sampling to gradually restore image information. Corresponding to the encoder part, the decoder part has performed up-sampling 4 times. Each up-sampling will expand the size of the feature map and reduce the number of input channels by half. By concatenating the corresponding feature maps of the encoder and the decoder, a shallow network can be used to save better detailed position information to assist the segmentation. Finally, the segmentation result is obtained through the sigmoid activation function.

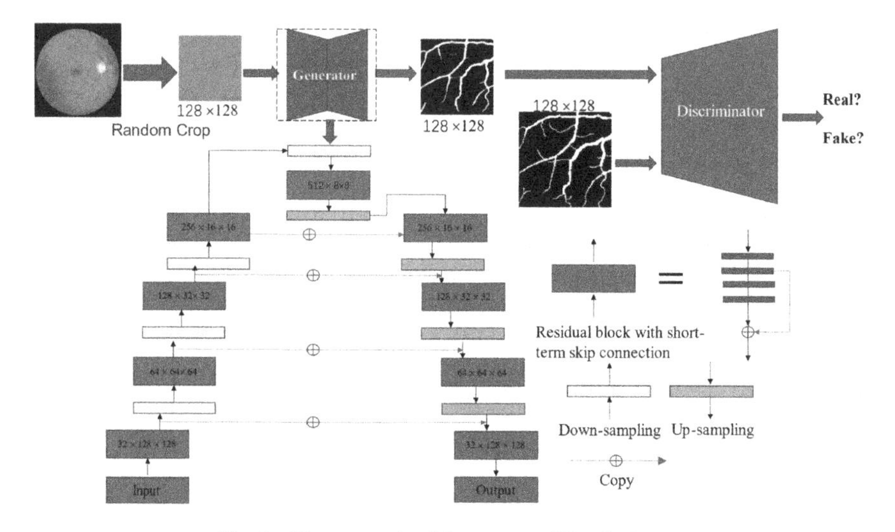

Fig. 1. The network of the proposed Patch-GAN.

In the discriminator part, the architecture of the discriminator is shown in Fig. 3, which consists of two dense connection blocks, three convolutional blocks and two compression layers. The features of the input images are extracted through the convolutional blocks and dense connection blocks. There are three IN-Leaky ReLU-Conv layers in each dense connection block. The dense connection block structure is shown in Fig. 4, in simple terms, the input of each layer comes from the output of all previous layers. If the output of the n_{th} layer neural network is x_n, then the output of the dense connection at the n_{th} layer can be expressed as:

$$y_n = H_n([x_0 \ldots, x_{n-1}]). \tag{1}$$

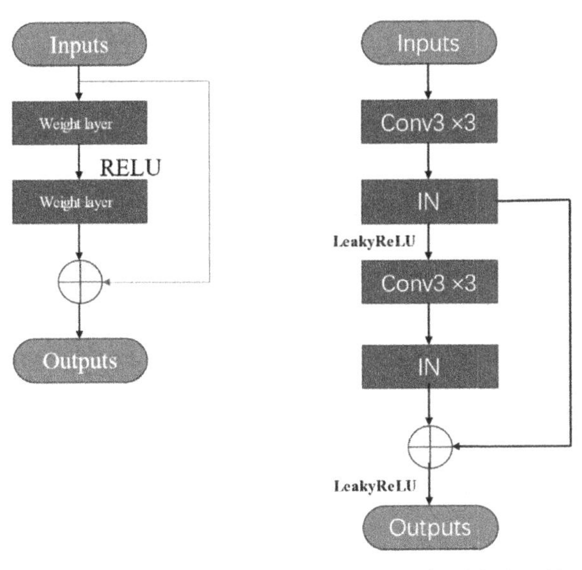

Fig. 2. Original residual architecture and improved residual architecture.

where $[x_0 \ldots, x_{n-1}]$ represents that the output feature map from layer 0 to $n - 1$ is stitched together. The 1×1 convolution layer is added into the discriminator to reduce the number of input feature layers, which not only reduces the dimension of calculation, but also integrates the features of each channel. Finally, the fused features are passed through two multi-layer convolutional layers to obtain the probability prediction of the output.

Fig. 3. Discriminator architecture.

2.2 Loss Function

WGAN-GP loss function is used in this paper. The WGAN-GP loss function adds a gradient penalty term on the basis of the WGAN [17] loss function. The introduced

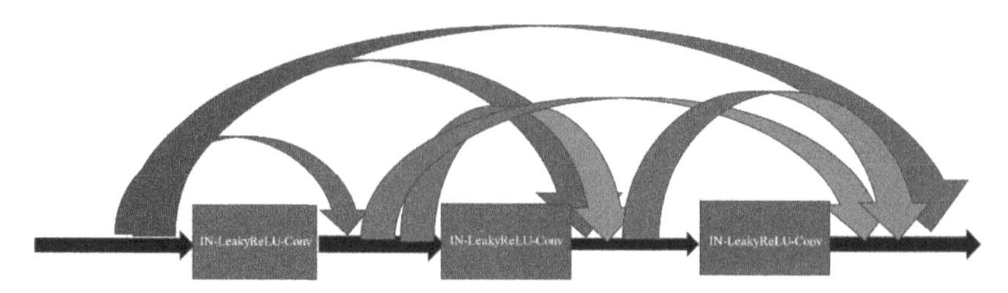

Fig. 4. Dense connection blocks.

gradient penalty term improves the disadvantages of gradient explosion while ensuring that the gradient does not disappear, so that the generator can update the parameters smoothly. The overall Patch-GAN loss function is defined as:

$$L(G, D) = E_{x\sim P_{data(x)}}[D(G(x))] - E_{y\sim P_{data(y)}}[D(y)] + \lambda E_{\hat{x}\sim P_{\hat{x}}}\left[\left(\left\| \nabla_{\hat{x}} D(\hat{x})\right\|_2 - 1\right)^2\right]. \quad (2)$$

where x represents the input retinal fundus image, and y stands for the ground truth. The value of λ is set to 10. $\hat{x}$ means random interpolation sampling on the line of y and $G(x)$, that is, $\hat{x}$ is defined by:

$$\hat{x} = \in (y) + (1- \in)(G(x)). \quad (3)$$

where ϵ is between 0 and 1. GAN framework solves the optimization problem of formula (4) by adversarial training of generator and discriminator.

$$\min_G \left[\max_D E_{x\sim P_{data(x)}}[D(G(x))] - E_{y\sim P_{data(y)}}[D(y)] + \lambda E_{\hat{x}\sim P_{\hat{x}}}\left[\left(\left\| \nabla_{\hat{x}} D(\hat{x})\right\|_2 - 1\right)^2\right]\right]. \quad (4)$$

In order to make D judge more accurately, D(y) should be maximized and D(G(x)) should be minimized. At the same time, the generator should prevent the discriminator from making correct judgments by generating new fake samples. Therefore, the final loss function is defined as the target minimax value. The loss function of the discriminator can be expressed as formula (5). The loss in the generator includes the pixel-level loss between the generated vessel tree and ground truth and the loss against the discriminator. L_1 loss is chosen as pixel-level loss. The loss of the generator is defined as formula (6).

$$\min_{\theta_D} L_{GAN}(D) = E_{x\sim P_{data(x)}}[D(G(x))] - E_{y\sim P_{data(y)}}[D(y)] + \lambda E_{\hat{x}\sim P_{\hat{x}}}\left[\left(\left\| \nabla_{\hat{x}} D(\hat{x})\right\|_2 - 1\right)^2\right]. \quad (5)$$

$$\min_{\theta_G}\left[\lambda L_1 - E_{x\sim P_{data(x)}}[D(G(x))]\right]. \quad (6)$$

λ is a hyperparameter that can be used to balance the two losses, which is set at 10 in the experiment.

3 Experimental Results

3.1 Datasets

There are 40 retinal images with 565 × 584 pixels in the DRIVE dataset. In the training stage, 20 pictures are selected for training and the rest for testing. The STARE dataset has 20 retinal fundus images with 605 × 700 pixels. In this paper, ten of them are selected for training and the rest for testing. The first expert's annotation serves as the ground truth. Since there are few pictures in the training set, data enhancement is used to expand the dataset in order to prevent overfitting. Mainly by randomly cropping the input image, and then randomly flipping the cropped image. The training epoch is set to 20,000 with a mini-batch of 1, and the learning rate is set to 0.0002. The experiment is test on a computer equipped with GeForce RTX 3090 GPU. The experiment is based on Anconda+Pytorch and the training time is about 5 h.

3.2 Evaluation Criteria

In order to verify the effectiveness and generalization of Patch-GAN on the task of retinal vessel segmentation, two publicly available retinal image datasets are used for comparative experiments. The Otsu in [18] is used to binarize the probability map and compare the probability map with the ground truth. In this paper, five evaluation indexes are used to evaluate the effectiveness of the proposed retinal vessel segmentation method. They are Specificity (Sp), Precision (Pr), Accuracy (Acc), Sensitivity (Se), and F_1 score.

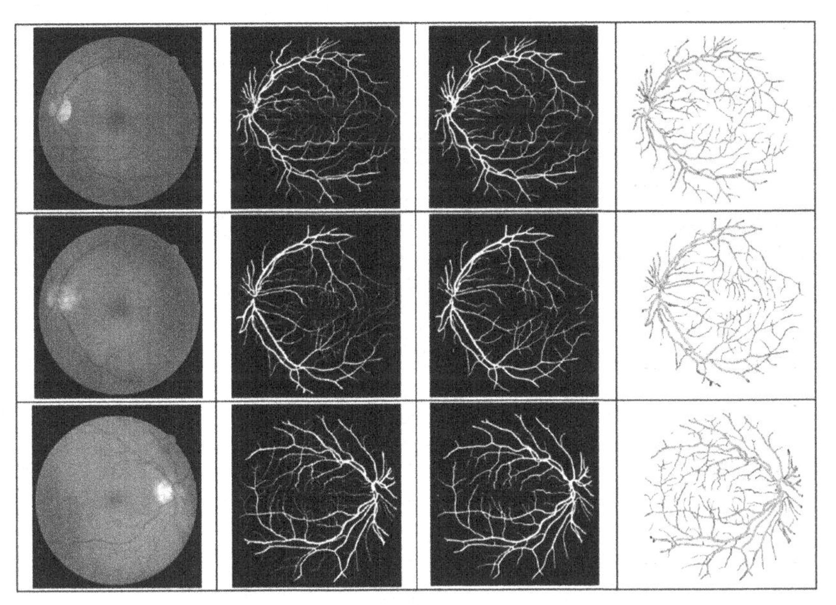

Fig. 5. Some results on the DRIVE dataset. From left to right: the original retinal images, ground truths, results of segmentation and visualization results. (Color figure online)

The evaluation criteria are given by the following formulas.

$$A_{CC} = \frac{TP + TN}{TP + FP + FN + TN}. \quad (7)$$

$$S_e = \frac{TP}{FN + TP}. \quad (8)$$

$$S_P = \frac{TN}{TN + FP}. \quad (9)$$

$$P_r = \frac{TP}{TP + FP}. \quad (10)$$

$$F_1 = 2 \times \frac{P_r \times S_e}{P_r + S_e}. \quad (11)$$

Where the number of pixels correctly marked as blood vessels is called true positive (TP), the number of pixels incorrectly marked as blood vessels is called false positive (FP), the number of pixels incorrectly marked as background is called false negative (FN), the number of pixels that are correctly marked as background is called true negative (TN). Some previous research results are compared with the proposed Patch-GAN. Table 1 describes the comparison of various segmentation methods using the DRIVE dataset. From Table 1, we can see that the average accuracy of segmentation results in the DRIVE dataset is 0.9658, which is 0.98% higher than SUD-GAN. Table 2 indicates the comparison results of the STARE dataset. As shown in Table 2 that the Patch-GAN

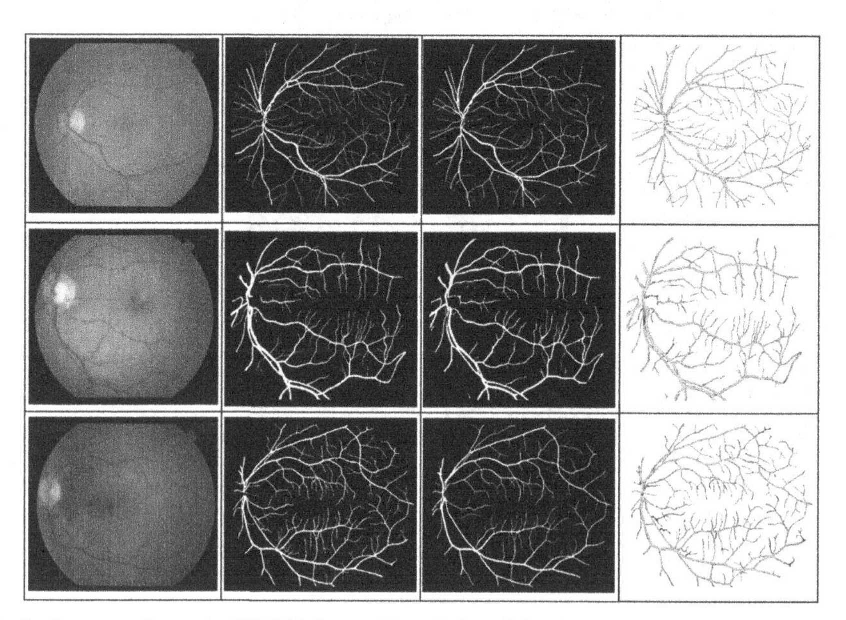

Fig. 6. Some results on the STARE dataset. From left to right: the original retinal images, ground truths, results of segmentation and visualization results. (Color figure online)

reaches the highest F_1 score on the STARE dataset. The average accuracy is 0.9745. We can see it is 0.82% higher than SUD-GAN. The segmentation results of the DRIVE and STARE datasets are shown in Fig. 5 and 6. Where green, white, red and blue represent TP, TN, FP and FN, respectively. The figures show that the retinal blood vessels segmented by the Patch-GAN method have few red and blue areas, indicating that there are few false positives and false negatives. In addition, there are smoother blood vessels and blood vessel edges, The segmentation results are also closer to the actual annotation results of the experts.

Table 1. Comparison results of different methods on DRIVE dataset.

Methods	A_{cc}	S_e	S_p	P_r	F_1
Soares [8]	0.9524	0.5258	0.9933	0.8829	0.6591
Maninis [9]	0.9411	0.9477	0.9405	0.6046	0.7382
Orlando [10]	–	0.7897	0.9684	-	-
Son [11]	0.9703	0.8014	0.9865	0.8509	0.8254
Hu [12]	0.9632	0.7543	0.9814	–	–
Guo [13]	0.9551	0.7800	0.9806	–	–
Yang [14]	0.9560	0.8340	0.9820	–	–
Patch-GAN	0.9658	0.8154	0.9806	0.8054	0.8104

Table 2. Comparison results of different methods on STARE dataset.

Methods	A_{cc}	S_e	S_p	P_r	F_1
Soares [8]	0.9659	0.6850	0.9892	0.8409	0.7550
Maninis [9]	0.9562	0.9403	0.9575	0.6474	0.7668
Orlando [10]	–	0.7680	0.9738	–	–
Son [11]	0.9754	0.7841	0.9913	0.8823	0.8303
Hu [12]	0.9632	0.7543	0.9814	–	–
Guo [13]	0.9660	0.8201	0.9828	–	–
Yang [14]	0.9663	0.8334	0.9897	–	–
Patch-GAN	0.9745	0.8126	0.9885	0.8600	0.8356

4 Conclusion

By checking the blood vessels in the color fundus images, many diseases can be reflected, such as diabetes, hypertension, etc. Automatically and accurately segmenting the blood vessel regions can assist doctors in disease diagnosis. However, the background color of the retina and the diseased area will affect the segmentation of the blood vessel area. The retinal vessel segmentation network called Patch-GAN proposed in this paper can achieve end-to-end segmentation of retinal vessels. A generator with a short connection structure is used and a discriminator with a deep dense connection to achieve precise segmentation of retinal vessels through adversarial training. In particular, the addition of the WGAN-GP loss function can help better optimize network parameters, and avoids the occurrence of over-fitting by randomly cropping and flipping the input image. Comparative experiments on the two datasets have achieved good results, verifying the effectiveness and generalization of the method in this paper. However, the small blood vessels segmented by Patch-GAN are not very good. Therefore, in the next work, we can introduce some of the latest loss functions and preprocessing methods to focus on the segmentation of tiny blood vessels.

References

1. Kocur, I., Resnikoff, S.: Visual impairment and blindness in Europe and their prevention. Brit. J. Ophthalmol. **86**(7), 716–722 (2002)
2. Wang, S.B., et al.: A spectrum of retinal vasculature measures and coronary artery disease. Atherosclerosis **268**, 215–224 (2018)
3. Abràmoff, M.D., Garvin, M.K., Sonka, M.: Retinal imaging and image analysis. IEEE Rev. Biomed. Eng. **3**, 169–208 (2010)
4. Fraz, M.M., et al.: An ensemble classification-based approach applied to retinal blood vessel segmentation. IEEE Trans. Biomed. Eng. **59**(9), 2538–2548 (2012)
5. Awad, M., Khanna, R.: Deep learning. In: Efficient Learning Machines, pp. 167–184. Apress, Berkeley, CA (2015). https://doi.org/10.1007/978-1-4302-5990-9_9
6. Long, J., Shelhamer, E., Darrell, T.: Fully convolutional networks for semantic segmentation. In: Proceeding CVPR, pp. 3431−3440. IEEE Press (2015)
7. Ronneberger, O., Fischer, P., Brox, T.: U-Net: convolutional networks for biomedical image segmentation. In: Navab, N., Hornegger, J., Wells, W.M., Frangi, A.F. (eds.) MICCAI 2015. LNCS, vol. 9351, pp. 234–241. Springer, Cham (2015). https://doi.org/10.1007/978-3-319-24574-4_28
8. Soares, J.V.B., Leandro, J.J.G., Cesar, R.M., Jelinek, H.F., Cree, M.J.: Retinal vessel segmentation using the 2-D Gabor wavelet and supervised classification. IEEE Trans. Med. Imaging **25**(9), 1214–1222 (2006)
9. Maninis, K.-K., Pont-Tuset, J., Arbeláez, P., Van Gool, L.: Deep retinal image understanding. In: Ourselin, S., Joskowicz, L., Sabuncu, M.R., Unal, G., Wells, W. (eds.) MICCAI 2016. LNCS, vol. 9901, pp. 140–148. Springer, Cham (2016). https://doi.org/10.1007/978-3-319-46723-8_17
10. Orlando, J.I., Prokofyeva, E., Blaschko, M.B.: A discriminatively trained fully connected conditional random field model for blood vessel segmentation in fundus images. IEEE Trans. Biomed. Eng. **64**, 16–27 (2017)
11. Son, J., Park, S.J., Jung, K.H.: Retinal Vessel Segmentation in Fundoscopic Images with Generative Adversarial Networks (2017). http://arxiv.org/abs/1706.09318
12. Hu, K., et al.: Retinal vessel segmentation of color fundus images using multiscale convolutional neural network with an improved cross-entropy loss function. Neurocomputing **309**, 179–191 (2018)
13. Guo, S., Wang, K., Kang, H., Zhang, Y.J., Gao, Y.Q., Li, T.: BTS-DSN: deeply supervised neural network with short connections for retinal vessel segmentation. J. Med. Inform. **126**, 105–113 (2019)

14. Yang, T.J., Wu, T.T., Li, L., Zhu, C.H.: SUD-GAN: deep convolution generative adversarial network combined with short connection and dense block for retinal vessel segmentation. J. Digit Imaging **33**(4), 946–957 (2020)
15. Gulrajani, I., Ahmed, F., Arjovsky, M., Dumoulin, V., Courville, A.: Improved training of Wasserstein GANs. In: Advances In Neural Information Processing Systems 30 (NIPS 2017) (2017)
16. Goodfellow, I.J., et al.: Generative adversarial nets. In: Advances In Neural Information Processing Systems 27 (NIPS 2014), pp. 2672−2680 (2014)
17. Arjovsky, M., Chintala, S., Bottou, L.: Wasserstein GAN. arXiv preprint ar Xiv:1701.07875 (2017)
18. Otsu, N.: A threshold selection method from gray-level histograms. IEEE Trans. Syst. Man Cybern. **9**(1), 62–66 (1979)

Research of Medical Image Registration Based on Characteristic Ball Constraint in Conformal Geometric Algebra

Tianyu Cheng[1], Juping Gu[1,2](✉), Liang Hua[2], Jianhong Zhu[2], Fengshen Zhao[2], and Yong Cao[2]

[1] Nanjing University of Science and Technology Nanjing, Jiangsu 210094, China
gu.jp@ntu.edu.cn

[2] School of Electrical Engineering, Nantong University Nantong, Jiangsu, China

Abstract. Multimodal 3D medical images have the characteristics of huge operational data and multi degrees of freedom in geometric transformation. However, there are some difficulties in multimodal medical image registration, such as low registration efficiency and speed. In order to meet the clinical needs, a feature sphere constrained registration algorithm based on conformal geometric algebra is proposed in this paper. Firstly, 3D contour point clouds are extracted from multimodal medical images. Then, the spatial conformal sphere is constructed, and the feature points are calculated based on the spatial projection constraint from the point cloud to the conformal sphere. Finally, rotation operators are constructed by feature points to realize fast 3D medical image registration. Experimental results indicate that the registration algorithm based on geometric feature constraints in this paper is more effective for registration of multimodal 3D medical images, with high accuracy, anti-noise ability, rapid calculation, and strong universality.

Keywords: Multimodal medical images · 3D registration · Conformal geometric algebra · Geometric feature constraints

1 Introduction

Multimodal medical image registration is a hot topic in the field of the combination of engineering with medicine, which is of great significance for clinical diagnosis and treatment. According to the imaging content, medical images can be divided into anatomical images (CT, MRI, etc.) and functional images (SPECT, PET, etc.). The former provides anatomical morphological information of human internal organs, while the latter reflects the metabolic function. The registration of the two types of images can provide more abundant lesion information, improve the diagnostic accuracy of doctors, and play a role in guiding clinical surgical treatment.

From the dimension of registration, medical image registration can be divided into 2D registration [1], 2D/3D registration [2] and 3D registration [3]. Compared with 2D registration, 3D skull registration has a great improvement in dimension and space complexity. At this stage, 3D registration has a certain research foundation. For example,

M. Fei et al. (Eds.): LSMS 2021/ICSEE 2021, CCIS 1467, pp. 54–63, 2021.
https://doi.org/10.1007/978-981-16-7207-1_6

the early registration method based on markers [4]. The method of implantable and non-implantable marked points is to complete registration by minimizing the distance of marked points. This method has high registration accuracy, but it requires manual intervention, and it is easy to be disturbed by external factors, and cannot achieve retrospective registration research. The subsequent development of registration methods based on internal features, such as singular value decomposition [5], nearest point iteration [6], B-spline interpolation [7] and other algorithms, have achieved non-invasive registration, but the registration accuracy is not high. In recent years, the introduction of mutual information algorithm [1] is a breakthrough of non-contact registration algorithm, which has the advantages of independent of imaging equipment, strong anti-noise ability and high registration accuracy. However, the way that global voxel information participates in calculations greatly increases the amount of data calculations. This makes the registration time-consuming and poor real-time clinical application. In order to improve the registration speed under the premise of ensuring the registration accuracy, literature [8] proposed a registration method combining geometric algebra and Iterative Closest Point (ICP), which has a corresponding improvement in registration accuracy and speed. References [9] and [10] continue to explore the advantages of geometric algebra in geometric feature registration, which successfully improve the speed of registration, and the registration accuracy is satisfactory.

On the basis of previous research work, this paper deeply studies the key factors that affect the registration speed, and proposes a new registration algorithm based on conformal geometric algebraic characteristic sphere constraints. In the framework of conformal geometric algebra, a free sphere in conformal space is constructed, and the spatial position of the free ball is determined based on the value of local vertical projection. Finally, the registration is completed by using the conformal space sphere to calculate the rotation operators. The process of point gathering in the sphere participating in the construction of invariants weakens the interference of noise. At the same time, the conformal operation has the advantages of reducing the operation process and intuitive geometric meaning. The invariant numerical solution method based on the geometric feature constraints of the outer contour greatly reduces the amount of operational data, which not only improves the registration accuracy and anti-noise ability, but also significantly reduces the registration time consumption to meet the clinical practical needs.

2 Basic Theory and Calculation of Conformal Geometric Algebra

Geometric Algebra, which was first established by Grassmann and Clifford, is a form of algebraic operation with complete geometric meaning. Geometric algebra, also known as Clifford algebra, is considered to be a perfect combination of algebra and geometry. Geometric algebra has developed many branches so far. The conformal geometric algebra (Conformal geometric algebra, CGA) created by Li Hongbo is a new type of Clifford algebra in five-dimensional space. Its geometric expression which is not constrained by spatial coordinates can better represent the geometric characteristics of geometry [11]. The algorithm of this paper is also carried out under this framework. This section mainly introduces the calculation and related theories involved in this paper. More knowledge about Geometric Algebra can be found in reference [12].

Inner product, outer product and geometric product are the core of geometric representation and operation in supporting geometric algebraic space. Inner product is a kind of dimensionality reduction operation, which can express geometric topological attributes such as distance and angle between multi-dimensional vectors. Outer product is a kind of dimension enhancement operation, which can be used for dimension expansion and shape construction. Geometric product is a mixed operation of joint inner product and outer product, which is similar to the combination of real and imaginary parts in complex numbers. For example, the geometric product expression of multiple vectors $A=\sum_{i=1}^{n}\langle A\rangle_i$ and $B=\sum_{i=1}^{m}\langle B\rangle_i$ is

$$AB = A \cdot B + A \wedge B. \tag{1}$$

Based on the geometric product, this paper briefly describes the geometric inverse of multiple vectors, the geometric inverse, the rotation operator and the geometric representation of conformal space [11, 12].

a. Geometric inversion

The geometric inverse is equivalent to the reverse sort, marked as $A^{\dagger}$. If $A = a_1 \wedge a_2 \wedge \cdots \wedge a_n$, then $A^{\dagger} = a_n \wedge a_{n-1} \wedge \cdots \wedge a_1$. The geometric inverse of the multiple vector $A=\sum_{i=1}^{n}\langle A\rangle_i$ is expressed as:

$$A^{\dagger} = \sum_{i=0}^{n} (-1)^{i(i-1)/2} \langle A\rangle_i. \tag{2}$$

b. Geometric inverse

The geometric inverse is the inverse operation of the geometric product, denoted as A^{-1}. Its calculation formula is

$$A^{-1} = \frac{A^{\dagger}}{\|A\|^2}. \tag{3}$$

c. Rotation operator

The rotation operator can be seen as a vector that performs two reflection operations, as shown in Fig. 1.

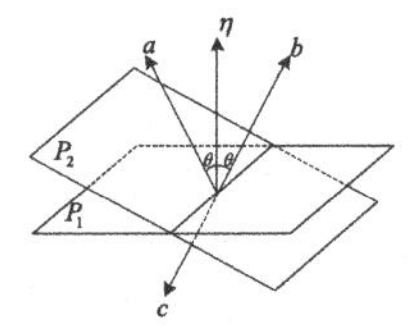

Fig. 1. The geometric meaning of rotation operator.

In the figure above, if you want to rotate vector a to vector b, you can first construct the angular bisector η of vectors a and b. Then, vector a is reflected by the vertical plane

P_1 of η to get vector c. Based on the geometric relation, it can be demonstrated that the vector c is also perpendicular to the vertical plane P_2 of vector b. So the vector c can be reflected by P_2 to get vector b. This means that vector a rotates to the position of vector b, after two reflection motions. The above description is expressed as:

$$b = -b(-hah)b = bhahb = TaT^{-1}. \quad (4)$$

Where $\eta = \frac{a+b}{\|a+b\|}$. The rotation operator can realize the overall transformation of geometry, which is very suitable for the motion transformation of multi-dimensional geometry.

d. Geometric representation of conformal space

In Conformal Geometric Algebraic space, the representation of geometry is shown in the following Table 1.

Table 1. The expression of basic geometry in conformal geometric algebra space.

Geometry	Standard form	Direct expression	Order
Point	$X = x + \frac{1}{2}x^2 e_\infty + e_0$	$S^* = X_1 \wedge X_2 \wedge X_3 \wedge X_4$	4
Sphere	$S = c + \frac{1}{2}(c^2 - r^2)e_\infty + e_0$	$X^* = (-I_M x - \frac{1}{2}x^2 e_\infty + e_0)$	4
Line	$L = \pi_1 \wedge \pi_2$	$L^* = P_1 \wedge P_2 \wedge e_\infty$	3
Plane	$\pi = n + de_\infty$	$\pi^* = P_1 \wedge P_2 \wedge P_3 \wedge e_\infty$	4

3 Geometric Feature Constraints Based on Conformal Geometric Algebraic Space

Although there are differences in intracranial tissue images in multimodal medical imaging, there is still a high similarity in the outer contours of the skull protected by the skull. As shown in Fig. 2, the geometric features of the face are more prominent. The "concave spot" of the bridge of the nose and the corner of the eye are the most prominent. In view of this characteristic, a facial feature point location method based on local feature constraints of conformal geometric algebra is proposed in this paper.

The point cloud data X_i of facial contours is expressed in conformal geometric algebraic space as:

$$X_i = x_i e_1 + y_i e_2 + z_i e_3 + \frac{1}{2} w_i e_\infty + e_0 \quad \left(w = \sqrt{x_i^2 + y_i^2 + z_i^2}\right). \quad (5)$$

First of all, a free plane P_i passing through the point cloud X_i is constructed, which is expressed as:

$$P_i = v_i^P + d_i e_\infty. \quad (6)$$

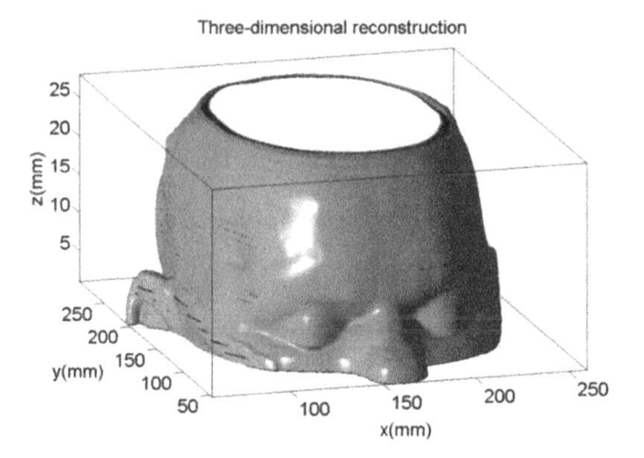

Fig. 2. 3D reconstruction of skull medical image.

Where v_i^P is the unit normal vector of plane P_i in conformal space. d_i is the projection distance from the plane to the origin. Then, a sphere S_i with radius r is constructed by taking the point cloud X_i as the center of the sphere. The point cloud set that falls inside S is marked as $\{X_{i,j}\}_{j=1}^{n_i}$. The idea of spatial geometric constraint is that when the sum of the projection distance from the point cloud set $\{X_{i,j}\}_{j=1}^{n_i}$ to the plane P_i is the minimum, the angle P_i of the normal vector v_i is determined. The above process is described with a formula below.

In the framework of conformal geometric algebra, the projection metric from point to plane can be expressed by inner product. Therefore, the norm form of the sum of the projection distance from the point cloud set $\{X_{i,j}\}_{j=1}^{n_i}$ to the plane P_i is expressed as:

$$D(P_i) = \sum_{j=1}^{n_i} \left\| X_{i,j} \cdot P_i \right\|^2$$
$$= \sum_{j=1}^{n_i} \left\| (x_j e_1 + y_j e_2 + z_j e_3 + \frac{1}{2}\sqrt{x_j^2 + y_j^2 + z_j^2} e_\infty + e_0) \cdot (x_i^P e_1 + y_i^P e_2 + z_i^P e_3 + d_i e_\infty) \right\|^2. \tag{7}$$

When solving the normal vector of each P_i, X_i is translated to the origin. This means that $d_i = 0$, and (x_j, y_j, z_j) turns into $(x_j - x_i, y_j - y_i, z_j - z_i)$. Equation (7) can be expressed as:

$$\begin{aligned} D(P_i) &= \sum_{j=1}^{n_i} \left\| X_j \cdot v_i^P \right\|^2 \\ &= \sum_{j=1}^{n_i} \left[(x_j - x_i)x_i^P + (y_j - y_i)y_i^P + (z_j - z_i)z_i^P \right]^2 \\ &= \sum_{j=1}^{n_i} (x_j - x_i)^2 \left(x_i^P\right)^2 + (y_j - y_i)^2 \left(y_i^P\right)^2 + (z_j - z_i)^2 \left(z_i^P\right)^2 \\ &+ 2(x_j - x_i)(y_j - y_i)x_i^P y_i^P \\ &+ 2(x_j - x_i)(z_j - z_i)x_i^P z_i^P + 2(y_j - y_i)(z_j - z_i)y_i^P z_i^P \end{aligned} \tag{8}$$

Obviously, the above equation is an optimization problem for calculating the maximum value of the objective function. Here, a Lagrangian function with $\varphi(v_i^P) = \|v_i^P\|^2 - 1 = 0$ as a single constraint is constructed to solve the optimization problem, which is expressed as:

$$L_1(V_i, \lambda) = D(v_i^P) + \lambda_i \varphi(v_i^P). \tag{9}$$

Next, the partial derivatives of unknowns $x_i^P, y_i^P, z_i^P, \lambda_i$ are calculated respectively. After constructing the simultaneous equations, the solution of each unknown number can be obtained. Finally, by substituting the $v_{i,\min}^P$ value into equation (8), the minimum value of the projection can be calculated.

In order to reduce computational redundancy, the algorithm only needs to calculate the projection maximum $D_{i,\min}$ of facial point cloud in reference mode and floating mode. The calculated $\{D_{i,\min}\}_{i=1}^n$ values are arranged from large to small, in which the point with the largest value is the facial feature point. Verified by a large number of experiments, the maximum value of more than 100 feature points are concentrated in the "concave point" of the bridge of the nose. Here, the algorithm takes the first 50 feature points to form a feature set, and calculates the centroid of the set to finally locate the feature points. This calculation method can also enhance the robustness of the location algorithm.

After aligning the feature points of the floating mode and the reference mode, the facial orientation still needs to be corrected to make up for the angle deviation during imaging. The method of this paper is to use the feature points to search down the outer contour points and locate the straight line of the bridge of the nose. Then, the direction vector v^r and v^f of the two-mode straight line is calculated, and the rotation operator T_1 is constructed. The rotation operator is expressed as:

$$T_1 = v^r \overline{\eta}. \tag{10}$$

Where N is the unit angle bisector vector of A and B, namely:

$$\overline{\eta} = \frac{v^r + v^f}{\|v^r + v^f\|}. \tag{11}$$

The expression of floating mode point cluster $\{v_i^f\}_{i=1}^n$ rotated by T_1 is

$$v_i^r = T_1 v_i^f T_1^\dagger. \tag{12}$$

The "sandwich" operation structure of rotation operator can realize coordinate-independent rotation, translation and stretch transformation, and reconstruct the motion process of spatial geometry. It can realize the overall transformation of the point cluster under the premise that the structure and characteristics of the geometry remain unchanged [11]. After the rotation operator space correction, the registration is completed.

4 Experiment and Result Analysis

4.1 Data Source

The multi-modal medical image data used in this article comes from the "Retrospective Image Registration Evaluation Project (RIRE)" of Vanderbilt University. RIRE is a network database and provides a registration evaluation platform. Researchers can upload registration results to compare with its "gold standard", which is a more objective evaluation standard of algorithm accuracy and stability.

The cranial image data of 7 patients were selected. Each patient included 1 set of CT data and 4 sets of MR data (PD, PD_rectified, T1 and T1_rectified). In the experiment, CT is used as the floating image and MR as the reference image.

4.2 Image Preprocessing

Image preprocessing is an important part of feature registration. First of all, the preprocessing needs to unify the 3D spatial resolution according to the image specifications provided by RIRE data. Taking the CT image with 29 slices and the MR_PD image with 26 slices as examples, the comparison of the spatial resolution is as follows Table 2:

Table 2. Uniform spatial resolution.

Modal	Data scale			Spatial resolution (mm)			Total number of point clouds
	x	y	z	x	y	z	
CT	512	512	29	0.65	0.65	4.0	64827
PD	256	256	26	1.25	1.25	4.0	54793
CT-correct	334	334	116	1.0	1.0	1.0	62592
PD-correct	320	320	104	1.0	1.0	1.0	53664

The core of registration is based on the geometric features of the skull contour, so it is important to segment the external contours of the skull effectively and stably. The Canny algorithm based on the optimal edge detection algorithm is used in this paper, and its accuracy and efficiency can meet the requirements of this algorithm.

4.3 Locating Feature Points

As shown in Fig. 3, the set of marker points in the figure are the feature points calculated by this algorithm. The feature points spread from inside to outside, and taking the first 50 maximum values in the set of points can ensure the stability of the algorithm. At the same time, the anti-noise ability of the algorithm is enhanced by the neighborhood point cloud participating in the operation, which is also the guarantee of location accuracy.

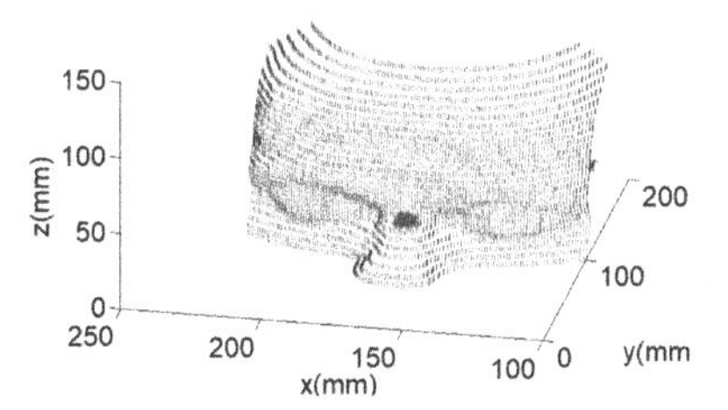

Fig. 3. Calculation of feature points.

In order to verify the validity of the feature point, the coordinates of the feature point are marked by hand, and then the coordinates of the "concave point" are located by the algorithm. Then, the positioning error is calculated according to the following formula:

$$E = \frac{1}{n}\sum_{i}^{n} \|p_i - p_s\|^2. \tag{13}$$

Where n is the total number of samples. Next, 35 sets of 3D image data from 7 patients are taken as samples. In order to simulate the possible situation of imaging, the 3D image is rotated by 0°, 15°and −15°. The experimental results are shown in Table 3.

Table 3. Average error of feature point location (mm).

Sample	Patient_001	Patient_002	Patient_003	Patient_005	Patient_006	Patient_007	Patient_009
0°	2.71	0.93	5.12	2.95	2.97	1.72	2.43
15°	2.71	1.01	5.14	2.95	2.98	1.74	2.43
−15°	2.69	1.02	5.14	2.95	2.98	1.72	2.43

As can be seen from Table 3, the feature point location algorithm in this paper has high accuracy. The algorithm benefits from the superiority of the constructed feature sphere and can avoid the potential error caused by rotation. At the same time, participating in the operation in the form of neighborhood point cloud can weaken the influence of special noise and enhance the robustness of the algorithm. In the feature extraction step, the radius of the sphere is based on the ratio of the eye to the whole face. The experimental results show that the effect of feature point location is better when the radius is 20 (mm).

4.4 Registration Result

Next, 1 set of CT and 4 sets of MR images of 7 patients were selected for registration. The result of registration was uploaded to RIRE database for evaluation. The experimental results were compared with the mutual information algorithm and the ICP algorithm. The results are shown in Table 4.

As can be seen from the above table, the registration accuracy of this algorithm is significantly improved compared with the ICP algorithm. And the accuracy level of this method is similar to that of the mutual information algorithm.

Table 4. Error evaluation and comparison.

Registration method	Average error/mm	Median error/mm	Maximum error/mm
ICP algorithm	2.75	4.22	6.58
Mutual information	1.40	1.53	3.72
The proposed method	2.61	1.83	3.79

4.5 Registration Speed

Here is a comparison of registration time, as shown in Table 5. This algorithm is written on Matlab R2010a, the computer configuration is: Intel(R) Core (TM) i7-6700HQ, CPU 2.60GHz, memory 12GB, Win10 64-bit operating system.

Table 5. Comparison of algorithm speed.

Registration method	Average time (s)
ICP algorithm	57.5
Mutual information	1470
The proposed method	2.5

As can be seen from the above table, compared with the mutual information algorithm, the timing consumption of this algorithm is greatly reduced, and the problem of low registration efficiency is improved. This algorithm is more time-saving than the nearest point iterative algorithm, at the same time the registration accuracy is higher, and the algorithm is more robust.

There are three main factors for the high speed of registration: one is that the conformal operation reduces the amount of calculation, and the traditional Euclidean distance "point-to-point" operation is transformed into "point-line" operation. The second is that the rotation operator has higher computational efficiency than the rotation matrix. The 3D point cloud is rotated by 15°, and the time consumption of the rotation matrix and the rotation operator is compared as shown in Table 6.

Table 6. The efficiency comparison of rotation operator and rotation matrix.

Method	Average time consuming (s)
Rotation matrix	89.5
Rotation operator	57.3

The third important factor of low time consumption is that it only uses the geometric characteristics of external contour point cloud, which greatly reduces the amount of data and the scale of operation.

5 Conclusion

In this paper, under the framework of conformal geometric algebra, a multimodal 3D medical registration algorithm based on conformal sphere spatial constraints is proposed. On the premise of ensuring the accuracy of registration, this algorithm can greatly reduce the time consumption of registration and meet the requirements of clinical diagnosis and treatment. This is mainly due to the advantages of geometric algebraic reduction calculation and geometric feature constraints, and also proves that it has the potential to study and popularize in spatial geometry and the construction of feature invariants.

References

1. Plattard, D., Soret, M., Troccaz, J., et al.: Patient set-up using portal images: 2D/2D image registration using mutual information. Comput. Aided Surg. **5**(4), 246–262 (2015)
2. Miao, S., Wang, Z.J., Rui, L.: A CNN regression approach for real-time 2D/3D registration. IEEE Trans. Med. Imaging. **35**(5), 1 (2016)
3. Brockmeyer, D., Gruber, D.P., Haller, J., et al.: High-accuracy 3D image-based registration of endoscopic video to C-arm cone-beam CT for image-guided skull base surgery. Laryngoscope **122**(1), 1925–1932 (2018)
4. Wu, K., Kalvin, A.D., Williamson, B., et al.: Providing visual information to validate 2-D to 3-D registration. Med. Image Anal. **4**(4), 357–374 (2000)
5. Wu, J.: Rigid 3D registration: a simple method free of SVD and eigen-decomposition. IEEE Instrum. Meas. Magn. **69**(10), 8288–8303 (2020)
6. He, Y., Lee, C.H.: An improved ICP registration algorithm by combining PointNet++ and ICP algorithm. In: 2020 6th International Conference on Control, pp. 741–745. ICCAR, Singapore (2020)
7. Bricq, S., Kidane, H.L., Zavala-Bojorquez, J., et al.: Automatic deformable PET/MRI registration for preclinical studies based on B-splines and non-linear intensity transformation. Med. Biol. Eng. Comput. **56**(9), 1531–1539 (2018)
8. Cao, W.M., Liu, H., Xu, C.: 3D medical image registration based on conformal geometric algebra. Sci. Sinica Informationis. **43**(2), 254–274 (2013)
9. Hua, L., Yu, K., Ding, L., et al.: A methodology of three-dimensional medical image registration based on conformal geometric invariant. Math. Probl. Eng. 1–8 (2014)
10. Hua, L., Cheng, T.Y., Gu, J.P., et al.: 3D medical image registration based on Clifford relative invariant and region of interest. J. Graphics **38**(1), 90–96 (2017)
11. Li, H.B.: Conformal geometric algebra for motion and shape description. Comput. Aided Des. Comput. Graphics **18**(7), 895–903 (2006)
12. Sierra, S.J.: Geometric algebras on projective surfaces. J. Algebra **324**(7), 1687–1730 (2010)

Geometric Transformation Image Registration Based on Discrete Feature Point Fitting

Xiaodan Yu[1(✉)], Jianguo Wu[1], and Liang Hua[2]

[1] School of Electrical and Energy Engineering, Nantong Institute of Technology, Nantong, China

[2] College of Electrical Engineering, Nantong University, Nantong, China

Abstract. Image registration is the premise of image fusion, it is important to find a high-precision registration method to make the multi-modal medical images align between the coordinate points of the same particle in the space. Previous researchers studied multi-modal medical image registration, which is based on the three-dimensional contour points and the largest common area of the images. Although the final experimental results are acceptable, part of the original image data is lost. Instead of intercepting the maximum common region, all layers of the two modes are involved in the calculation of the registration algorithm in this paper. The feature points of each layer are extracted, and the feature axis is obtained by linear fitting. The rotation operator is constructed based on this characteristic axis, then the 3D models of the two modes are registered in space.

Keywords: Skull image registration · Clifford algebra · Mass moment of inertia · Geometric feature invariants

1 Introduction

Medical images can be roughly divided into two types according to the different description of human body information: one is the anatomical image used to describe tissue structure and anatomical information, the other is to describe the metabolic information in the body. Cranial image registration technology has many advantages in clinical medicine [1].

Image registration has been born very early, and has a long history in the field of medicine. In foreign countries, it was Anuta P.E. et al. That was the first to study a fast Fourier transform (FFT) method to register different images [2]. Then, the multi-mode image registration based on feature neighborhood mutual information method is proposed by P.A. Legg et al. [3]; The registration method of ICP was proposed by N. Senin et al. [4]. The domestic research on medical image registration started relatively late, but it has also made remarkable achievements. Shuqian Luo et al. Proposed multimodal medical image registration based on maximum mutual information [5]. In recent years, some experts have applied Clifford algebra [6] in the field of information processing and computer graphics. Hui Liu et al. Used Clifford differential and Clifford gradient to calculate the gradient norm of a pixel in a multispectral image to judge whether the

M. Fei et al. (Eds.): LSMS 2021/ICSEE 2021, CCIS 1467, pp. 64–72, 2021.
https://doi.org/10.1007/978-981-16-7207-1_7

pixel is an edge point [7]. Liang Hua et al. applied Clifford algebra to the registration of 3D multimodal medical images, and transferred the classical ICP algorithm to Clifford algebra space for calculation [8].

In this paper, the geometric invariants of each set of images are found and established in the geometric algebra and analytic geometry space by combining the characteristics of the skull contour, the overall structure and the differences of the human posture in the actual image acquisition. High precision translation, rotation, scaling and other operations are implemented for the images to be registered, and the images with fixed modes are registered and proofread.

2 Mathematical Theory and Calculation Method

2.1 Mass Moment of Inertia

Mass moment of inertia refers to the measure of inertia of a rigid body when it rotates around a fixed axis (a unique property, that is, a rotating object keeps its uniform circular motion or static), and is represented by the letter I or J. In classical mechanics, the unit of mass moment of inertia is $Kg \cdot m^2$. For a particle, its mass moment of inertia is $I = mr^2$, where m is the mass of the particle and R is the vertical distance from the particle to the rotation axis.

The expression of mass moment of inertia is as follows:

$$I = \sum_i m_i r_i^2. \tag{1}$$

If the mass of a rigid body is continuously distributed, the formula for calculating the mass moment of inertia can be written as follows:

$$I = \int\int_V r^2 dm = \int\int_V r^2 \rho dV. \tag{2}$$

2.2 Geometric Algebra

Clifford algebra, also known as geometric algebra [6], integrates two operations of inner product and outer product.

1) Geometric product

The geometric product is the sum of outer product and inner product. For any two vectors a and b, the geometric product can be expressed as follows:

$$ab = a \cdot b + a \wedge b. \tag{3}$$

Where, the dot product $a \cdot b$ is a scalar and the outer product $a \wedge b$ is a double vector, then ab is a multiple vector (Fig. 1).

4) Reflection

The reflection vector a' of vector a in plane B can be expressed as (Fig. 2):

$$a' = a_{\perp B} - a_{\parallel B} = -\eta a \eta^{-1}. \tag{4}$$

Note: when η is a unit vector, $\eta^{-1} = \eta$.

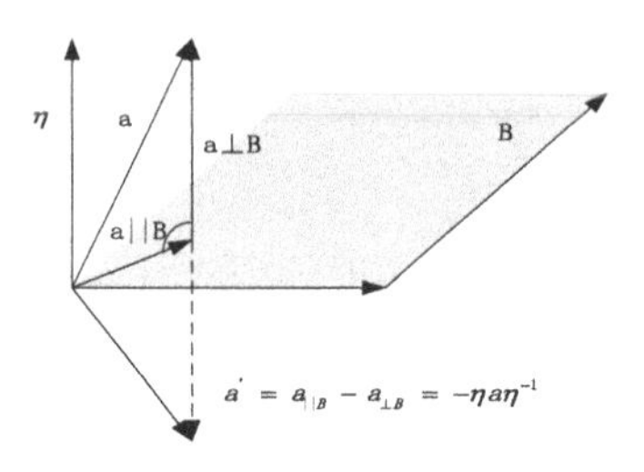

Fig. 1. Reflection of vector on plane

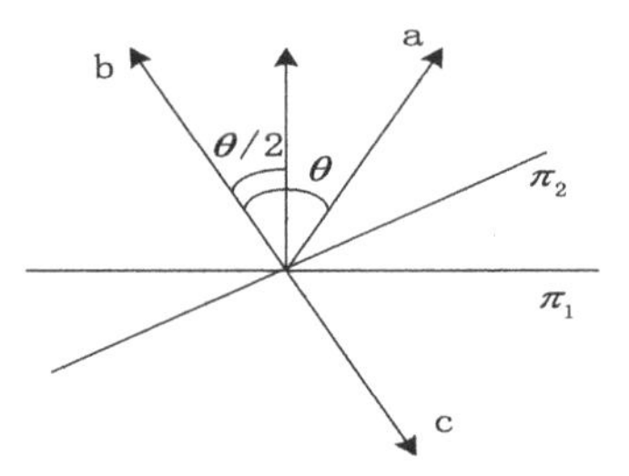

Fig. 2. A rotation to b

5) Rotation

Rotate unit vector a at any angle θ to get unit vector b, make the unit angle bisector η of vectors a and b in the plane a ∧ b, and make the vertical plane π1 of the vector η, then make a reflection vector about the plane π1 for vector a, the reflection vector c of a can be expressed as:

$$c = -\eta a \eta. \tag{5}$$

Then making the vertical plane π2 of vector b, we can get that vectors b and c are perpendicular to the plane π2, Then b can be expressed as:

$$b = -bcb = -b(-\eta a \eta)b = b\eta a \eta b = RaR^{-1}. \tag{6}$$

The bisection vector of unit angle is $\eta = \frac{a+b}{\|a+b\|}$, and the rotation operator is $R = b\eta$.

3 Image Preprocessing

3.1 Two Dimensional Image Edge Extraction

First of all, Canny operator is used to extract the edge of the two-dimensional image. Finally, only the outermost contour of the image is obtained, and then the outermost contour of each layer is extracted by horizontal and vertical scanning method. So far, the edge of the image has been extracted to prepare for the three-dimensional modeling of the image (Fig. 3).

3.2 3D Modeling of Contour

From a physical point of view, due to the protection of the skull, the soft tissue inside the skull is not easy to deform due to internal and external factors, thus maintaining the clarity of the contour, so we can treat the skull image as a rigid body (Fig. 4).

The edge points are stacked in the Z direction to get the three-dimensional contour of the skull. Get the three-dimensional outline of the skull, as shown in Fig. 5.

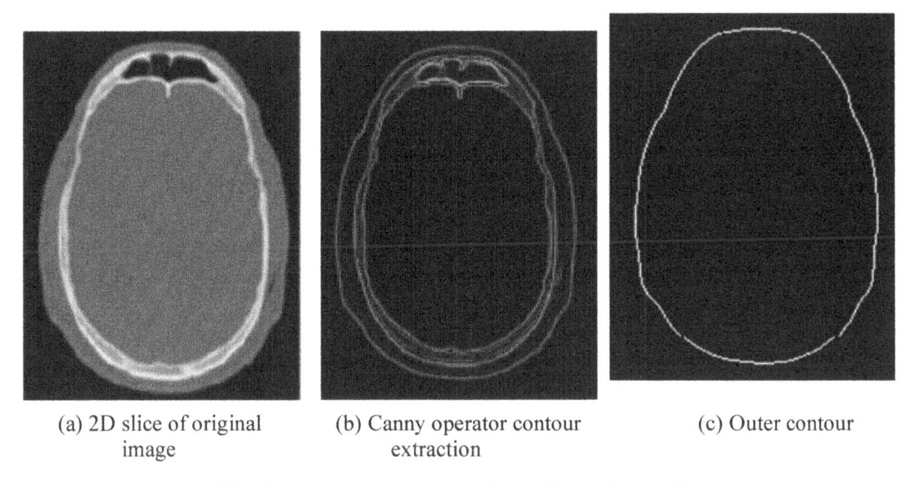

Fig. 3. Contour extraction of two-dimensional slice

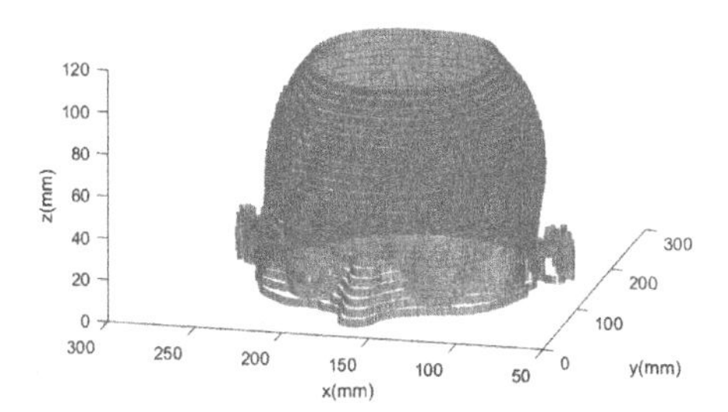

Fig. 4. Three dimensional outline of skull

4 Discrete Feature Points Extraction and Linear Fitting

After unifying the spatial resolution of the image groups of the two modes, Canny operator is used to extract contour points, the images of the two modes are stacked in the Z direction to obtain the 3D model of the image group, the centroids of the three-dimensional models of the two modes are translated to the center of the plane with the average height of the left and right ear tips as the height.

The characteristic points of each layer are obtained by using the principle of minimum mass moment of inertia, the formula of mass moment of inertia for discrete points is as follows:

$$I = \sum_i m_i r_i^2. \tag{7}$$

For the convenience of calculation, the mass of discrete points is recorded as unit 1, let the characteristic points of each layer be O_Z, where Z is the height of each slice in the coordinate system. The contour point of the registration mode 3D model whose ordinate is $Z = 0$ is taken as an example: firstly, all the contour points in $Z = 0$ layer of the registered modal 3D model are extracted $P_{Z=0}$, let the characteristic point of this layer be $O_{z=0}$ (x,y,0); then the sum of mass moment of inertia of each contour point around this point is calculated; finally, when the sum of mass moment of inertia of all contour points to this point is minimum, the corresponding values of X and y are obtained, the obtained points $O_{z=0}$ is the feature points of this layer. The specific calculation process is as follows:

Firstly, the mass moment of inertia of each contour point around this point is calculated:

$$I = m_i r_i^2 = (x_i - x)^2 + (y_i - y)^2. \tag{8}$$

Then the sum of mass moment of inertia of all contour points around this point in this layer is:

$$I = \sum_i m_i r_i^2 = \sum_i \left[(x_i - x)^2 + (y_i - y)^2\right]. \tag{9}$$

The partial derivatives of x and y are obtained from the above formula:

$$\begin{cases} \frac{\partial I}{\partial x} = \sum_i (2x - 2x_i) = 0 \\ \frac{\partial I}{\partial y} = \sum_i (2y - 2y_i) = 0 \end{cases}. \tag{10}$$

By solving the partial derivative equation, the values of x and y are obtained, then the feature points of the registered modal 3D model in $Z = 0$ layer is obtained, and the sum of the mass moment of inertia from all the contour points to this point is minimum. The feature points of other layers of the registration mode are solved by the same method, then the eigenvectors of registration modes are obtained by linear fitting of these discrete points h_f, The process of linear fitting is as follows:

Suppose that the linear equation of the feature discrete points of the registered modal three-dimensional model is as follows:

$$\frac{x - x_f}{X} = \frac{x - y_f}{Y} = \frac{x - z_f}{Z}. \tag{11}$$

The equations of x, y with respect to Z are obtained:

$$\begin{cases} x = \frac{x}{z}(z - z_f) + x_0 = az + b \\ y = \frac{y}{z}(z - z_0 + y_0 = cz + d \end{cases}. \tag{12}$$

The least square method is used to list the equations:

$$\begin{cases} Q_x = \sum_{f=1}^{m} [x_f - (az_f + b)]^2 \\ Q_y = \sum_{f=1}^{m} [y_f - (cz_f + d)]^2 \end{cases}. \tag{13}$$

The partial derivatives of equations for a, b, c and d are obtained as follows:

$$\begin{cases} b\sum_{f=1}^{m} z_f + a\sum_{f=1}^{m} z_f^2 = \sum_{f=1}^{m} x_f z_f \\ bm + a\sum_{f=1}^{m} z_f = \sum_{f=1}^{m} x_f \\ d\sum_{f=1}^{m} z_f + c\sum_{f=1}^{m} z_f^2 = \sum_{f=1}^{m} y_f z_f \\ dm + c\sum_{f=1}^{m} z_f = \sum_{f=1}^{m} y_f \end{cases}. \tag{14}$$

If $F = \begin{bmatrix} z_1\ z_2 \cdots z_m \\ 1\ \ 1 \cdots 1 \end{bmatrix}$, $A = [a\ b]'$, $X = [x_1 \cdots x_m]'$, $B = [c\ d]'$, $Y = [y_1 \cdots y_m]'$, then the solution of the above equations (8) can be transformed into the solution of the following matrix:

$$\begin{cases} FF'A = FX \\ FF'B = FY \end{cases}. \tag{15}$$

The solutions of vectors A and B can be obtained by taking all of the m discrete characteristic points into the above matrix equation, the solutions of a, b, c and d are obtained, then the equation of the line where the discrete feature points are located is obtained.

The eigenvector h_f is the unit vector in the direction of (X, Y, Z). Using the same method, the characteristic points of each layer of the fixed mode are obtained, when the sum of the mass moment of inertia of the discrete points of the layer contour to this characteristic point is minimum. The eigenvectors of the fixed modes are obtained by linear fitting of these discrete eigenpoints h_r, the eigenvectors h_f and h_r of two different modes are used to construct the rotation operator, the eigenvector of the registration mode h_f is rotated to the position coincident with the reference mode according to the rotation operator, the rotation operator R is as follow:

$$R = h_r \frac{h_f + h_r}{\left\| h_f + h_r \right\|}. \tag{16}$$

5 Experimental Results and Analysis

The medical images used in this experiment are from the "retrospective image registration evaluation" project library of Vanderbilt University in the United States, and this study was based on patient _ 001, the fixed CT image and PD image group are selected for registration. The 3D model of CT and PD image group is established before registration, the three-dimensional model is translated to the center of the plane with the average height of the left and right ear tips as the height as the origin. Figure 5 shows the multimodal 3D modeling before registration, Fig. 6 shows the 3D model of multimode after registration.

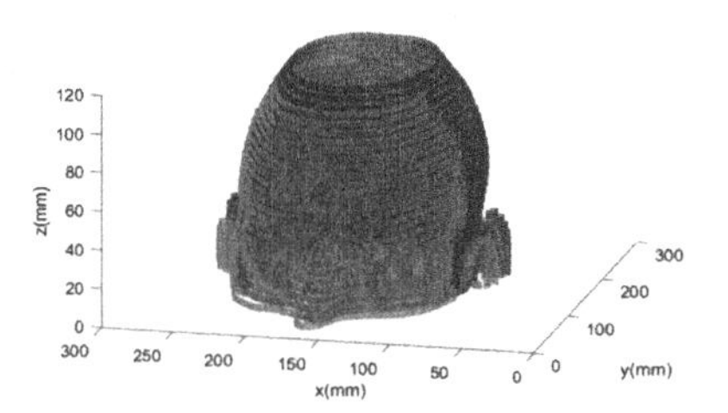

Fig. 5. 3D modeling before registration

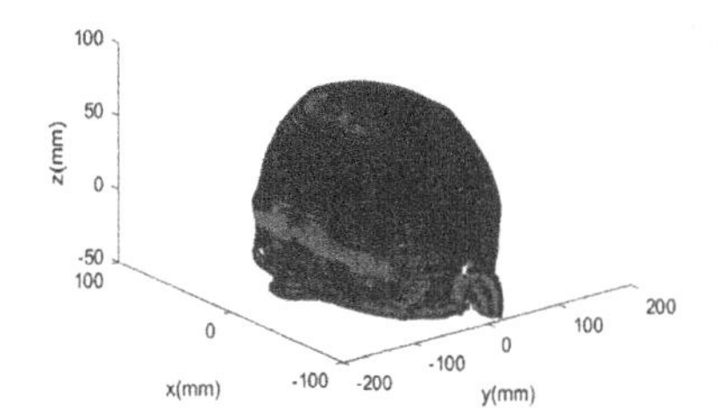

Fig. 6. 3D modeling after registration

Table 1. Registration parameters

Modality	Geometric feature vector
Registration mode (CT)	$h_f = -0.134e_1 - 0.193e_2 + 0.972e_3$
Fixed mode (PD)	$h_r = -0.079e_1 - 0.188e_2 + 0.979e_3$
Rotation operator R	R = 0.9997-0.0050e_{12}+0.0031e_{23} -0.0272e_{31}

Table 2. Error between experimental results and gold standard

Patient	VOI1	VOI2	VOI3	VOI4	VOI5	VOI6	VOI7	VOI8
Patient_001	1.935	3.364	3.319	2.394	2.498	2.625	0.496	1.073
Mean error:1.112 mm			Max error:3.364 mm			Mean error:2.394 mm		

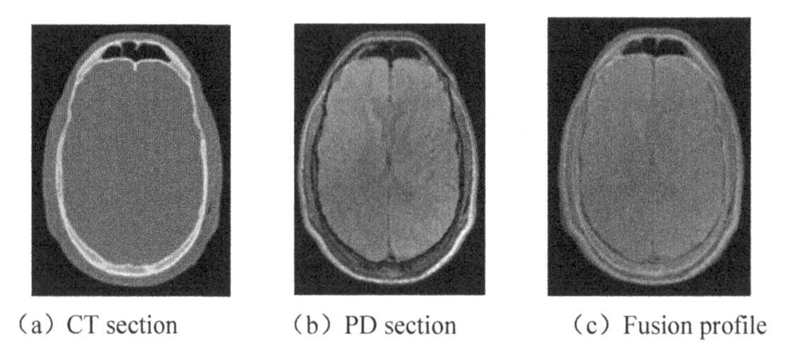

(a) CT section (b) PD section (c) Fusion profile

Fig. 7. Cross section parallel to xoy plane

The experimental registration parameters are shown in the following table (Table 1):

The evaluation results are shown in Table 2:

The fusion effect is shown in Figs. 7:

The effectiveness of the method in this chapter can be well verified by three-dimensional model, gold standard and fusion renderings, the experimental algorithm also achieves good results in other image groups of patient 1 and other image groups of patients. It is proved that the method in this chapter also has strong robustness.

The error of this algorithm is compared with the previous methods of other scholars in this field. The error comparison of various algorithms is shown in Table 3.

In this paper, all the contour points are involved in the calculation, and combined with the geometric invariance of mass moment of inertia, the feature points are constructed, the line where the feature points are fitted linearly is taken as the feature vector, and the rotation operator is constructed. This method has the advantages of intuitive geometric meaning, simple calculation and reliable experimental results.

Operation efficiency is also an important index to judge the experimental method, this experiment is carried out on MATLAB r2017a, and the computer configuration is

Table 3. Error comparison of various algorithms

Registration method	Error (mm)		
	Average error	Median error	Maximum error
Nearest point iteration	2.75	4.22	6.58
Maximum mutual information	1.40	1.53	3.72
Algorithm based only on Clifford algebraic space	1.201	2.578	3.815
Algorithm in this paper	1.133	1.847	3.563

CPU: Intel CeleronG5400 3.7 GHz, Memory: 8.00 GB, operating system: Windows 10 64 bit.

Comparing the data in Table 4, we can find that, the average time of this algorithm is much smaller than that of other algorithms. These algorithms are ICP registration methods proposed by N. Senin et al., Improved ICP image registration algorithm proposed by Weimin Li et al., and the mutual information maximization method proposed by Viola P. Because the geometric invariants based on Clifford algebraic space reduce the operation scale and complexity to a certain extent, so the time of registration is greatly shortened.

Table 4. Comparison of registration speed

Registration method	Average time (s)
Basic ICP	57.5
LM ICP	37.3
Maximum mutual information	1470
Algorithm in this paper	3.6

6 Conclusion

This experiment is based on the skull contour points, the accuracy of contour points extraction will directly affect the accuracy of the experiment, although the experimental error is within the acceptable range. Therefore, to find more accurate and optimized edge extraction method will be a new breakthrough in this paper. And due to the protection of the head, human brain tissue is generally fixed shape and not easy to deform, so the research of registration is rigid registration, in order to better grasp the medical image registration methods, flexible image registration is also an important research topic, and will become a mainstream direction of image registration.

References

1. Shen, Y.P.: Medical image registration technology. Chinese J. Med. Phys. **30**(01), 3885–3889 (2013)
2. Anuta, P.E.: Spatial registration of multispectral and multitemporal digital imagery using fast fourier transform techniques. Geosci. Electron. IEEE Trans. **8**(4), 353–368 (1970)
3. Legg, P.A., Rosin, P.L., Marshall, D.: Feature neighbourhood mutual information for multi-modal image registration: an application to eye fundus imaging. Pattern Recogn. **48**(6), 1937–1946 (2015)
4. Senin, N., Colosimo, B.M., Pacella, M.: Point set augmentation through fitting for enhanced ICP registration of point clouds in multisensor coordinate metrology. Robot. Comput. Integr. Manuf. **29**(1), 39–52 (2013)
5. Luo, S.Q., Li, X.: Multimodal medical image registration based on maximum mutual information. Chinese J. Image Graph. **5**(7), 551–558 (2000)
6. Clifford, W.K.: On the classification of geometric algebras. In: William Kingdon Clifford (ed.) Mathematical Papers, pp. 397—401 (1882)
7. Liu, H., Xu, C., Cao, W.M.: Multispectral image edge detection based on Clifford algebra. J. Southeast Univ. (Natural Science edition) **42**(02), 244–248 (2012)
8. Hua, L., Huang, Y., Ding, L.J., Feng, H., Gu, J.P.: 3D multimode medical image registration on Clifford algebraic space. Optoelectron. Eng. **41**(01), 65–72 (2017)

Biomedical Signal Processing, Imaging, Visualization and Surgical Robotics

The Feasibility of Prosthesis Design with Gradient Porous Structural

Qiguo Rong(✉) and Danmei Luo

Department of Mechanics and Engineering Science, College of Engineering, Peking University, Beijing 100871, China
qrong@pku.edu.cn

Abstract. The purpose of this study was to develop a design method for prosthesis with gradient porous tetrahedral structure, which combines the material optimization of continuum body in the macro level and the element size control of porous structure in the micro level. The internal density of the porous prosthesis can change with the structural stress and achieve the maximum porosity satisfying the strength and stiffness requirements. Compared with the regular porous structure, the optimum design can have a more uniform stress distribution, and the stiffness of the prosthesis is closer to the surrounding bone tissue.

Keywords: Prosthesis · Topology optimization · Gradient porous structure · 3D printing

1 Introduction

Because of biocompatibility, high strength, low stiffness and other advantages, titanium alloys are widely used in artificial prostheses. However, the elastic modulus of titanium alloy is still much higher than that of human bone tissue, and often causes the effect of stress shielding. Porous design can reduce the elastic modulus improve bone tissue adhesion, ingrowth and osseointegration. In addition, it can be easily realized by the additive manufacturing technology. Through precise design of external form and internal pore structure, the geometric and mechanical properties of the prosthesis can be adjusted to match the surrounding bone tissue, and the goal of the optimum biological performance and the lightest weight design of prosthesis can be achieved.

The micro-structures of porous prosthesis can be roughly divided into two types: regular porous structure and random porous structure. Compared with regular porous structure, irregular porous structure has more flexibility in design and can better realize the bionic design idea. Many researchers have made explorations in this field and proposed many modeling methods for irregular porous structures [1–3]. For example, some scholars proposed a hexahedral mesh generation method based on solid model, some scholars proposed truss structure based on principal stress trajectory [4], and some scholars developed optimization method for porous structure [5].

The purpose of this paper was to develop a parametric modeling method of porous prosthesis based on tetrahedral structure. The mechanical behavior of porous structures

M. Fei et al. (Eds.): LSMS 2021/ICSEE 2021, CCIS 1467, pp. 75–83, 2021.
https://doi.org/10.1007/978-981-16-7207-1_8

was calculated by FEM based on the representative element volume (unit cell). The relationship between geometric parameters and equivalent elastic modulus of porous tetrahedral structures was obtained by fitting the calculated results. Finally, an example is given to validate the proposed method.

2 Design of Non-uniform Porous Femoral Prosthesis

The femur is the largest long bone in the human body. It consists of cortical bone, cancellous bone and marrow. In this study, a 40 mm segmental bone defect in femoral shaft was assumed. The locking titanium plate combined with porous tetrahedral prosthesis was applied to repair the defect. What kind of porous design can make the overall stiffness the most appropriate, and at the same time the prosthesis has the lightest weight and the largest porosity?

The femur is composed of two distinct materials in cross-section. The outer layer is cortical bone, and the central area is medullary cavity or cancellous bone. The elastic modulus of the inner and outer layers varies greatly. It is not appropriate to adopt uniform porous structure, but to design porous prosthesis with non-uniform modulus. Firstly, the optimal distribution of elastic mudulus in the design domain was obtained by structural topology optimization method SKO. Then, the pore size distribution was determined according to the relationship between elastic modulus and pore size. Finally, the porous structure prosthesis was generated. The porosity of porous structures should be within the range of 50%–90% to meet the requirements of printability and mechanical requirements. The equivalent elastic modulus of elasticity should be within the range of 3 GPa and 15 GPa. Therefore, the upper and lower bounds of the elastic modulus were set to be 15 GPa and 3 GPa respectively during the optimization process.

2.1 FE Modeling

CT scan of the femur of a healthy female volunteer (23 years old, 60 kg) was performed to obtain the required femoral CT data. The CT data were imported into the medical image processing software Mimics. After processes thresholding, tissue segmentation, Boolean operation and three-dimensional reconstruction, a three-dimensional geometric model of the right femur, including the marrow cavity, was obtained in STL format (Fig. 1a). The noise and errors of the geometric model were removed in the inverse CAD software Geomagic Studio, and the smooth NURBS model of cortical bone (Fig. 1b) and cancellous bone (Fig. 1c) was obtained.

A three-dimensional solid model of locking plate and bilayer cortical screw was established in CAD software Solidworks. The locking plate was made of Ti6Al4V and placed on the anterolateral femur (tension band). The internal surface of the plate was basically identical with the anatomical shape of the bone surface. The proximal end attached tightly to the femoral shaft, and the distal end was fitted to the femoral lateral condyle. The thickness of the plate was 5 mm and the width was 11.5 mm. Four bicortical locking screws were fixed to the femoral shaft at the proximal end of the plate, and four cancellous locking screws at the distal end of the plate were fixed to the metaphyseal end of the femur. The length of the screws penetrated both sides of the cortical bone.

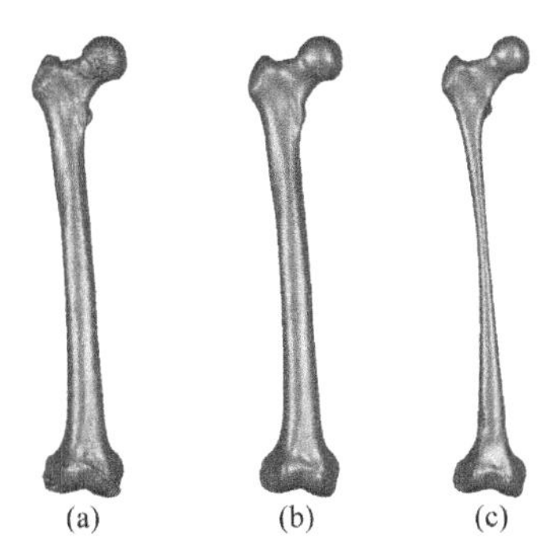

Fig. 1. Three-dimensional model of femur: (a) geometric model; (b) cortical bone; (c) cancellous bone.

The geometric models of cortical bone, cancellous bone, medullary cavity, titanium plate and titanium screws were assembled in FE software ANSYS. The complete finite element model of locking plate system and femur was then established (Fig. 2a). The model was meshed into linear tetrahedral elements (335899 nodes and 175530 elements).

Based on the complete femoral geometric model, the femoral shaft defect with 40 mm length was simulated. The finite element model of defective femoral joint plate (Fig. 2b) consists of 323757 nodes and 1743666 elements. Furthermore, a linear hexahedral element was used to divide the defect area and simulate the effect of porous prosthesis repair. A finite element model with "porous prosthesis" was established. The porous prosthesis part was an equivalent entity model based on homogeneity method. In the model, the interface between the titanium plate and the titanium nail was set as binding contact.

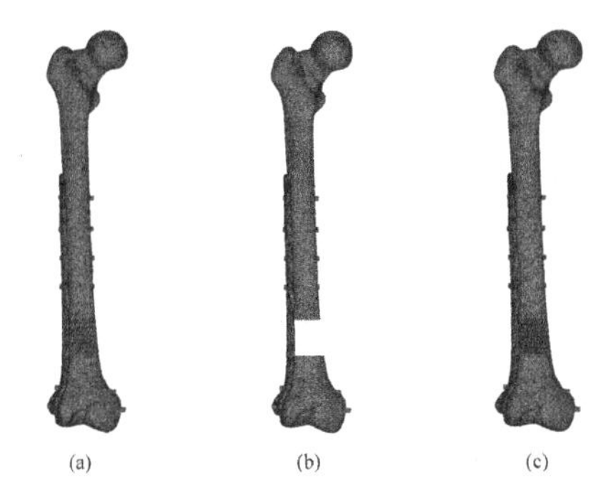

Fig. 2. Three-dimensional FE model of the locking plate system and femur: (a) FE model with intact bone; (b) FE model with bone defect; (c) FE model with prosthesis.

Considering that the prosthesis is implanted in the middle of femur and surrounded by cortical bone and medullary cavity in this study, it can be considered that the simplified method of material attribute assignment only considers three material attributes of bone, namely cortical bone, cancellous bone and medullary cavity. This simplified model has sufficient calculation accuracy. All materials in the model are homogeneous, isotropic and linear elastic. The elastic modulus and Poisson's ratio of materials are listed in Table 1.

Simulate the normal physiological state of human bipedal standing position (abduction 15°, femoral shaft axis and vertical direction 15° angle), at this time the force acting on each side of the hip joint is 1/2 of the upper body gravity, the size is about 200 N, joint force through the center of the femoral head vertical downward. All the degrees of freedom of all the nodes on the surface of the distal femoral condyle are restrained, and the 200 N joint force is evenly distributed to the nodes on the surface of the femoral head.

Table 1. Material parameters for FE analysis.

Material	Modulus (MPa)	Poisson's ratio
Cortical bone	167000	0.3
Cancellous bone	137	0.3
Mark	1	0.3
Ti6Al4V	114000	0.3

2.2 Stress Analysis and Optmization

Five models are analyzed in this study:

1) Intact femoral joint with titanium plate/defective femoral joint with titanium plate.

Firstly, the mechanical properties of the complete femoral joint titanium plate model under standing loads are studied, which will be an important reference for the follow-up prosthesis design. Then, the mechanical properties of the defective femoral joint titanium plate model under standing load were compared.

2) Porous prosthesis with equivalent modulus of 3 GPa/15 GPa.

Homogeneous porous prosthesis was used to repair bone defect. According to the requirements of biological properties, porous prosthesis can achieve satisfactory bone growth effect when its porosity is 50%~90%. According to the relationship between the porosity of porous tetrahedral structure and the equivalent modulus of elasticity obtained in Chapter 2, when the porosity is in the range of 50%~90%, the corresponding elastic modulus range of the structure is 2.26 GPa~14.99 GPa. Considering the structural strength, 3 GPa~15 GPa is used as the range of equivalent modulus of elasticity of porous prosthesis, and the corresponding porosity is 49.51%~85.13%. In order to simplify the calculation, the homogenization method is used for finite element analysis of porous prosthesis. The equivalent modulus of elasticity is directly defined as the material property, which greatly simplifies the finite element calculation. Considering two extremes, one is to use porous prosthesis with equivalent modulus of elasticity of 3 GPa, the other is to use porous prosthesis with equivalent modulus of elasticity of 15 GPa to compare their biomechanical properties.

3) Porous prosthesis (beam element) with equivalent modulus of 3 GPa/15 GPa.

According to the relationship between the length-thickness ratio of porous tetrahedral structure and the equivalent elastic modulus obtained in Chapter 2, when the range of elastic modulus of porous tetrahedral structure is 3 GPa~15 GPa, the corresponding length-thickness ratio is 7.41~3.36, the diameter of rod is 0.4 mm, and the corresponding unit size is 2.964 mm~1.344 mm. According to element size and rod diameter, the finite element models of porous tetrahedral beam elements with equivalent modulus of elasticity of 3 GPa and 15 GPa are established in defect geometry model. Each rod is divided into three beam elements. The biomechanical properties under standing loads were calculated respectively.

4) SKO topology optimization on the model with equivalent modulus of 15 GPa established by homogenization method.

The range of elastic modulus varies from 3 GPa to 15 GPa, and SKO topology optimization is carried out. The constraint conditions are that the difference between the total displacement of the structure and the total displacement of the whole bone calculated is less than 0.1 mm, and the end of iteration condition is that the volume change ratio is less than 0.1%. After 16 iterations, the final optimized material distribution model was obtained. Computational optimization model biomechanical properties.

5) Optimal non-uniform porous tetrahedral prosthesis (beam element).

The optimal material distribution is transformed into the corresponding non-uniform porous tetrahedral prosthesis beam element model, and its biomechanical properties are analyzed.

2.3 Results

To determine the optimal stiffness of the prosthesis, we first analyze the condition of the complete bone combined with titanium plate, and use the results of the complete bone as the optimization criterion to guide the prosthesis design.

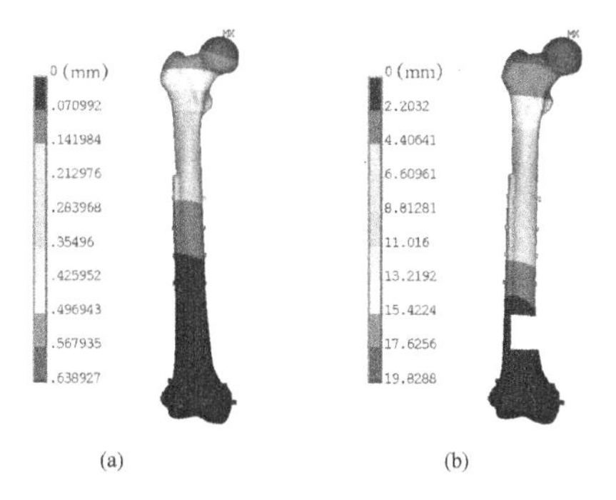

Fig. 3. Displacement.

Figure 3a and Fig. 3b are displacement distributions of complete bone and defective bone repaired with titanium plate under standing load, respectively. It can be seen that the maximum displacement occurs at the large rotor under standing load. For the complete

Table 2. FE analysis results.

Model	Displacement of greater trochanter	Maximum stress of titanium plate (MPa)	Maximum stress of prosthesis (MPa)	Maximum contact stress of upper contact area (MPa)	Maximum contact stress of lower contact area (MPa)	Pososity (%)
Intact bone	0.638	95.604	/	2.735	1.836	/
Defect bone	19.828	1196.52	/	0.205	0.700	/
3 GPa EP	0.789	95.666	2.020	2.841	14.854	85.13
3 GPa PP	0.800	95.649	81.622	45.963	38.648	85.13
15 GPa EP	0.631	95.603	2.278	2.860	2.060	49.51
15 GPa PP	0.632	95.600	30.767	12.014	8.977	49.51
Optimum EP	0.635	95.611	3.044	2.957	2.153	68.43
Optimum PP	0.639	95.592	24.392	10.424	10.962	68.43

[1]EP: equivalent prosthesis; PP: porous prosthesis

femoral joint titanium plate model, the maximum displacement is 0.64 mm. For the defective femoral joint titanium plate model, the maximum displacement is 19.83 mm, which is much larger than the normal situation. This shows that the titanium plate alone cannot provide sufficient stiffness for repairing large bone defects.

Table 2 lists all the finite element computational statistics. Analyzing the stress value of titanium plate and nail, the maximum stress is 95.604 MPa for the complete femur model, and 1196.52 MPa for the defective femur model, which exceeds the ultimate strength of titanium alloy material and may cause structural damage. This indicates that the strength of titanium plate is far from enough for repairing bone defect by using titanium plate alone. To sum up, it is not feasible to repair large bone defect with titanium plate alone. It needs to be done with bone prosthesis.

The range of elastic modulus of tetrahedral porous prostheses which meet the requirements of biology and stiffness has been defined to be 3 GPa–15 GPa. In this study, the equivalent modulus of elasticity of porous prostheses is used as the design interval, and the corresponding porosity is 49.51%–85.13%. Two extreme cases were analyzed, namely, the equivalent elastic modulus of the prosthesis with 3 GPa and 5 GPa.

Firstly, the equivalent model based on homogenization method is used to calculate. In terms of structural stiffness, the displacements of the large rotor corresponding to the complete bone, the equivalent prosthesis of 3 GPa and the equivalent prosthesis of 15 GPa are 0.638 mm, 0.789 mm and 0.632 mm, respectively. The overall deformation of the bone structure after 3 GPa equivalent prosthesis repair is larger than that under the condition of complete bone, which indicates that the stiffness is insufficient. The overall deformation of the bone structure after 15 GPa equivalent prosthesis repair is slightly less than that of the complete bone, which indicates that the stiffness compatibility of 15 GPa prosthesis repair can be basically satisfied.

To verify the accuracy of homogenization method, finite element models of beam elements with equivalent elastic modulus of 3 GPa (rod diameter is 0.4 mm, element size is 2.964 mm) and 5 GPa (rod diameter is 0.4 mm, element size is 1.344 mm) were analyzed. Compared with the 3 GPa equivalent prosthesis and the 3 GPa beam element prosthesis, the displacement difference of the two large rotors is 0.011 mm; compared with the 15 GPa equivalent prosthesis and the 15 GPa porous beam element prosthesis, the displacement difference of the two large rotors is 0.001 mm. The calculated results of the equivalent model and the beam element model are highly consistent, which shows that the proposed homogenization method can effectively calculate the structural stiffness.

There is an order of magnitude difference between the stress level of the prosthesis obtained by homogenization method and that of the real prosthesis structure, which is caused by the discontinuity of the micro-model structure. In this paper, the real stress of the prosthesis is not evaluated by homogenization method. The maximum stress value of 3 GPa porous prosthesis is 81.62 MPa, while that of 15 GPa porous prosthesis is 30.77 MPa, which indicates that the strength of 3 GPa prosthesis is lower than that of 15 GPa prosthesis. It can be roughly judged that the service life of 3 GPa prosthesis under fatigue load is shorter than that of 15 GPa prosthesis.

Similarly, the stress difference between the equivalent model and the beam element model on the bone contact surface is very large, which is also caused by the discontinuity of the prosthesis. The contact position of the thin rod of the porous prosthesis at the bone contact surface will form a star-like stress concentration. In this paper, the homogenization method is used to evaluate the real stress of bone tissue around the prosthesis. The maximum stress value of the upper end-to-bone interface of 3 GPa porous prosthesis is 45.96 MPa, while that of the lower end-to-bone interface is 38.65 GPa. The upper end-to-bone interface is slightly larger than the lower end-to-bone interface. The maximum stress value of the upper end-to-bone interface of 15 GPa porous prosthesis is 12.014 MPa, and that of the lower end-to-bone interface is 8.977 GPa, which is still slightly larger than that of the lower end-to-bone interface. The stress value of the end-to-bone interface of the 15 GPa porous prosthesis is significantly lower than that of the 3 GPa porous prosthesis, and the risk of bone tissue damage after implantation is also lower.

In conclusion, the smaller the pore size of the unit, the greater the strength and stiffness of the prosthesis, the smaller the stress on the surrounding bone tissue, but the smaller the porosity and the heavier the weight of the prosthesis. Although the strength and stiffness matching of the prosthesis with the equivalent modulus of elasticity of 15 GPa is better than that of the prosthesis with the equivalent modulus of elasticity of 3 GPa, the porosity of 49.51% is much lower than that of the 3 GPa prosthesis (85.13), which means that the weight of the prosthesis is 3.4 times of that of the 3 GPa prosthesis. Neither of the two prostheses is the best prosthesis for bone repair. We need to optimize the prosthesis structure to meet the requirements of strength and stiffness as well as self-weight.

The prosthesis was optimized by optimization method. SKO method was used to optimize the structure of the 15 GPa equivalent model. After 18 iterations, the results converged and the optimal material distribution was obtained. At this time, the porosity of the structure increased from 49.51% to 68.43%. The displacement of the large rotor of the optimized prosthesis was 0.639 mm, very close to the 0.638 mm of the complete bone, and the stiffness coordination was highly consistent. Meanwhile, the maximum stress of the optimized prosthesis was 24.39 MPa, which was lower than that of the 3 GPa homogeneous prosthesis (81.62 MPa) and the 15 GPa homogeneous prosthesis (30.77 MPa), indicating that the strength of the optimized prosthesis was enhanced. The maximal stress values of the optimal prosthesis were 10.424 MPa and 10.962 MPa, respectively. The risk of bone tissue destruction was lower after implantation.

In Fig. 4, the process of converting the optimal material distribution to the corresponding non-uniform pore distribution is given. The modulus distribution of the prosthesis element is optimized according to the stress distribution of the 15 GPa equivalent prosthesis. The outer layer of the prosthesis is the main bearing area, while the inner layer bears less, which is closely related to the structural characteristics of the outer cortical bone and the inner medullary cavity of the femur. The optimum material distribution is mainly composed of 3 GPa and 15 GPa. There are hardly any elements with intermediate modulus. According to the geometric contour of the element, the design model is fitted to the surface. In the design model, the tetrahedral structure is filled according to the corresponding unit size. Finally, the porous tetrahedral structure with the optimum non-uniform pore distribution is obtained. It is worth noting that the intersection surface of the two elements. On the other hand, the unit forms a smooth transition.

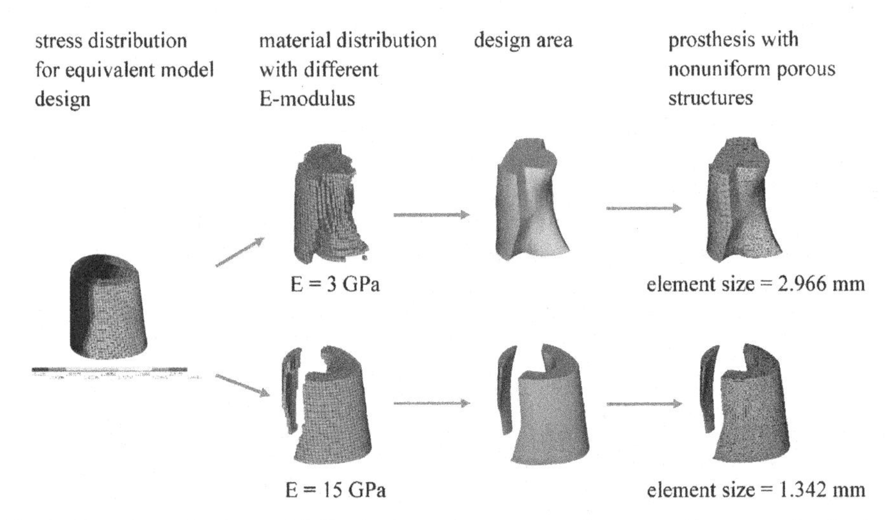

Fig. 4. From the optimal material distribution to the corresponding non-uniform pore distribution diagram.

3 Conclusions

In this chapter, a design method of porous tetrahedral prosthesis with maximum porosity and non-uniform pore distribution is proposed based on macro-structure topology optimization algorithm and micro-element size control algorithm. Firstly, the equivalent solid model of porous structure is established by homogenization method. Then, the stress distribution of the model is analyzed by finite element method. Then, according to the stress distribution of the model, SKO bionic structure topology optimization method is used to optimize the material distribution in the structure, and the maximum porosity of the structure is achieved under the constraint conditions. Finally, according to the optimized material elastic modulus, based on the function relationship between the geometric parameters of porous structure and the equivalent elastic modulus proposed in this paper, the volume fraction of the corresponding partially filled porous structure is determined. Within the model, the tetrahedral frame filled with different densities is generated. In the area with large stress, the distribution of the frame joints is dense, while in the area with small stress, the distribution of the frame joints is dense. It's sparse. Taking the design of non-uniform porous femoral prosthesis as an example, the results show that the structural stiffness of the optimized model is very close to that of the intact bone, and the strength of the prosthesis is better than that of the homogeneous structure prosthesis, and the weight of the prosthesis is greatly reduced. Based on the stress distribution, the porous tetrahedral structure with the largest non-uniform porosity distribution can be constructed under the constraint of strength and stiffness.

Acknowledgements. This work was supported by National Natural Science Foundation of China (No. 11872074).

References

1. Lal, P., Sun, W.: Computer modeling approach for microsphere-packed bone scaffold. Comput. Aided Design. **36**(5), 487–497 (2004)
2. Cai, S., Xi, J.: A control approach for pore size distribution in the bone scaffold based on the hexahedral mesh refinement. Comput. Aided Design. **40**(10–11), 1040–1050 (2008)
3. Rainer, A., Giannitelli, S.M., Accoto, D., et al.: Load-adaptive scaffold architecturing: a bioinspired approach to the design of porous additively manufactured scaffolds with optimized mechanical properties. Ann. Biomed. Eng. **40**(4), 966–975 (2012)
4. Reinhart, G., Teufelhart, S.: Optimization of mechanical loaded lattice structures by orientating their struts along the flux of force. Procedia CIRP **12**, 175–180 (2013)
5. Wang, W., Wang, T.Y., Yang, Z., et al.: Cost-effective printing of 3D objects with skin-frame structures. ACM Trans. Graph. **32**(6), 177 (2013)

Evaluation of Fatigue Based on Reaction Time and fNIRS of RSVP Small Target Detection

Yusong Zhou[1], Banghua Yang[1,2](✉), Linfeng Yan[3], and Wen Wang[3](✉)

[1] School of Mechanical Engineering and Automation, Shanghai University, Shanghai 200444, China
{zhouyusong,yangbanghua}@shu.edu.cn

[2] Research Center of Brain Computer Engineering, Shanghai University, Shanghai 200444, China

[3] Department of Radiology, Functional and Molecular Imaging Key Lab of Shaanxi Province, Tangdu Hospital, Fourth Military Medical University, Xi'an 710038, China
wangwen@fmmu.edu.cn

Abstract. Fatigue has an impact on the target recognition accuracy in small target detection task based on the rapid serial visual presentation (RSVP) paradigm. It is necessary to evaluate brain fatigue objectively to adjust RSVP parameters, so as to prolong the effective task execution time. This paper proposes a method to evaluate the degree of brain fatigue based on reaction time and Oxyhemoglobin (HBO) index of functional near-infrared spectroscopy (fNIRS). A 100-min RSVP task was designed to induce the fatigue state. The fNIRS data of 17 channels in the prefrontal cortex (PFC) region and 440 groups of auditory key reaction time data of 11 subjects were collected and analyzed. The degree of brain fatigue can be divided into conscious state and fatigue state by combining the subjective evaluation scale, probability density distribution and cumulative distribution of reaction time, and the concentration change of HBO. The results show that, in the RSVP task, with the increase of brain fatigue, the reaction time of human gradually increases and the overall distribution is lognormal, the reaction time of HBO reduces and the concentration increases in PFC at the same time. Therefore, the reaction time index and fNIRS index can be used as potential indicators to evaluate the brain fatigue model in RSVP small target detection task, and provide the basis for the later research on the monitoring and regulation of brain fatigue.

Keywords: RSVP · Small target detection · Reaction time · fNIRS

1 Introduction

RSVP small target detection task requires subjects to watch high frequency picture flow and judge it. Long-term task will induce mental fatigue. Mental fatigue is a very complex psychological and physiological phenomenon and there is no unified definition of it at present. Grandjean believes that mental fatigue is a transitional state between wakefulness and sleep [1, 2]. From the perspective of clinical medicine, Layzer believes that mental fatigue refers to the inability to continue to concentrate and maintain efficiency

M. Fei et al. (Eds.): LSMS 2021/ICSEE 2021, CCIS 1467, pp. 84–93, 2021.
https://doi.org/10.1007/978-981-16-7207-1_9

in the process of work [3, 4]. Mental fatigue is a gradual and cumulative process, and is often accompanied by dizziness, inattention, slow reaction, etc. Resulting in damage to memory, learning, mathematical operation, logical reasoning and other cognitive functions [5]. Therefore, it has great significance to have research on mental fatigue in some specific working environment or occupation.

In terms of the research on relationship between fatigue and reaction time, Langner et al. [6] found that after long-term task, people's overall reaction time increased, indicating that long-term task will produce mental fatigue, and reduce cognitive efficiency. Guo M ZH et al. [7] found that driving for a long time can lead to a fluctuating increase in fatigue of driver, which leads to an increase of reaction time; Foong R et al. [8] used the reaction time parameter as the basis for dividing the fatigue level. Sonnleitner A et al. [9] found that reaction time and α-band would increase with the increase of task time. Verdonck S et al. [10] emphasized the validity of the selective reaction time test in psychology and neuroscience. Li SH W et al. [11] collected driver's eye movement data, reaction time and execution time, and quantified driving fatigue using entropy weight method.

In addition, in the study of fNIRS for fatigue assessment, the advantages of fNIRS technology are mainly reflected in advanced cognitive activities [12, 13]. At the same time, fNIRS can also measure muscle oxygen content to achieve the purpose of measuring body fatigue [14, 15]. At present, relevant studies have found that the activity of HBO in PFC is increased, while the activity of Deoxyhemoglobin (HBR) is decreased and Blood oxygen saturation (SaO2) is decreased under fatigue state [16–18], and the activity of HBO in the left and right sides of the PFC is asymmetric under fatigue state [19, 20].

In conclusion, the changes of reaction time index and fNIRS index are closely related to fatigue degree. The fatigue assessment based on reaction time is simple, direct and real-time, while fNIRS is real and objective. At present, in the small target detection task based on RSVP paradigm, there are few studies on the evaluation of brain fatigue by combining reaction time index and fNIRS index. This paper analyzes the reaction time and fNIRS data of the subjects in the task, explores the relationship between these two indexes and mental fatigue, and gives an objective evaluation and quantification of brain fatigue.

2 Materials and Methods

2.1 Subjects

11 male right-handed subjects of Shanghai University were recruited in the study, age: 22–28 (24.2 $\pm$ 1.4). All subjects had no history of psychology, spirit, nerve, color blindness, drug abuse, coffee, tea, alcohol, smoking and other habits, and had good sleep quality. The subjects volunteered to participate in the experiment and fill in written consent form, and we pay certain remuneration to them after the experiment. Before the experiment, the subjects were not exposed to this type of experimental operation.

2.2 Experimental Paradigm

The paradigm consists of four 25-min Sessions. Each Session includes one Preparation, ten Blocks and one Rest. The target images are small ship images based on complex background environment, and the non-target images are small non-ship images with high similarity to the target image.

The ratio of target to non-target image is 1/9 in the paradigm. Preparation includes 1-min eye opening rest, 1-min eye closing rest, Adaptation1 (frequency: 1 Hz, 10 pictures), Adaptation2 (frequency: 2 Hz, 10 pictures), Adaptation3 (frequency: 4 Hz, 10 pictures), Adaptation4 (frequency: 10 Hz, 10 pictures, cycles: 3). Each Block contains 10 Experiments, each Experiment includes 10 s of picture presentation (frequency: 10 Hz, 100 pictures) and 5 s of rest. The subjects need to identify the target picture in the 10 s' picture stream. In the random position of each Block will play a sound, and the subjects need to press the keyboard immediately after hearing it. The Rest required the subjects to close their eyes for one minute.

The fatigue scales were filled out before and after each Session for 8 times. The fatigue scale includes four items: thinking clarity, energy concentration, sleepiness and comprehensive evaluation of fatigue. The items are extracted from the fatigue self-conscious symptom questionnaire of Fatigue Research Association of Japan Industry Association, and the scale has a 10-point scale [21]. Before the experiment, the subjects were trained to be familiar with the task requirements. The structure diagram of paradigm is shown in Fig. 1, and the flow chart is shown in Fig. 2.

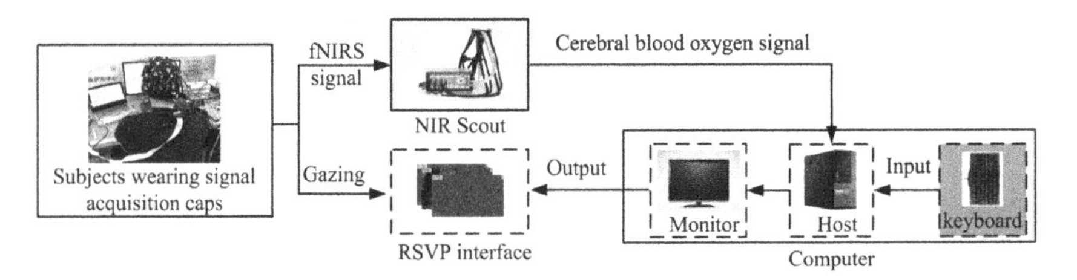

Fig. 1. The structure diagram of small target detection task based on RSVP paradigm.

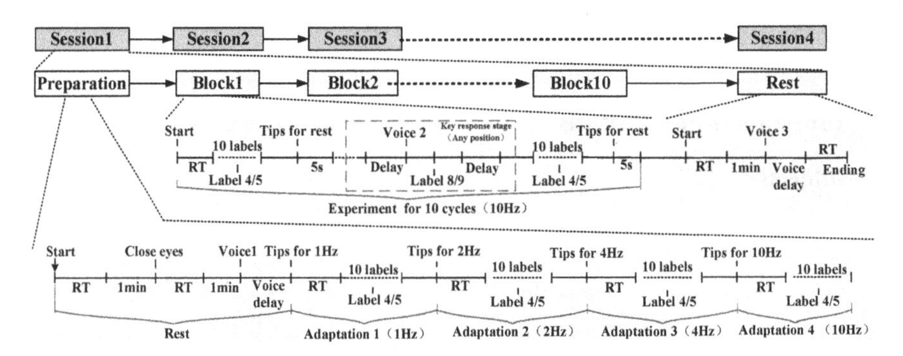

Fig. 2. The flow chart of the RSVP paradigm.

2.3 Data Acquisition

The presentation of the pictures is based on E-Prime 3.0 software, the computer is based on Windows 10 operating system, the main frequency is 3.2 GHz, and the keyboard is Dell kb216t Wired Keyboard.

fNIRS data acquisition is based on NIRScout functional NIRS equipment of NIRX company, with a total of 17 channels. The measuring is on PFC, including 8 light sources (TX) and 8 detectors (Rx). The sampling rate is set at 7.81 Hz. The layout of 17 channels is shown in Fig. 3.

The picture of data collection in the experiment site is shown in Fig. 4.

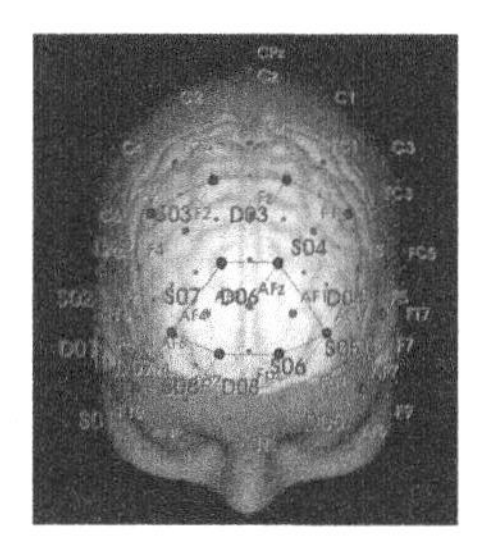

Fig. 3. The layout of fNIRS.

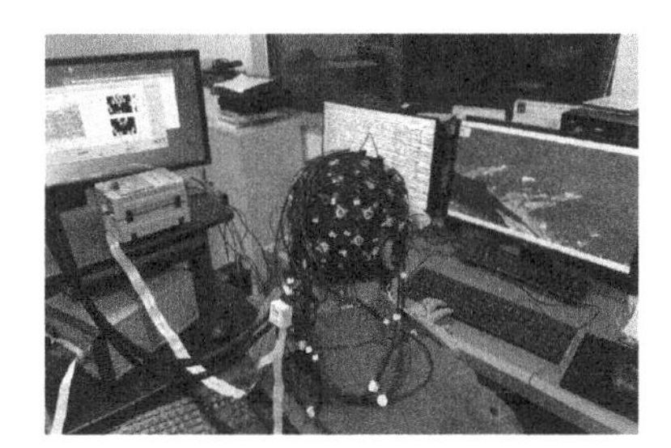

Fig. 4. The experiment site.

2.4 Data Processing

2.4.1 Data Fitting of Reaction Time

The probability density function and cumulative distribution function can be used to analyze the reaction time data. According to the relevant literature, the probability density of reaction time under fatigue task often conforms to the Log-Normal Distribution.

Assuming that the reaction time is represented by a random variable x, $x > 0$, the probability density function of lognormal distribution of x can be described by the following formula:

$$f(x, \mu, \sigma) = \frac{1}{\sqrt{2\pi} x\sigma} e^{-\frac{(\ln x - \mu)^2}{2\sigma^2}} \tag{1}$$

Recorded as: $\ln X \sim N(\mu, \sigma^2)$. Where μ is the average of ln$x$ and σ^2 is the standard deviation of ln$x$.

The cumulative distribution function and mathematical expectation of lognormal distribution of x can be described as follows:

$$F(x) = \frac{1}{2} + \frac{1}{2} erf\left(\frac{\ln(x) - \mu}{\sqrt{2}\sigma}\right) \tag{2}$$

$$E(x) = e^{\mu + \frac{\sigma^2}{2}} \tag{3}$$

2.4.2 fNIRS Data Processing

In this study, firstly, the standard deviation is used to detect the sudden change of fNIRS waveform, and the fluctuation data caused by pseudo trace is removed according to the threshold (STD threshold = 5). Secondly, the band-pass filter (0.01 Hz–0.2 Hz) is used to remove physiological interference and high-frequency noise components. Finally, according to the Mod. Beer-Lambert-Law (MBLL) to concentrate hemodynamic state. The comparison of the near infrared data before and after the processing is shown in Fig. 5.

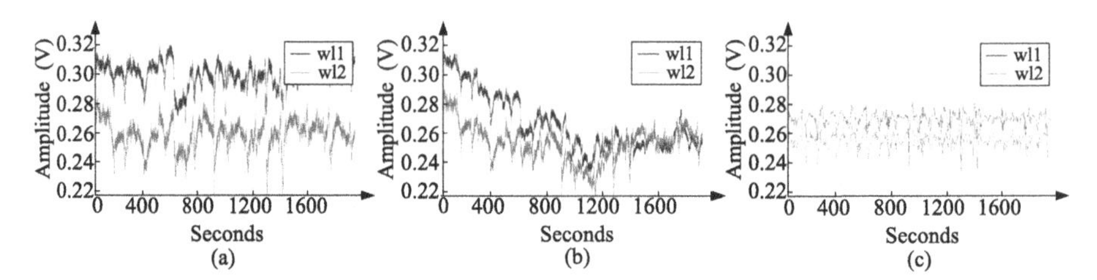

Fig. 5. The comparison of the near infrared data before and after the processing: (a) represents raw data, (b) represents the data after Removing artifacts, (c) represents the data after filtering.

3 Result

3.1 Subjective Evaluation Result

The subjective fatigue degree was the average of the total score of the fatigue scale. After repeated measurement analysis of variance (ANOVA) test, the specific changes are shown in Fig. 6, and the results are shown in Table 1. The results showed that there was a significant difference in subjective fatigue before and after the fatigue task of each Session ($P < 0.05$). With the increase of the number of Sessions, subjective fatigue also increased. Therefore, from the subjective evaluation, RSVP small target detection task successfully induced mental fatigue.

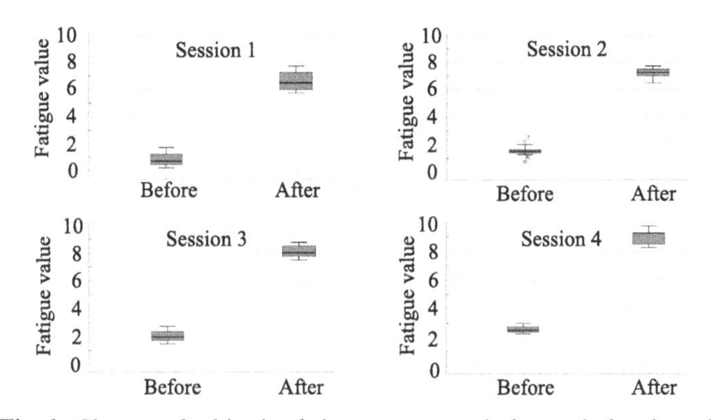

Fig. 6. Changes of subjective fatigue assessment before and after the task.

Table 1. Analysis results of subjective evaluation data before and after the task.

Group	Parameter	Session1	Session2	Session3	Session4
Before the experiment	*M*	0.86	1.48	2.04	2.55
	SD	0.48	0.45	0.42	0.25
After the experiment	*M*	6.64	7.23	8.09	8.95
	SD	0.71	0.45	0.48	0.51
	F	605.22	1163.80	3275.74	1325.40
	P	0.000	0.000	0.000	0.000

3.2 Behavioral Result

440 groups of reaction time data of 11 subjects in 4 Sessions were obtained. As shown in Fig. 7, in each Session, with the increase of Blocks, the reaction time of each subject increased gradually. After observing and fitting the reaction time data of each Session and all Sessions, it is found that the reaction time of the subjects in the RSVP small target detection task conforms to the lognormal distribution, as shown in Fig. 8, which are the probability density distribution and cumulative distribution curve of the subjects' reaction time respectively. Among them, the reaction time of all Sessions conforms to the lognormal distribution of $\mu = 6.5354$, $\sigma^2 = 0.2888$, then the probability density function of reaction time is as follows:

$$f(x) = \frac{0.7423}{x} e^{-\frac{(\ln x - 6.5354)^2}{0.5776}} \tag{4}$$

The mean value of lognormal distribution is as follows:

$$E(x) = 796.16 \tag{5}$$

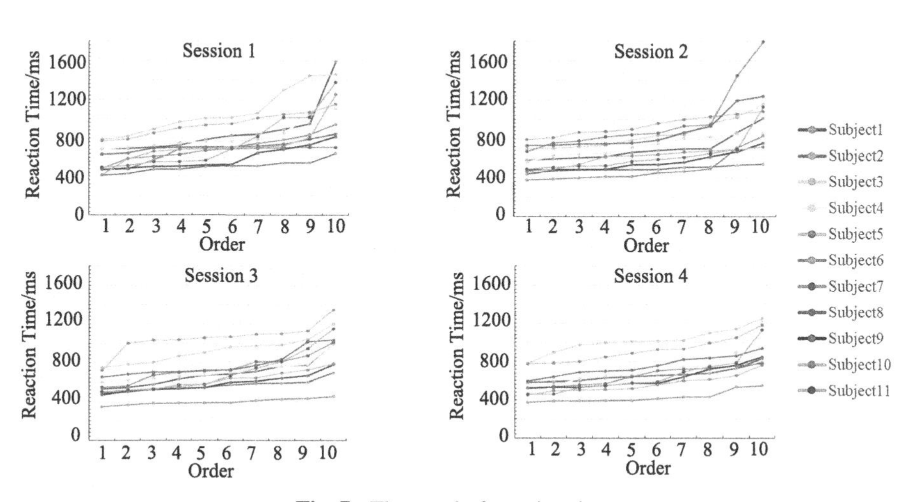

Fig. 7. The trend of reaction time.

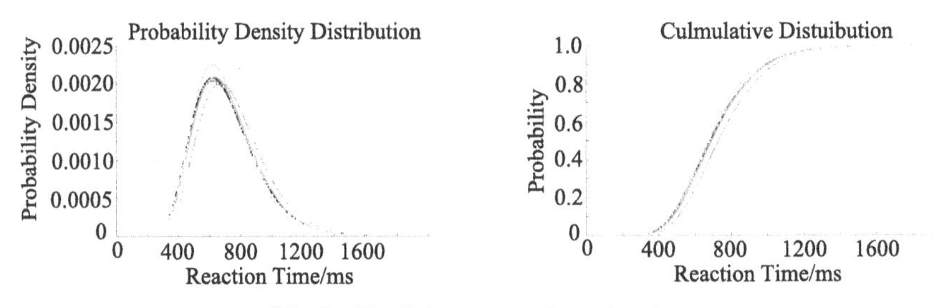

Fig. 8. The fitting curve of reaction time.

That is to say, the expected reaction time of the subjects in the RSVP small target detection task is 796.16 ms.

3.3 Analysis Result of fNIRS Data

Taking the mathematical expectation of reaction time as the segmentation point, the HBO concentration in the corresponding Block of all the subjects is selected to average (reaction time < 796.16 ms, and reaction time > 796.16 ms). The results of each fNIRS channel are drawn through the brain topographic map, as shown in Fig. 9. The HBO concentration data of all the blocks in each Session is averaged, and the results of each fNIRS channel are plotted by brain topographic map, as shown in Fig. 10. In the figure, the darker the red is, the higher the HBO concentration is, and the darker the blue is, the lower the HBO concentration is. The data of each Session and all Sessions is overlaid, and the concentration of HBO and HBR of all the first 5 s and the last 20 s of the time when the target image appears is selected to average. The curves of HBO and HBR concentration data is plotted, as shown in Fig. 11. In the figure, the red curve shows the change of HBO concentration, and the blue curve shows the change of HBR concentration.

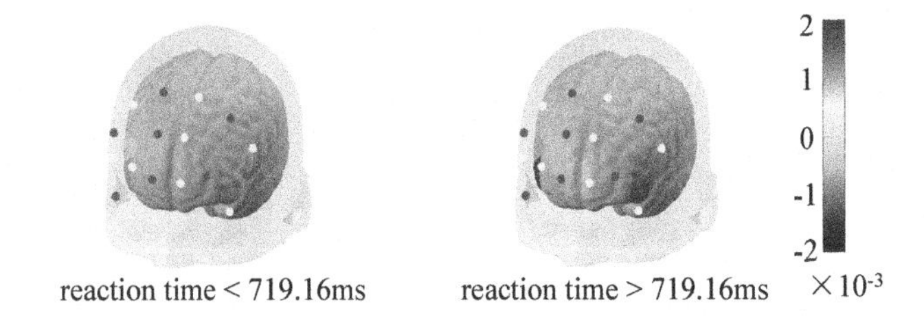

Fig. 9. The topographic maps of average HBO concentration in different stages. (Color figure online)

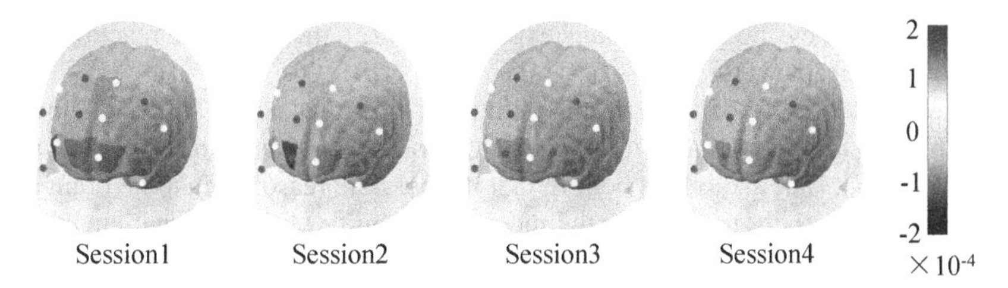

Fig. 10. The topographic map of average concentration of HBO in different sessions. (Color figure online)

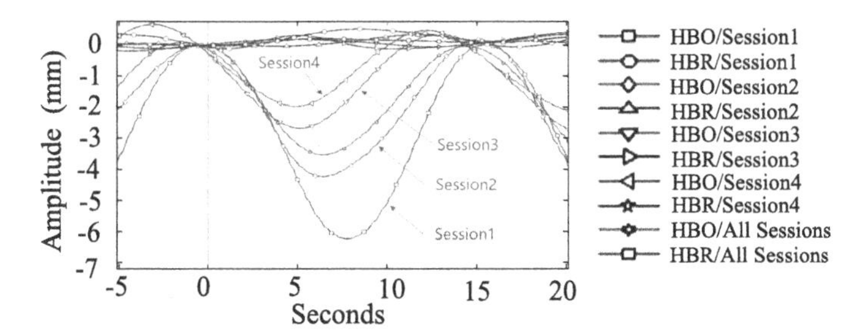

Fig. 11. The curve of average concentration of HBO and HBR in different sessions.

4 Discussion

Subjective scale can reflect the real subjective feelings of the subjects from the psychological level. In the RSVP small target detection task, the subjective fatigue feelings of the subjects have a great impact on the completion of the task. Therefore, the subjective scale evaluation results are meaningful. From the result of subjective scale evaluation, each Session successfully induced fatigue, and with the increase of the number of Sessions, the fatigue degree is also increasing after the completion of the task. This is due to the continuous long-term task leading to the accumulation of fatigue. The subjective fatigue evaluation value before the task in Session4 is higher than that of Session3. It may be that the subjects know that the experiment is going to end, and they are psychologically excited, which alleviates the feeling of fatigue to a certain extent.

Reaction time can objectively reflect the subject's reaction ability to specific tasks under the current body state. Relevant studies show that under the state of brain fatigue, people's reaction time will increase [6, 7], which is consistent with the results of this study. This study attempts to study the fatigue state of the subjects from the probability density distribution and cumulative distribution of reaction time. According to the mathematical expectation of the fitting curve, the awake and fatigue stages of the subjects in the RSVP small target detection task are divided, and the result is verified by the near infrared data. In addition, from the distribution of reaction time of four Sessions, we can see that with the increase of Session times, the peak value of probability density curve increases, the reaction time corresponding to the peak arrival point decreases, and the reaction time corresponding to the point where the cumulative distribution curve tends to be stable decreases. This may be that people gradually form conditioned reflex to the task of pressing keys because of many experiments [22], therefore, the reaction time is reduced to a certain extent.

Relevant studies have shown that the activity of prefrontal cortex (PFC) will increase under the state of brain fatigue [16, 17], which is consistent with the results of this study. Khan M.J. et al. [20] and others believe that the possible reason is that the brain needs to increase the activity to enhance the monitoring ability of the external environment under the state of fatigue. It can be seen from Fig. 9 that in the brain state divided by the mathematical expectation of reaction time, the HBO concentration in the PFC is lower when in awake stage, while the HBO concentration in the PFC is higher when in fatigue stage. This indicating that the brain needs to activate the PFC to a greater extent to ensure the completion of the task in the fatigue state. At the same time, it is observed that the HBO concentration on both sides of the PFC is significantly higher than that in other regions. This indicating that the region for the execution of RSVP small target detection task of the brain is concentrated on both sides of PFC. The results verify the effectiveness of fatigue assessment based on reaction time. Combined with Fig. 10 and Fig. 11, we can see that with the increase of Session times, the HBO concentration in PFC is increasing at the moment of picture appearance. Meanwhile, the reaction speed of HBO is getting faster and faster. It shows that with the repetition of the tasks, short-term rest can not alleviate the mental fatigue of the subjects. Fatigue is accumulating, and the brain needs a longer rest time to recover from fatigue.

5 Conclusion

In this paper, the RSVP small target detection task is designed to induce mental fatigue, and exploring the changes of reaction time and fNIRS index in the fatigue state of human body. The mental fatigue degree is divided into awake and fatigue state by mathematical expectation of reaction time, and the fNIRS data is used to verify the result. The result shows that the reaction time index and fNIRS index can be used as potential indicators to evaluate the brain fatigue model in RSVP small target detection task, and this provide the basis for the later research on the monitoring and regulation of brain fatigue.

Acknowledgments. This work is supported by Key Research & Development Project of National Science and Technique Ministry of China (No. 2018YFC0807405, No. 2018YFC1312903), National Natural Science Foundation of China (No. 61976133), and Major scientific and technological innovation projects of Shan Dong Province (2019JZZY021010).

References

1. Grandjean, E.: Fatigue in industry. J. Occup. Environ. Med. **36**(3), 175–186 (1979)
2. Grandjean, E.: Fitting the Task to the Human: A Textbook of Occupational Ergonomics, 5th edn. Taylor & Francis, London (1997)
3. Layzer, R.B.: Asthenia and the chronic fatigue syndrome. J. Muscle Nerve. **21**(12), 1609–1611 (1998)
4. Campagnolo, N., Johnston, S., Collatz, A., et al.: Dietary and nutrition interventions for the therapeutic treatment of chronic fatigue syndrome/myalgic encephalomyelitis: a systematic review. J. Hum. Nutr. Diet. **30**(3), 247–259 (2017)

5. Miao, D.M., Wang, J.S.H., Liu, L.: Handbook of Military Psychology. China Light Industry Press, Beijing (2004)
6. Langner, R., Steinborn, M.B., Chatterjee, A., et al.: Mental fatigue and temporal preparation in simple reaction-time performance. J. Acta Psychol. **133**(1), 64–72 (2009)
7. Guo, M.Z., Li, S.W.: Driving fatigue quantification based on driver's reaction time. J. Jilin Univ. (Eng. Technol. Edn.) **50**(03), 951–955 (2020)
8. Foong, R., Ang, K.K., Quek, C., et al.: An analysis on driver drowsiness based on reaction time and EEG band power. In: 2015 37th Annual International Conference of the IEEE Engineering in Medicine and Biology Society (EMBC), Milano, Italy, pp. 7982–7985 (2015)
9. Sonnleitner, A., Treder, M.S., Simon, M., et al.: EEG alpha spindles and prolonged brake reaction times during auditory distraction in an on-road driving study. Accid. Anal. Prev. **62**, 110–118 (2014)
10. Verdonck, S., Tuerlinckx, F.: Factoring out nondecision time in choice reaction time data: theory and implications. J. Psychol. Rev. **23**(1), 1–6 (2016)
11. Li, S.W., Yin, Y.N., Wang, L.H., et al.: Driving fatigue quantization based on entropy weight method. J. South Chin. Univ. Technol. (Nat. Sci. Edn.) **45**(8), 50–56 (2017)
12. David, A.B., Clare, E.E., Marco, F., et al.: Twenty years of functional near-infrared spectroscopy: introduction for the special issue. J. NeuroImage. **85**(11), 1–5 (2014)
13. Marco, F., Valentina, Q.: A brief review on the history of human functional near-infrared spectroscopy (fNIRS) development and fields of application. J. NeuroImage. **63**(2), 921–935 (2012)
14. Jöbsis, F.F.: Non-invasive, infra-red monitoring of cerebral O_2 sufficiency, bloodvolume, HbO_2-Hb shifts and bloodflow. Acta Neurol. Scand. Suppl. **64**(1), 452–453 (1977)
15. Zhang, L., Zhang, C., Feng, H.E., et al.: Research progress on the interaction effects and its neural mechanisms between physical fatigue and mental fatigue. J. Biomed. Eng. **32**(5), 1135–1140 (2015)
16. Madsen, S.J., Yang, V., Jansen, E.D., et al.: Shed a light in fatigue detection with near-infrared spectroscopy during long-lasting driving. In: SPIE BIOS, San Francisco, California, USA, pp. 96901T (2016)
17. Muthalib, M., Kan, B., Nosaka, K., Perrey, S.: Effects of Transcranial direct current stimulation of the motor cortex on prefrontal cortex activation during a neuromuscular fatigue task: an fNIRS study. In: Van Huffel, S., Naulaers, G., Caicedo, A., Bruley, D.F., Harrison, D.K. (eds.) Oxygen Transport to Tissue XXXV. AEMB, vol. 789, pp. 73–79. Springer, New York (2013). https://doi.org/10.1007/978-1-4614-7411-1_11
18. Li, Z.Y., Dai, S.H.X., Zhang, X.Y., et al.: Assessment of Cerebral oxygen saturation using near infrared spectroscopy under driver fatigue state. J. Spectro. Spectral Anal. **30**(1), 58–61 (2010)
19. Mehta, R.K., Parasuraman, R.: Effects of mental fatigue on the development of physical fatigue: a neuroergonomic approach. Hum. Fact. J. Hum. Fact. Ergon. Soc. **56**(4), 12 (2014)
20. Khan, M.J., Hong, K.S.: Passive BCI based on drowsiness detection: an fNIRS study. Biomed. Optics Exp. **6**(10), 4063–4078 (2015)
21. He, H., Dong, G.: Application study on Japanese "subjective fatigue symptoms" (2002 version) in a Chinese manufacture. Chin. J. Ergon. **15**(03), 26–28 (2009)
22. Shang, Y.C.: Learning behavior of classical and operational conditioning in animals. Bull. Biol. **40**(012), 7–9 (2005)

Multi-label Diagnosis Algorithm for Arrhythmia Diseases Based on Improved Classifier Chains

Hao Huang, Jintao Lv, Yu Pu, Yuxuan Wang, and Junjiang Zhu(✉)

College of Mechanical and Electrical Engineering, China Jiliang University, Hangzhou, China

Abstract. Electrocardiogram (ECG) has been proved to be the most common and effective approach to investigate arrhythmia. In clinical, a segment of ECG signal often indicates several arrhythmia diseases. Therefore, the automatic diagnosis algorithm of arrhythmia can be seen as a multi-label classification problem. In order to improve the classification accuracy, a method based on improved classifier chains is proposed in this paper. First, a deep neural network is pre-trained to extract the features of the ECG, and then multiple extreme random forest classifiers are used to construct a classifier chain in line with the process of clinical diagnosis, thereby completing the multi-label diagnosis of arrhythmia diseases. The experiment results show that compared with the method based on neural network, the subset accuracy of proposed method is improved from 76.62% to 83.94%, while other indicators are also improved.

Keywords: Deep learning · Classifier chains · Attention mechanism · Extra random forest

1 Introduction

Arrhythmia, also known as cardiac arrhythmia or heart arrhythmia, is a group of conditions in which the heartbeat is irregular, too fast, or too slow [1]. There are four main groups of arrhythmia: extra beats, supraventricular tachycardias, ventricular arrhythmias and bradyarrhythmia [2]. The types of arrhythmia are various, and some of them can exist at the same time or alone. Clinical diagnosis of arrhythmia usually depends on resting electrocardiogram (ECG). In practice, one ECG signal usually embraces several arrhythmia diseases at the same time. Arrhythmia affects millions of people, but the number of professional cardiologists who can understand the ECG is particularly scarce relative to the number of patients. Therefore, it is significant to develop a computer-aided ECG diagnosis algorithm.

Traditional ECG diagnosis algorithm extracts the feature of the ECG signal and then classifiers it to complete the diagnosis. For example, based on the morphological characteristics and the statistical analysis of the ECG signal, scholars have used support vector machines (SVM) [3], random forest (RF) [4], and multilayer perceptron (MLP) [5] to achieve the diagnosis of arrhythmia. However, traditional methods rely on expert knowledge and are complicated in operation.

M. Fei et al. (Eds.): LSMS 2021/ICSEE 2021, CCIS 1467, pp. 94–103, 2021.
https://doi.org/10.1007/978-981-16-7207-1_10

As an end-to-end method, deep neural network can automatically extract ECG data features. Multi-label classification algorithms based on deep learning can be divided into two types. One is the problem transformation method, such as Binary Relevance [6], Label Powerset [7] and Classifier Chains [8]. The basic idea of Binary Relevance algorithm is to decompose the multi-label problem to several binary classification problem, which ignores the correlation between the label. The Label Powerset algorithm maps each combination to a unique combination id number, which leads to label sparsity and difficult model training. The basic idea of Classifier Chains is to transform the multi-label learning problem into a chain of binary classification problems, where subsequent binary classifiers in the chain is built upon the predictions of preceding ones. This algorithm has the advantage of exploiting label correlations while has a huge computational burden. Another mothed is algorithm adaptation. A deep neural network can tackle multi-class learning problem by changing the form of the output layer to deal with multi-label problem directly [9–11]. It essentially splits multiple multi-label problems into multiple binary classification problems and does not consider the correlation between labels.

Therefore, a multi-label arrhythmia diagnosis algorithm based on classifier chains is proposed to utilize the ECG features extracted by deep neural networks. In this paper, a deep neural network is pre-trained to extract the features of the ECG signal at first, and then multiple Extreme Random Forest (ERF) [12] classifiers are used to construct a classifier chain according to the clinical diagnosis idea, thereby completing the multi-label diagnosis of arrhythmia diseases. This algorithm reduces the computational burden by using the ECG feature rather than raw ECG signal as the input for classifier chain. Due its chaining property, this method also mines the associations between labels and improves the accuracy of multi-labels classification obviously.

2 Method

After completing the collection and labeling of ECG data, this paper pre-trained a deep neural network which is used to automatically extract ECG features. Subsequently, multiple extreme random forest classifiers are used to form a classifier chain to mines the associations between labels. Those extracted features are used as the input of the classifier chain, therefore classifier chains complete the multi-label diagnosis of arrhythmia diseases, as shown in Fig. 1.

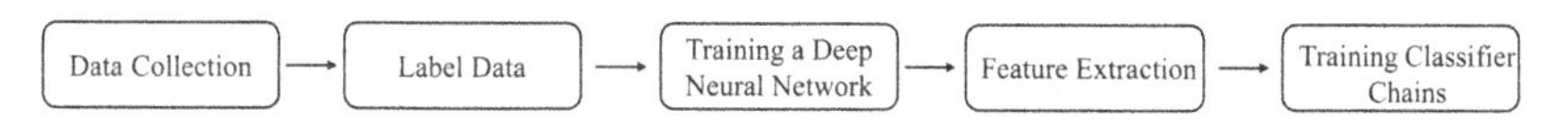

Fig. 1. The flow chart of method in this article.

2.1 Dataset and the Process of ECG

Clinically, the ECG used to analyze arrhythmia diseases is a 10-s 12-lead resting ECG signal. The multi-label arrhythmia diagnosis algorithm needs to be based this type of ECG signal to be applied to the clinical diagnosis. The commonly used ECG databases in the world include MIT-BIH Arrhythmia Database [13], American Heart Association

ECG Database [14], Common Standards for Electrocardiography Database [15] and European ST-T Database [16], but most of the ECG data of them are only 2-leads and few types of arrhythmia disease in the database among these databases, which do not meet the needs of algorithm.

Therefore, this paper utilizes the 12-lead clinical resting data collected by Shanghai. Shuchuang Medical Technology Company in several hospital in Shanghai. The ECG signal length is 10 s, and the sampling frequency is 500 Hz. In order to collect data accurately, the electrode sheet first amplifies the value of the ECG signal by 400 times when collecting the ECG signal and then uses discretization. Due to the power frequency interference when collecting the ECG signal, a trap filter is designed in the hardware circuit. The Butterworth band-pass filter [17] is used to eliminate high-frequency and low-frequency noise, and then the ECG signals of each lead after the filtered wave are rearranged together according to the lead order.

The several arrhythmia diseases are selected to label the ECG signal. These arrhythmia diseases are atrial fibrillation, atrial flutter, sinus tachycardia, sinus rhythm, sinus arrhythmia, sinus bradycardia, atrial premature beat (PAC), ventricular premature beat (PVC), A-V junction premature beats (JPBs) and First-degree atrioventricular block (I AVB). Two professional cardiologists label each ECG data. If cardiologists disagree with each other, the third cardiologist will determine the label. The distribution of arrhythmia diseases in different data sets is shown in Table 1.

Table 1. Overview of datasets.

	Normal	Atrial Fibrillation	Atrial Flutter	Sinus Tachycardia	Sinus Rhythm	Sinus Arrhythmia
Train	7056	11639	13628	10847	43980	4420
Validation	608	1002	1248	926	3805	393
Test	2343	2531	2221	2733	7811	858
	Sinus Bradycardia	PVC	PAC	JPBs	I AVB	All Num
Train	5486	9982	12057	9869	13516	90000
Validation	499	933	1022	856	1191	7873
Test	837	1748	2542	1229	1844	16991

2.2 Deep Neural Network

Models based on the deep residual neural network (ResNet) [18] series have shown excellent performance in classification tasks. This paper intends to use the ResNet-34 as the feature extractor. Subsequently, this paper introduces the long short-term memory (LSTM) [19] layer to mine the relationship between features. In the LSTM layer, a hidden state output is generated at every moment, but LSTM only outputs the output at the last time step, ignoring the hidden output at the intermediate time step. The attention mechanism [20] is used to assign different weights to the hidden outputs in the

LSTM that have different effects on the results. These features then are input into a fully connected layer classification. In the last layer, the sigmoid function is used to calculate the probability of arrhythmia. The overall structure of the model is shown in Fig. 2.

The model proposed in this paper can be divided into two parts: feature extraction part and feature decision based on fully connected layer. After the model is trained, the fully connected layer in the model is removed to retain the feature extraction part of the model, which can be used to extract the features of the ECG data and input the ECG features into the classifier chain. This method can avoid the significant computational burden caused by inputting the raw ECG signal into the classifier chain.

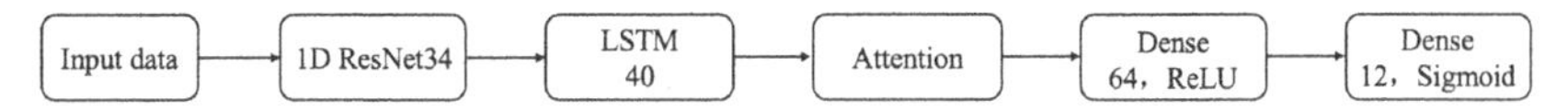

Fig. 2. The structure of the pre-trained model.

2.3 Classifier Chains

As a multi-label algorithm of problem transform, classifier chains combine multiple binary classifiers in a certain order. In classifier chains, each classifier is used to predict the corresponding label. Each classifier uses all available features provided to the model and the labels used for the earlier model in the chain as training data. In this way, classifier chains mine the association between labels.

In the classifier chains, the order of each classifier has a great impact on the model results, and it is necessary to design the order of tags. Therefore, following the clinical diagnosis of arrhythmia diseases, six model is designed to predict the several label of ECG signal: normal, dominant rhythms (including atrial fibrillation, atrial flutter, sinus tachycardia, sinus rhythm, sinus arrhythmia, sinus bradycardia), atrial premature beat (PAC), ventricular premature beat (PVC), A-V junction premature beats (JPBs) and First-degree atrioventricular block (I AVB).

As a typical classification algorithm in machine learning, extreme random forest is used to complete the classification of each label. In the training phase, after the ECG

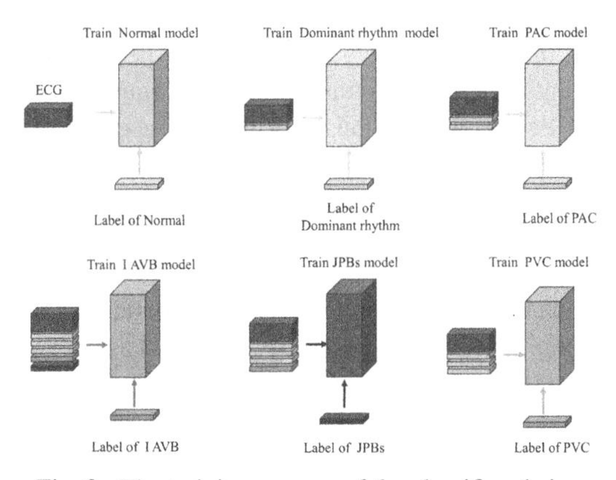

Fig. 3. The training process of the classifier chains.

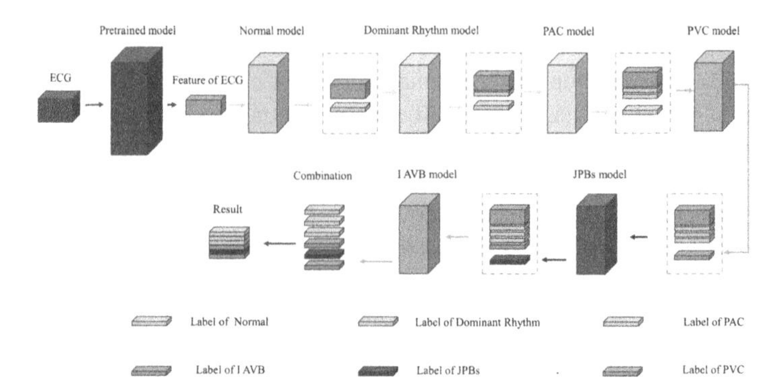

Fig. 4. The test process of the classifier chain.

signal is extracted, these models are trained in order. As shown in Fig. 3, when training the model, the ECG features and all the label previously used are used as input for training. In the testing phase, the feature of ECG in test set is extracted by the pre-trained model firstly and then the feature of ECG will be input to the classifier chains next. As is shown in Fig. 4, the previous model's prediction results and the ECG features are utilized as the input to the next classifier for prediction. Finally, each classifier's prediction results in the classifier chains are combined to the final multi-label of ECG signals.

3 Experiment

The experimental environment is to build a model based on the convolutional neural network proposed above and then use the pre-processed ECG data and labels to input into the model and pre-train the model. Adam is selected as an optimizer. The loss function is the binary cross-entropy loss function. The initial learning rate is 0.001. The weight attenuation is L2 regularization, and the attenuation coefficient is set as 0.0005, and the momentum is 0.9. The epoch of training sessions is 100. The early stopping mechanism is introduced during model training to detect the model's performance on the validation set. If the loss does not decrease after 10 consecutive training, it means that the model has converged and then stopped training. After the pre-training model training is completed, the fully connected layer part of the model is removed, and the feature extraction part of the model is retained. The pre-trained model is used to extract the feature of ECG. The feature and the label will be next used to train six extreme random forest classifiers.

3.1 Evaluation Metrics

There are several evaluation indictors in multi-label classification problem as shown in formulas 1 to 6. Among them, Subset Accuracy represents the ratio that the diagnosis of arrhythmia is wholly correct, which is the most important indicator to measure the clinical diagnosis algorithm for arrhythmia. $\boldsymbol{X}$ represents the overall instance $\boldsymbol{x}$ set. $\boldsymbol{y}$ represents the label corresponding to the instance $\boldsymbol{x}$, $\boldsymbol{h}_{(\boldsymbol{x})}$ represents the classification result of the multi-label classifier for $\boldsymbol{x}$, $\boldsymbol{L}_j$ represents the jth label in the label vector corresponding to the instance $\boldsymbol{x}$, $\otimes$ is the logical AND symbol, $\boldsymbol{TP_j}$ (Ture Positive) represents the number of positive samples correctly predicted by the model in the $\boldsymbol{j}$th label, and $\boldsymbol{FP_j}$ (False Positive) and $\boldsymbol{FN_j}$ (False Native) respectively represents the number of positive and Native samples incorrectly predicted by the model in the $\boldsymbol{j}$th label.

$$Subset\ Accuracy(h) = \frac{1}{|X|}\sum_{x \in x}[h(x) = y]. \tag{1}$$

$$Hamming\ Loss(h) = \frac{1}{|X|}\sum_{x \in X}\frac{1}{l}\sum_{j=1}^{l}[(L_j \in h_{(x)}) \otimes (L_j \in y)]. \tag{2}$$

$$Jacccard\ Index(h) = \frac{1}{|X|}\sum\nolimits_{x \in X}\frac{h(x) \cap y}{h(x) \cup y}. \tag{3}$$

$$precison = \frac{\sum_{j=1}^{j=l} TP_j}{\sum_{j=1}^{j=l} TP_j + FP_j}. \tag{4}$$

$$recall = \frac{\sum_{j=1}^{j=l} TP_j}{\sum_{j=1}^{j=l} TP_j + FN_j}. \tag{5}$$

$$F1 = \frac{2 * precision * recall}{precsion + recall}. \tag{6}$$

3.2 Result of Pre-trained Model

After training, the pre-training model is tested on the test set, and the results are shown in Table 2. On subset accuracy which is the most important indicator for measuring ECG data, it can be seen that the subset accuracy of the pre-trained model is only 76.62%. It means that the deep neural network has certain limitations in dealing with the multi-label problem of ECG. However, considering that the pre-trained model is only used to extract ECG features, it is necessary to analyze the feature-extracting ability of the pre-trained model.

As shown in the Table 3, it can be found that the pre-trained model has good performance for each label. On the index of accuracy, the pre-trained model is suitable for atrial fibrillation, sinus tachycardia, sinus arrhythmia, sinus bradycardia and first-degree atrioventricular Block. These arrhythmia diseases are all greater than 97%, and

Table 2. Result of the pre-trained model in test set.

Methods	Subset accuracy	Hamming loss	Jaccard index	Precision	Recall	F1 score
Pre-trained model	76.62%	0.0326	0.8641	89.71%	85.80%	0.8760

the accuracy of other labels are also higher than 94%. The model can effectively extract the features which are used to distinguish different arrhythmia diseases. However, the pre-trained model cannot effectively complete the multi-label classification only relying on the fully connected layer. These features require a better multi-label classification algorithm to be used.

Table 3. Result of pre-trained model in different label.

	Normal	Atrial Fibrillation	Atrial Flutter	Sinus Tachycardia	Sinus Rhythm	Sinus Arrhythmia
Accuracy	97.16%	97.17%	96.75%	98.56%	94.62%	97.32%
Precision	91.67%	92.43%	87.07%	95.77%	93.36%	74.85%
Recall	87.37%	88.47%	88.25%	95.28%	94.37%	70.75%
	Sinus Bradycardia	PVC	PAC	JPBs	I AVB	
Accuracy	98.36%	95.17%	96.95%	96.67%	97.14%	
Pression	87.42%	81.94%	92.95%	78.14%	88.15%	
Recall	78.02%	68.02%	86.11%	75.02%	85.09%	

3.3 Compare with the Other Classifier

In order to effectively use features, this paper construct classifier chains by several Extra Random Forest (ERF) model to complete the multi-label diagnosis of arrhythmia diseases. In addition, this article also compare ERF with Support Vector Classification (SVC) [21], Random Forest (RF) [12], k-nearest neighbors classifier (kNN) [22], and Gradient Boosting Decision Tree (GBDT) [23]. The kernel function of SVM is RBF kernel function. The number of neighbors is set to 5 in kNN. In the extreme random forest, the estimators selected are 10 and there is no restriction on the maximum depth of the subtree. In the GBDT, the loss method is deviance, the learning rate is set to 0.1, and the depth of the subtree is selected as 100.

As shown in Table 4, it can be seen that the ERF has the highest accuracy in the most critical subset accuracy, and the result is 83.94%. Followed by ERF, subset accuracy of RF is 83.86%. The SVM, KNN and gradient boosting tree from high to low are 81.95%, 81.69% and 81.31%. Except for the subset accuracy, the ERF is better than other classifiers on Hamming Loss, Jaccard Index, Precision, Recall and F1 score. Besides, the effects of RF and ERF are better than SVC, KNN, and GBDT. Since using deep neural networks to extract features is a non-linear change, non-linear classifiers such as random forests have better performance when processing such features. The ERF uses all the samples in the training set during training and uses a random division method to divide the feature values. Therefore, the ERF obtains better generalization ability.

Table 4. Result of difficult classifier.

Methods	Subset accuracy	Hamming loss	Jaccard index	Precision	Recall	F1 score
ERF	**83.94%**	**0.0236**	**0.8954**	**92.67%**	**90.59%**	**0.9159**
SVC	81.95%	0.0263	0.8852	91.82%	89.44%	0.9054
RF	83.86%	0.0238	0.8944	92.69%	90.41%	0.9148
KNN	81.69%	0.0274	0.8827	91.05%	89.56%	0.9026
GBDT	81.31%	0.0272	0.8808	91.39%	89.19%	0.9022

3.4 Compared with Other Method

In order to study whether the model combination using classifier chain can effectively improve the performance of the multi-label algorithm. After using the pre-training model to extract the features, this paper compares serval classifier for each label, and then adapt the algorithm into a multi-label diagnosis algorithm by binary relevance. After using the pre-trained model to extract the ECG features, this paper trains the Multi-label Random Forest (ML-RF), Multi-Label SVM (ML-SVC), Multi-Label KNN (ML-kNN) and Multi-label gradient boosting tree (ML-GBDT). The results are shown in Table 5.

From Table 5, it can be found that the extreme random forest has a good performance, with the subset accuracy is 82.74%, followed by a multi-label random forest with a subset accuracy of 82.40%. It can be found that the method of using binary relevance has a significant drop compared with the method of using the classifier chain. The reason is that when using binary relevance to construct a classifier for each type of ECG label, the classifiers are trained independently of each other, so the correlation between the label and the label cannot be considered, thereby reducing the overall prediction accuracy.

Table 5. Result of pre-trained model in different label.

Methods	Subset accuracy	Hamming loss	Jaccard index	Precision	Recall	F1 score
Our method	**83.94%**	**0.0236**	**0.8954**	**92.67%**	**90.59%**	**0.9159**
ML-kNN	81.16%	0.0311	0.8697	89.87%	88.11%	0.8896
ML-RF	82.40%	0.0271	0.8762	91.94%	88.74%	0.9027
ML-ERF	82.74%	0.0269	0.8781	91.95%	88.89%	0.9036
ML-SVC	80.88%	0.0297	0.8693	91.03%	87.65%	0.8925
ML-GBDT	79.47%	0.0310	0.8627	90.32%	87.59%	0.8890

4 Summary

In order to improve the classification accuracy, a method based on improved classifier chains is proposed in this paper. First, a deep neural network is pre-trained to extract the features of the ECG signal, and then multiple extreme random forest classifiers are used to construct a classifier chain according to the clinical diagnosis idea, thereby completing the multi-label diagnosis of arrhythmia diseases. This paper also compared with the direct use of the deep neural network for multi-label prediction and the use of binary relevance for multi-label prediction, which demonstrated the effectiveness of the proposed method based on the classifier chains. In addition, to select the best classifier, this article compares with the SVC, RF, kNN, and GBDT classifiers. Experimental results show that extreme random forest has a good performance in processing ECG characteristics.

Acknowledgments. The research is supported by Supported by Natural Science Foundation (No: 61801454).

References

1. Antzelevitch, C., Burashnikov, A.: Overview of basic mechanisms of cardiac arrhythmia. J. Card. Electrophysiol. Clin. **3**, 23–45 (2011)
2. My Publications: Types of Arrhythmia - NHLBI, NIH. My Publications (2011)
3. Faziludeen, S., Sabiq, P.: ECG beat classification using wavelets and SVM. In: 2013 IEEE Conference on Information & Communication Technologies, pp. 815–818. IEEE (2013)
4. Emanet, N.: ECG beat classification by using discrete wavelet transform and Random Forest algorithm. In: 2009 5th International Conference on Soft Computing, Computing with Words and Perceptions in System Analysis, Decision and Control, pp. 1–4. IEEE (2009)
5. Peláez, J.I., Doña, J.M., Fornari, J.F., Serra, G.: Ischemia classification via ECG using MLP neural networks. Int. J. Comput. Intell. Syst. **7**(2), 344 (2014). https://doi.org/10.1080/18756891.2014.889498
6. Luaces, O., Díez, J., Barranquero, J., Coz, J.J., Bahamonde, A.: Binary relevance efficacy for multilabel classification. J. Prog. Artif. Intell. **1**, 303–313 (2012)
7. Senge, R., Coz, J.J., Hüllermeier, E.: On the problem of error propagation in classifier chains for multi-label classification. In: Spiliopoulou, M., Schmidt-Thieme, L., Janning, R. (eds.) Data Analysis, Machine Learning and Knowledge Discovery. SCDAKO, pp. 163–170. Springer, Cham (2014). https://doi.org/10.1007/978-3-319-01595-8_18
8. Read, J., Pfahringer, B., Holmes, G., Frank, E.: Classifier chains for multi-label classification. Mach. Learn. **85**, 333 (2011)
9. Zhu, J., Xin, K., Zhao, Q., Zhang, Y.: A multi-label learning method to detect arrhythmia based on 12-lead ECGs. In: Liao, H., et al. (eds.) MLMECH/CVII-STENT -2019. LNCS, vol. 11794, pp. 11–19. Springer, Cham (2019). https://doi.org/10.1007/978-3-030-33327-0_2
10. Huang, C., Zhao, R., Chen, W., Li, H.: Arrhythmia classification with attention-based res-BiLSTM-Net. In: Liao, H., et al. (eds.) MLMECH/CVII-STENT -2019. LNCS, vol. 11794, pp. 3–10. Springer, Cham (2019). https://doi.org/10.1007/978-3-030-33327-0_1
11. Cai, J., Sun, W., Guan, J., You, I.: Multi-ECGNet for ECG Arrythmia multi-label classification. IEEE Access **8**, 110848–110858 (2020). https://doi.org/10.1109/ACCESS.2020.3001284
12. Ahmad, I., Basheri, M., Iqbal, M.J., Rahim, A.: Performance comparison of support vector machine, random forest, and extreme learning machine for intrusion detection. J. IEEE Access **6**, 33789–33795 (2018)
13. Moody, G.B., Mark, R.G.: The impact of the MIT-BIH arrhythmia database. IEEE Eng. Med. Biol. Mag. **20**(3), 45–50 (2001). https://doi.org/10.1109/51.932724

14. Monina Klevens, R., et al.: Estimating health care-associated infections and deaths in U.S. hospitals, 2002. Pub. Health Rep. **122**(2), 160–166 (2016). https://doi.org/10.1177/003335490712200205
15. Laguna, P., Jané, R., Caminal, P.: Automatic detection of wave boundaries in multilead ECG signals: validation with the CSE database. J. Comput. Biomed. Res. **27**, 45–60 (1994)
16. Taddei, A., et al.: The European ST-T database: standard for evaluating systems for the analysis of ST-T changes in ambulatory electrocardiography. J. Eur. Heart J. **13**, 1164–1172 (1992)
17. Gomez, V.: The use of Butterworth filters for trend and cycle estimation in economic time series. J. Bus. Econ. Stat. **19**, 365–373 (2001)
18. Wu, Z., Shen, C., Van Den Hengel, A.: Wider or deeper: revisiting the ResNet model for visual recognition. J. Pattern Recogn. **90**, 119–133 (2019)
19. Graves, A.: Long short-term memory. In: Graves, A. (ed.) Supervised Sequence Labelling with Recurrent Neural Networks, pp. 37–45. Springer, Heidelberg (2012). https://doi.org/10.1007/978-3-642-24797-2_4
20. Yang, Z., Yang, D., Dyer, C., He, X., Smola, A., Hovy, E.: Hierarchical attention networks for document classification. In: Proceedings of the 2016 Conference of the North American Chapter of the Association for Computational Linguistics: Human Language Technologies, pp. 1480–1489(2016)
21. Brereton, R.G., Lloyd, G.R.: Support vector machines for classification and regression. Analyst **135**, 230–267 (2010)
22. Zhang, M.-L., Zhou, Z.-H.: ML-KNN: a lazy learning approach to multi-label learning. Pattern Recogn. **40**, 2038–2048 (2007)
23. Ke, G., et al.: LightGBM: a highly efficient gradient boosting decision tree. J. Adv. Neural Inf. Process. Syst. **30**, 3146–3154 (2017)

Support Vector Machine Classification Method Based on Convex Hull Clipping

Yaqin Guo(✉) and Jianguo Wu

Electrical and Energy Engineering College, Nantong Institute of Technology, Nantong, China

Abstract. Convex hull is one of the basic means to describe the shape of objects, which is used in many fields of computer graphics and images. This paper proposes a support vector machine classification method based on convex hull clipping. Principle component analysis (PCA) is used for pretreatment on the planar point set. The end point of convex hull is determined by the PCA fitting line, then convex hull is obtained by the rapid sorting method. During classification, the approximating line of the boundary is constructed by translating the fitting line to the class boundary. Samples in the region surrounded by the approximating line and convex hull are boundary samples, which are used to the train support vector machine (SVM). Experiments on two artificial data sets and the UCI standard data set show that the proposed method can improve classification accuracy and reduce training time, especially when the training set is large.

Keywords: Convex hull · Support vector machine · Principle component analysis · Planar point set

1 Introduction

In pattern recognition, support vector machine is as an efficient classifier. It has been widely used in various fields, such as handwriting recognition [1], character recognition [2] and face recognition [3] and so on. However, in large sample classification, although SVM has high classification accuracy, the slow classification speed restricted the development of the technology, and many researchers have being paid more and more attention to the research. Support vectors classification method based on matrix exponent boundary Fisher projection is proposed in [4]. By boundary Fisher thought, the original training samples are projected to the low dimensional space, selecting new samples are used to train support vector machine (SVM). Support vectors classification method based on projection vector boundary feature is proposed in [5]. According to statistical theory and normal distribution characteristics in one-dimensional space, determine projection line in two-dimensional space, the training samples are projected to the line, than construct boundary vector sets in two-dimensional space, which are used to train support vector machine (SVM).

The research shows that the main function is support vector in support vector machine classification, and the support vector is mainly located at the boundary of the sample. In

M. Fei et al. (Eds.): LSMS 2021/ICSEE 2021, CCIS 1467, pp. 104–113, 2021.
https://doi.org/10.1007/978-981-16-7207-1_11

computational geometry, convex hull can describe the shape of the object by using geometrical method [6]. The feature of convex hull is describe the boundary of the object, then the boundary is related to the support vector.

Combined with the above ideas, support vector machine classification method based on convex hull clipping is proposed. In this paper, the construction method of convex hull and the selection method of boundary vector are given. In the constructing convex hull, principle component analysis (PCA) is used to do the pretreatment on the planar point set [7]. The end point of convex hull is determined by PCA fitting line, then convex hull is obtained by rapid sorting method. In the classification, the approximating line of the boundary is constructed by translating the fitting line to the class boundary. The samples in the region surrounded by approximating line and convex hull are boundary samples, which are used to train support vector machine (SVM).

2 Constructing Convex Hull Method

The constructing method of convex hull mainly includes three parts, namely, pretreating the sample set, determining convex hull edge end point and constructing the convex hull.

2.1 Pretreating the Sample Set

In pretreating the sample set, principal component analysis (PCA) was used to fit the plane sample points [8]. Given a set of training data $X = \{X_1, X_2, \cdots, X_n\}^T$, which $X_i = (x_i, y_i)$, $x = (x_1, x_2, \cdots, x_n)^T$, $y = (y_1, y_2, \cdots, y_n)^T$. Mean point $(\overline{x}, \overline{y})$ is calculated in Formula (1).

$$\overline{x} = \frac{1}{n}\sum_{i=1}^{n} x_i$$

$$\overline{y} = \frac{1}{n}\sum\nolimits_{i=1}^{n} x_i. \tag{1}$$

Constructing covariance matrix in Formula (2).

$$C = \begin{bmatrix} cov(x, x) & cov(x, y) \\ cov(y, x) & cov(y, y) \end{bmatrix}. \tag{2}$$

Which $\mathrm{cov}(x, y) = \frac{1}{n-1}\sum_{i=1}^{n} (x - \overline{x})^T (y - \overline{y})$.

The eigenvalue and eigenvector of covariance matrix C are solved, and the eigenvector corresponding to the largest eigenvalue is found, which is the normal vector $n(a, b)$ of fitting line. Using the mean point $(\overline{x}, \overline{y})$ as the point on fitting line, the linear equation can be determined.

2.2 Determining Convex Hull Edge End Point Method

In Sect. 2.1, the fitting line can reflect the sample points distribution. Next, the two end points of the convex hull are determined, and the sample point set is divided into two

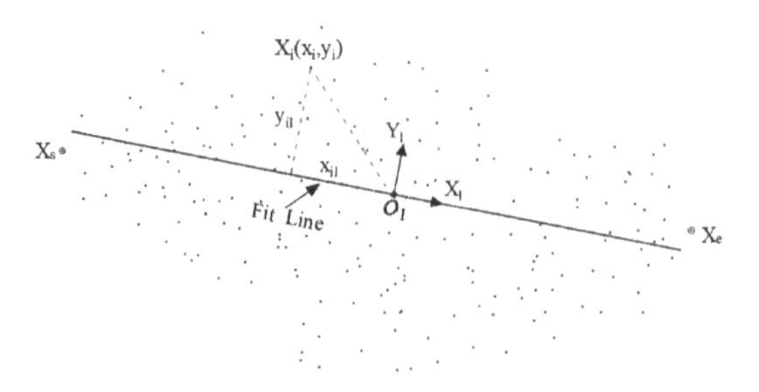

Fig. 1. Determining the convex hull edge end point.

parts by line, which is connected by two end points. The method for determining the convex hull edge end point is shown in Fig. 1.

New coordinate system $O_lX_lY_l$ is constructed by using the mean point $O_l(\overline{x}, \overline{y})$ as the origin, and taking the normal vector $n(a, b)$ of the fitting line as the axis O_lX_l direction. The new coordinates $X_{il}(x_{il}, y_{il})$ can be calculated by Formula (3) in the coordinate system $O_lX_lY_l$.

$$\begin{cases} x_{il} = \overrightarrow{X_iO_l} \cdot n \\ y_{il} = (x_i - \overline{x}) \times b + (y_i - \overline{y}) \times a \end{cases} . \tag{3}$$

According to the Formula (3), all points coordinates in the original sample set are calculated to obtain the new point set $X_l = \{X_{1l}, X_{2l}, \cdots, X_{nl}\}^T$, which$X_{il} = (x_{il}, y_{il})$. Sort the point set from small to large according to the size of x_{il}, the first point is used as the first end point X_s of the convex hull, and the last point is used as the ending point X_e of the convex hull in the sorted sequence. By connecting X_s and X_e, the dividing line is constructed. The point set is divided into two parts. It is shown in Fig. 2.

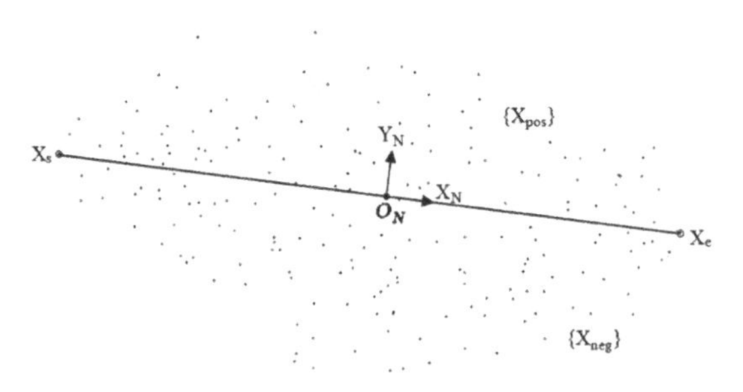

Fig. 2. Segmentation of plane point set.

Using the midpoint O_N of the line X_sX_e as the coordinate origin, taking the vector $\overrightarrow{X_sX_e}$ as the axis O_NX_N direction, new coordinate system is constructed. Using the above method, the sample points are converted to coordinate plane $O_NX_NY_N$, and get new sample point set $X_N = \{X_{1N}, X_{2N}, \cdots, X_{nN}\}^T$, which $X_{iN} = (x_{iN}, y_{iN})$. According to the positive and negative of y_{iN}, the sample point set is divided into two parts $\{X_{pos}\}$ and $\{X_{neg}\}$. In the process of searching convex hull, two parts can be searched separately, and the first and last points of the two parts are X_s and X_e. It is also the end point of the convex hull.

2.3 Constructing Convex Hull Method

Taking pointing set $\{X_{pos}\}$ as an example, constructing convex hull method is introduced. The point set $\{X_{pos}\}$ is sorted according to the size of x_{iN}. The sorting method is as follows.

(1) If $x_{iN} < x_{jN} (i \neq j)$, X_{jN} comes after X_{iN}.
(2) If $x_{iN} = x_{jN}$ and $y_{iN} < y_{jN} (i \neq j)$, X_{jN} comes after X_{iN}.

In the sorting process, the maximum value y_{iN} is found as the new convex hull edge point $X_{MN}(x_{MN}, y_{MN})$. The point set $\{X_{pos}\}$ is divided into left and right regions by the size of x_{MN}. It is shown in Fig. 3. The reference line $l_l(X_s X_{MN})$ is set in the left area, and the reference line $l_r(X_e X_{MN})$ is set in the right area. Taking the left region as an example, the search method of convex hull is introduced. Set normal vector of the reference line $l_l(X_s X_{MN})$ is $n'\left(a', b'\right)$, the distance is defined in Formula (4).

$$d_i = \overrightarrow{X_s X_{iN}} . n' \tag{4}$$

When the calculated distance d_i is negative, the point X_{iN} is removed from the point set, then when the distance d_i is positive, the point X_{iN} is retained. The new point set $\{X_{lpos}\}$ is constructed. Sort the point set $\{X_{lpos}\}$ from small to large according to the size of d_i, the last point is used as the convex hull. The decision principle of the right region is the same as that of the left region.

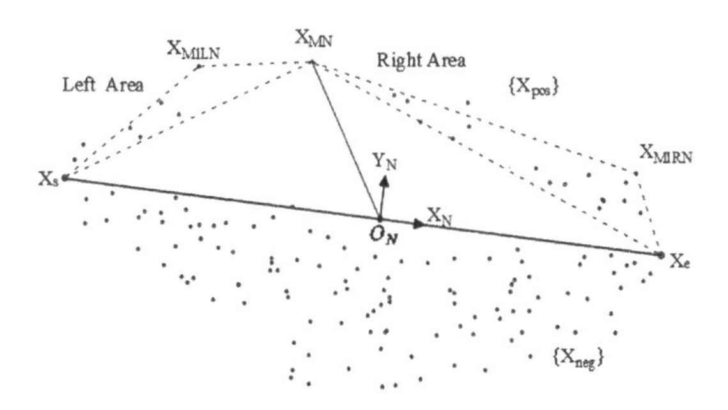

Fig. 3. Confirmation rule for the edge points of convex hull.

According to the above method, the left and right regions are subdivided by using the new edge point X_{M1LN} and X_{M1RN}. In the new region, the edge points of convex hull are searched continuously, which conform to the judgment principle. Multiple iterations are used until there are no convex hull edge points in the sample set. The convex hull is shown in Fig. 4.

3 Classification Method

3.1 Support Vector Machine

In the classification, for linear separable problems, optimal hyperplane is constructed in Formula (5) [9].

$$(w \cdot x) + b = 0 \tag{5}$$

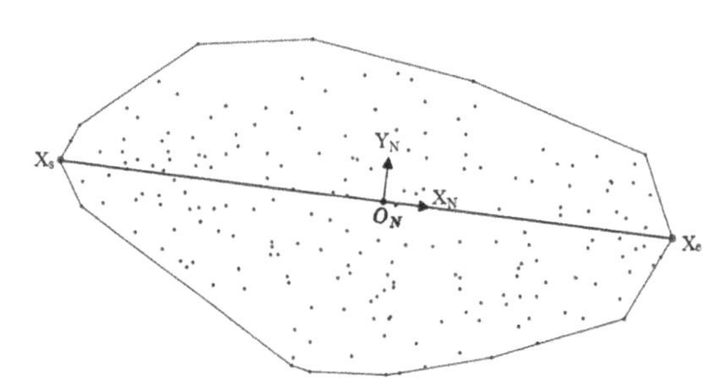

Fig. 4. Convex hull confirmed by the algorithm.

By Quadratic Programming method, optimal hyperplane equation is solved, support vector is obtained. For non-linear separable problems, the training samples are mapped to high-dimensional space by non-linear mapping, the solving method is the same as the linear separable problems. However, it takes a lot of time to train SVM.

The research shows that the main function is support vector in support vector machine classification, and the support vector is mainly located at the boundary of the sample. Therefore, the support vector can be preselected in the classification, then training speed will be accelerated.

3.2 Selecting Boundary Sample

According to the Sect. 2, the convex hull of the sample point set is constructed, and the convex hull is exactly the sample boundary. The support vector are adjacent boundary points of two kinds of samples, so the boundary constructed by convex hull will have redundant points. Selecting boundary sample method is shown in Fig. 5.

According to the above problems, the selecting boundary sample step is as follows.

(1) According to the Sect. 2, the convex hull of the sample point set is constructed.
(2) The approximating line of the boundary is constructed by translating the PCA fitting line to the class boundary. The samples in the region surrounded by approximating line and convex hull are boundary samples, which are used to train support vector machine (SVM).
(3) In the fitting line translation, the translation distance can be estimated by experimental method. In the experiment, the longest distance D_m between the fitting line and the sample point is calculated, the translation distance is $\frac{3}{4}D_m$ generally.

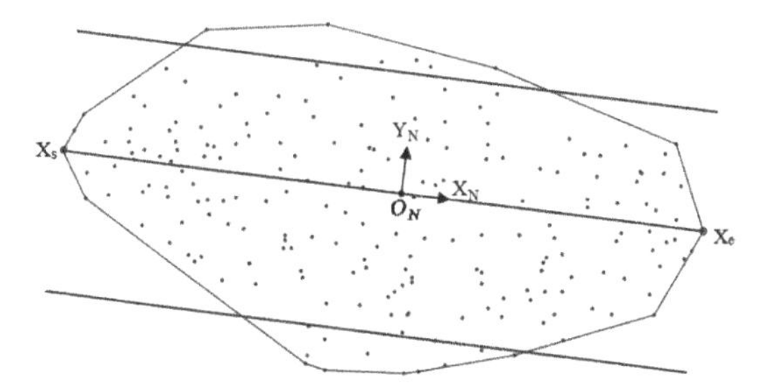

Fig. 5. Selecting boundary sample.

4 Experiment Result

In order to verify the effectiveness of the algorithm, linear separable and nonlinear separable data with different shapes are selected. In the practical application, UCI standard data set is selected [10].

4.1 Linear Separable Experiment

In the experiment, two linearly separable samples with different shapes are randomly generated. In the square sample, the Class 1 is $U([0, 2] \times [0, 0.95])$, and the Class 2 is $U([0, 2] \times [1.05, 2])$. The number of samples is 500 respectively. In the experiment, we set translation distance D is 0.36. The samples in the region surrounded by approximating line and convex hull are boundary samples. It is shown in Fig. 6. There are 400 samples in the training set, and there are 600 samples in the testing set. In the classification, Gaussian radial basis kernel function, namely. $K(x, y) = \exp[-\|x - y\|^2/2p^2]$ is used in the experiment.

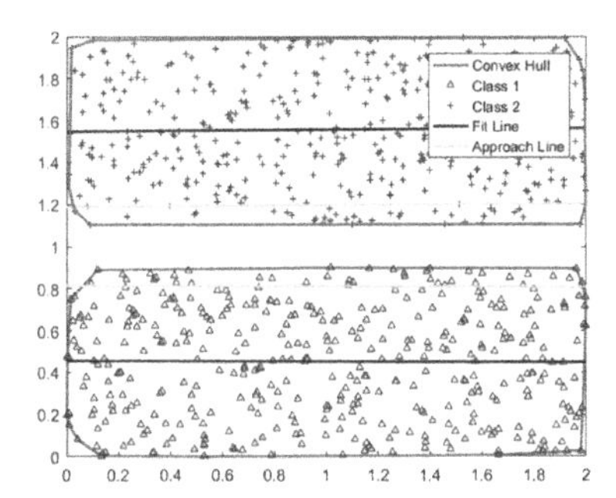

Fig. 6. Constructing boundary samples in linear the square data.

In the round sample, both classes samples radius r is 5. The center point is (3, 3) in Class 1, and the center point is (11, 11) in Class 2. The number of samples is 600 respectively. In the experiment, we set translation distance D is 3.9. The samples in the region surrounded by approximating line and convex hull are boundary samples. It is shown in Fig. 7. There are 600 samples in the training set, and there are 600 samples in the testing set. In the classification, kernel function and parameter are the same as the square samples.

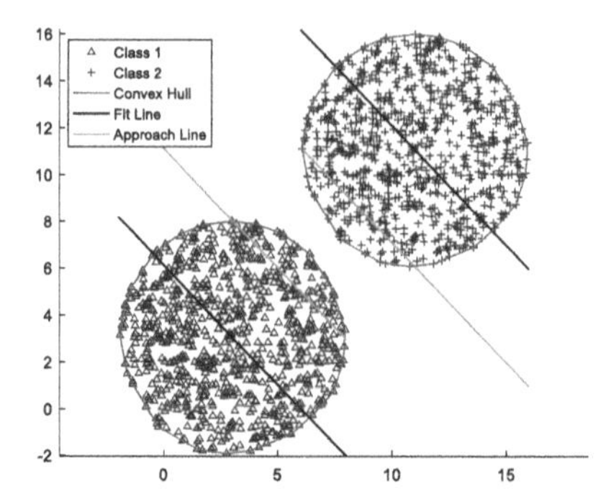

Fig. 7. Constructing boundary samples in the round data.

Table 1. Classification comparison linear separable algorithm.

Algorithm	Shape	Parameter	Boundary sample (number)	Training time (ms)	Classification accuracy (%)
Standard SVM	Square	–	–	184.2	100
	Round	–	–	201.3	100
The proposed algorithm	Square	0.36	72	100.3	100
	Round	3.9	102	110.6	100

Table 1 gives experiment results. It is noticed that training time includes pre-extracting time and SVM training time. Where applicable, the results are average value of 50 times experiments, which choose training samples and test samples randomly. From Table 1, whether it is square data or center data, the proposed method has the same recognition rate as the standard support vector machine, but the time is much shorter. The method has good generalization ability.

4.2 Non-linear Separable Experiment

In the experiment, two-dimensional cross data of different shapes are generated artificially. In the square sample, the Class 1 is $U([0, 2] \times [0, 1.1])$, and the Class 2 is $U([0, 2] \times [0.9, 2])$. The number of samples is 500 respectively. In the experiment, we set translation distance D is 0.41. It is shown in Fig. 8. In the classification, the testing method is the same as the linear square sample.

In the round sample, both classes samples radius r is 5. The center point is (5, 5) in Class 1, and the center point is (11, 11) in Class 2. The number of samples is 600 respectively. In the experiment, we set translation distance D is 3.9. The samples in the region surrounded by approximating line and convex hull are boundary samples. It is shown in Fig. 9. In the classification, the testing method is the same as the linear round sample.

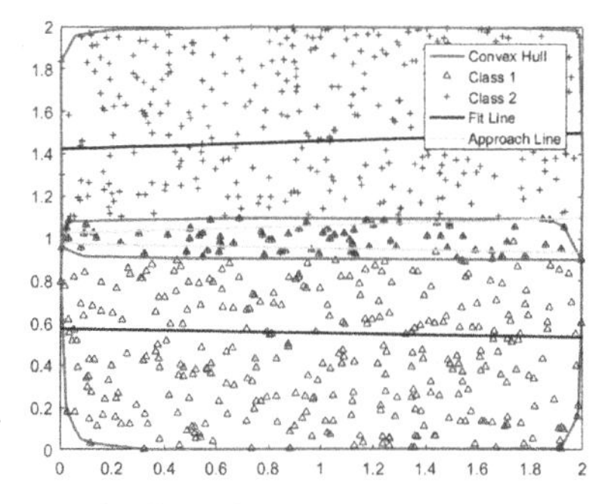

Fig. 8. Constructing boundary samples in non-linear square data.

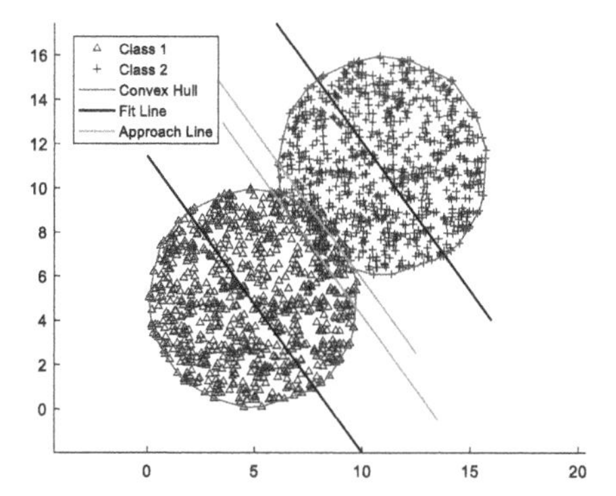

Fig. 9. Constructing boundary samples in non-linear round data.

Table 2 compare the proposed method and standard SVM. The testing method is the same as linear experiment. From Table 2, the proposed method can get a better performance.

Table 2. Classification comparison non-linear separable algorithm.

Algorithm	Shape	Parameter	Boundary sample (number)	Training time (ms)	Classification accuracy (%)
Standard SVM	Square	–	–	332	98.9
	Round	–	–	384	98.5
The proposed algorithm	Square	0.41	109	196	98.9
	Round	3.9	132	223	98.5

4.3 Practical Application Experiment

In order to further verify the actual effect of the method, waveform database are chosen from UCI databases. There are 5000 samples and 3 classes. Class 2 and 0 are used to training samples. In the classification, kernel function and parameter are the same as experiment 4.1. In order to provide a baseline for comparison, the results of standard SVM is added. In the experiment, we set different parameter to verify the algorithm. Table 3 gives the experiment results. From Table 3, the proposed method can get better performance and save more time.

Table 3. Classification comparison practical application algorithm.

Algorithm	Parameter	Boundary sample (number)	Training time (ms)	Classification accuracy (%)
Standard SVM	–	–	2123.2	86.9
The proposed algorithm	23	678	786.3	85.2
	27	632	743.2	84.6
	30	598	718.6	83.8

5 Conclusion

By analyzing the characteristics of convex hull, support vector machine classification method based on convex hull clipping is proposed. Firstly, convex hull construction steps are introduced, namely, pretreating the sample set, determining the convex hull edge end point and constructing the convex hull. Then, analyzing the characteristics of support vector machine (SVM) classification, and introducing the selection method of boundary samples. Experiments on two artificial data sets and UCI standard data set show that the proposed method can improve classification accuracy and reduce training time, especially in the large samples.

In the paper, the method of selecting boundary samples is used to train SVM. There are still several open questions need to study. For example, how to set parameter, how to improve training speed in high-dimensional space and large sample data, how to provide better method under different application environments.

Acknowledgments. This work was financially supported by Nantong Science and Technology Project (GCZ19048), and supported by Nantong Polytechnic College Young and Middle-aged Scientific Research Training Project (ZQNGG209), and supported by Nantong Key Laboratory of intelligent control and intelligent Computing, and by supported by Nantong Science and Technology Project (GC2019130).

References

1. Gan, J., Wang, W.Q., Lu, K.: In-air handwritten Chinese text recognition with temporal convolutional recurrent network. J. Pattern Recogn. **97**, 601–615 (2020)

2. Raymond, P., Felipe, P., Such, S.P., Frank, B.: Intelligent character recognition using fully convolutional neural networks. J. Pattern Recogn. **71**, 604–613 (2018)
3. Liao, Z.Y., Wang, Y.T., Xie, X.L., Liu, J.M.: Face Recognition by support vector machine based on particle swarm optimization. J. Comput. Eng. **12**, 248–254 (2017)
4. Guo, Y.Q.: Support vectors classification method based on matrix exponent boundary fisher projection. In: 2019 IEEE International Conference on Mechatronics and Automation, Tianjin, pp. 957–961 (2019)
5. Guo, Y.Q.: Support vectors classification method based on projection vector boundary feature. In: 2018 IEEE International Conference on Digital Image Processing, Shanghai, p. 1080 (2018)
6. Qian, H.B., Li, Y.L.: A SVM algorithm based on convex hull sparsity and genetic algorithm optimization. J. Chongqin Univ. **44**, 29–36 (2021)
7. Wang, H.Y., Zhou, Y.J.: Pattern recognition method of PCA-based statistical process control chart. J. Control Decis. Making **24**, 20–24 (2020)
8. Liu, B., Wang, T.: An efficient convex hull algorithm for planar point set based on recursive method. J. Acta Automatica Sinica **38**, 1376–1379 (2012)
9. Wu, Q., Zhao, X.: Incremental learning a algorithm of support vector machine based on vector projection of fisher discriminant. J. Xian Univ. Posts Telecommun. **23**, 79–84 (2016)
10. Frank A., Asuncion A.: UCI machine learning repository. http://www.ics.uni.edu

Surface EMG Real-Time Chinese Language Recognition Using Artificial Neural Networks

M. Majid Riaz[1,2] and Zhen Zhang[1,2(✉)]

[1] School of Mechatronic Engineering and Automation, Shanghai University, Shanghai 200444, China
zhangzhen_ta@shu.edu.cn

[2] School of Computer Engineering and Sciences, Shanghai University, Shanghai 200444, China

Abstract. Surface electromyography (sEMG) has recently been commonly used in sign language recognition owing to its numerous advantages (i.e., portability, lightweight, and ease of use) over cameras and inertial sensors. We proposed a real-time Chinese sign language recognition model based on sEMG signals in this research. sEMG data was collected using MYO armband on nine healthy volunteers who performed a series of 15 gestures. The signal was preprocessed using a sliding window method followed by a muscle detection operation. Five time-domain features were extracted from the preprocessed signal for feature extraction. We used a feed-forward Artificial Neural Network (ANN) to classify the signals with an activation time threshold to recognize the gestures. This model is 94.9% in accuracy and reacts in around 195.19 ms (real-time).

Keywords: Sign Language Recognition (SLR) · Armband · ANN · sEMG

1 Introduction

Surface electromyography (sEMG) is an electrophysiological signal associated with muscle activity, which aims to provide a more natural, convenient, and effective way of human-computer interaction, which has significant applications in prosthetic control, sign language recognition (SLR), electronic games, and other fields. According to the type of sensors used for data acquisition, the key methods for realizing SLR can be roughly divided into four types: 1) inertial sensors used in cyber gloves [1], 2) visual signal system [2] using cameras, 3) electromyography [3], and 4) radio frequency [5].

Because of its low system cost and faster response time, machine learning has recently emerged as the best technique for sign language recognition models based on sEMG. Support Vector Machine (SVM) [6], Artificial Neural Networks (ANN) [7], K-Nearest Neighbors (KNN) [8], and convolutional neural networks (CNN) [9] are the most widely used classification models for SLR.

Since SLR systems have a wide range of applications in virtual environmental control, robot control, smart homes, and game control, they must be implemented in real-time. An SLR device is called real-time if its recognition time is less than 300 ms [10], implying that the gesture should be recognized before it is performed. Kuo Yang et al.

M. Fei et al. (Eds.): LSMS 2021/ICSEE 2021, CCIS 1467, pp. 114–122, 2021.
https://doi.org/10.1007/978-981-16-7207-1_12

[11] proposed a real-time gesture recognition model with an average time response of 227.76 ms and an average accuracy of 98.7% for a set of five gestures. For 15 Chinese sign language (CSL) gestures, Ziyi Su et al. [12] proposed a real-time SLR model with a response time of around 300 ms and an average accuracy of 88.7%. [13] Proposed a real-time model based on an artificial feed-forward neural network that has an average response time of 11 ms and an average accuracy of 90.1 for a series of five gestures.

A real-time sEMG Chinese SLR system based on ANN is proposed in this paper. The raw sEMG signal is preprocessed with a low pass filter to remove noise and smooth it out. The preprocessed signal is passed through a sliding window for feature extraction, and a few features are extracted to create the feature vector. The signal is categorized using an ANN classifier with three layers, and the signal is then classified into various groups using a majority voting approach.

2 Sensors and Data Acquisition

2.1 Sensors and Dataset

The sEMG method was used to collect data in our experiment. We used the MYO armband shown in Fig. 1, a commercially available wearable device developed by Thalmic Labs Canada. An inertial measurement unit and eight sEMG sensors are included in the MYO armband. In human-computer interaction (HCI) research and production, the MYO armband is widely used.

Fig. 1. A MYO armband equipped with an IMU and eight EMG sensors

For data collection, the MYO armband was used with an IMU and eight EMG sensors, providing sEMG signal at 200 Hz. The MYO armband includes a Bluetooth interface that transfers the digital signal to the machine in real-time. In total, fifteen gestures are included in our dataset as shown in Fig. 2, with ten of them depicting CSL, namely one, two, three, four, five, six, seven, eight, nine, and ten. The remaining five gestures are composed of the letters A, B, C, D, and E.

The dataset includes nine right-handed stable subjects aged 22 to 35 years old, with six males and three females. The subjects were instructed to make the 15 movements depicted in Fig. 2. Each gesture was replicated 5 times for the training dataset and 30 times for the testing dataset, resulting in 270 samples of each gesture. An extra sample of "relax" was added to the dataset for the sake of comparison and classifying each gesture.

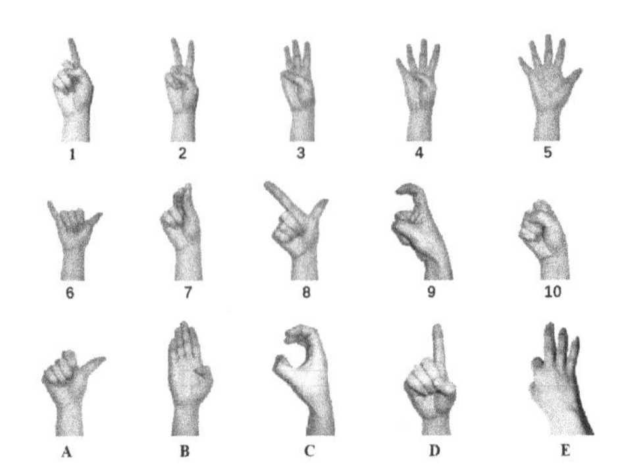

Fig. 2. Series of 15 gestures used in the experiment

3 Methodology

Based on the method proposed in [7], we created an SLR model. Pre-processing, feature extraction, classification, and post-processing are the four sub-sections of this section.

3.1 Pre-processing

As shown in Fig. 3, a sliding window is applied after data collection, with the main window N and various sub-windows S. The raw sEMG signal is represented by a matrix A of length $N * 8$, where N is the length of the main window and 8 is the number of sensors on the MYO armband. The raw sEMG signal is normalized to get data in the $[-1, 1]$ range, which is our matrix A's range.

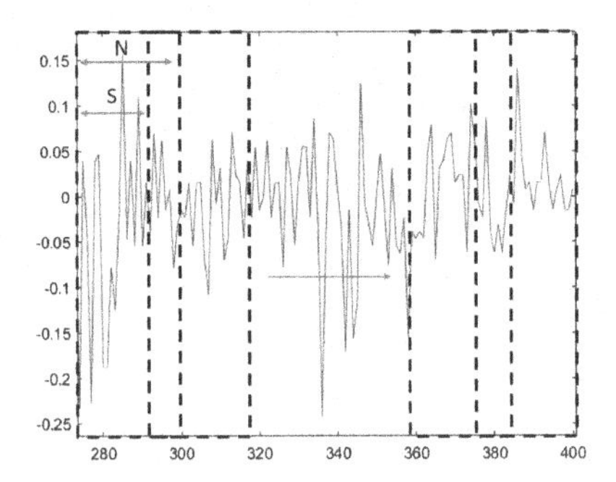

Fig. 3. Schematic diagram of sliding window

The raw sEMG signal must be pre-processed to isolate the features since it contains additional noise that must be eliminated, so the signal must be rectified and filtered. An absolute value function is used to rectify the signal, which is then passed through a 4th order low pass Butterworth filter with a logical cut-off frequency of 5 Hz discovered using the Fourier transform. Furthermore, some unnecessary parts of the signal, such as the head and tail, must be removed using the Kundu et al. muscle detect feature mentioned in [14].

3.2 Feature Extraction

The time domain knowledge of sEMG signals is investigated in this paper to extract effective features and identify 15 gestures. Since time-domain features are simple to compute and have high recognition rates, they are commonly used in sEMG signal feature classification. To extract the features from the pre-processed signal, a sliding window is used. The sliding window is added to the main window N to split it into smaller sub windows s, allowing for the extraction of more popular features by taking smaller pieces of the signal at a time. We identified 5 time-domain characteristics in this study as discussed in Table 1.

Table 1. Time domain features used in this paper

Mean Absolute Value (MAV)	$M = \frac{1}{N}\sum_{j=1}^{N} \lvert a(j)\rvert$
Slope Sign Change (SSC)	$S = \sum_{j=2}^{N-1} \lvert(a(j) - a(j-1)) \times ((a(j) - a(j+1))\rvert$
Root Mean Square (RMS)	$R = \sqrt{\frac{1}{N}\sum_{j=1}^{N} \lvert a((j)^2\rvert}$
Wave Length (WL)	$W = \sum_{j=2}^{N} \lvert(a(j) - a(j-1))\rvert$
Hjorth Activity Parameter (HP)	$A_{hp} = VAR(a(j)) = \frac{1}{N-1}\sum_{j=1}^{N} \lvert a((j)^2\rvert$
Hjorth Mobility Parameter (HP)	$M_{hp} = \sqrt{\frac{VAR((\frac{ds(j)}{dj}))}{VARa(j)}}$
Hjorth Complexity Parameter (HP)	$C_{hp} = \frac{Mob.((\frac{ds(j)}{dj}))}{Mob.a(j)}$

The features vector of the training dataset is visualized using T-Stochastic Neighbor Embedding (T-SNE), as shown in Fig. 4. T-SNE was visualized for a single user using four separate sliding window values: 200 ms, 300 ms, 400 ms, and 500 As the size of the window is increased, the features vectors of different groups get closer to each other. However, after reaching a certain threshold (400 ms), the number of features in the vector decreases while the duration increases.

To further observe effect of window size on accuracy, we took eight measurements of window size: 100 ms, 150 ms, 200 ms, 250 ms, 300 ms, 350 ms and 400 ms, as shown in Fig. 5. Accuracy keeps rising along with rise in window size, but after reaching a specific level (150 ms) it gradually starts reducing with the increase in window size. So we find out that window size = 30 points (150 ms) provides us the best recognition accuracy of 94.9%.

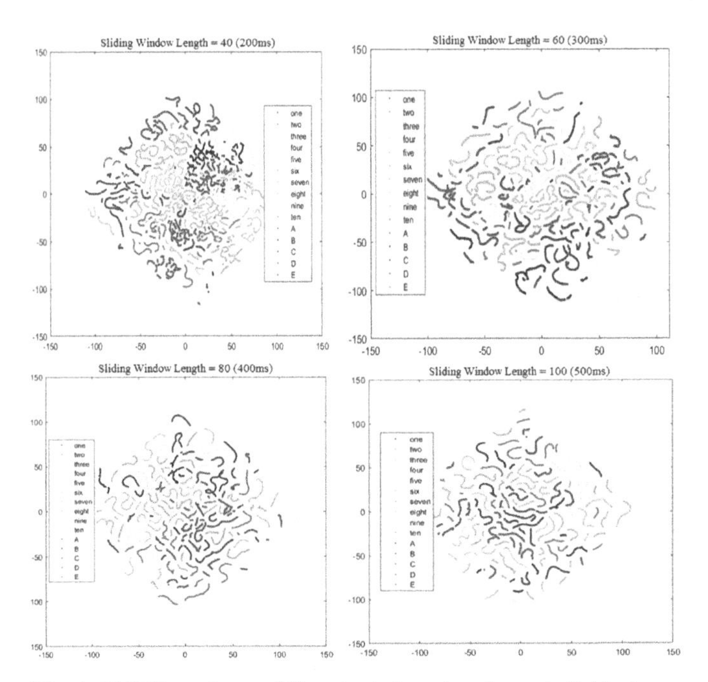

Fig. 4. T-SNE results on different window sizes for an individual user

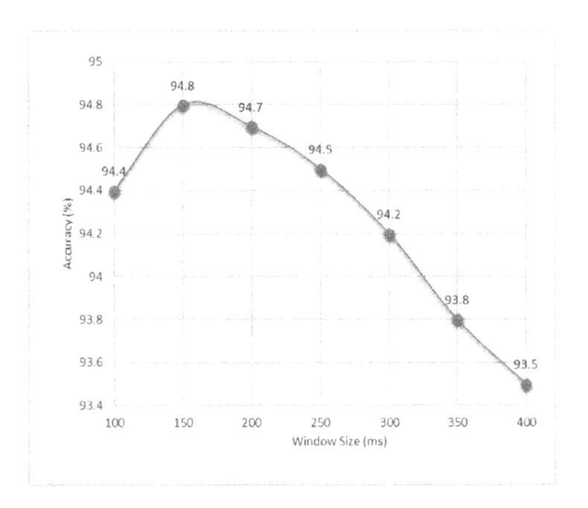

Fig. 5. Accuracy (%) for different values of window size

As we get the best accuracy value at window size = 30 points (150 ms), so we set the window size to 150 ms in our experiment. After defining the window size, we can move forward to find the feature vectors V_i of each sample signal. The feature vector consists of two parts; the preprocessed sEMG signal and five features from time-domain, which can be denoted as E_i and T_i respectively. The length of E_i is taken equal to *Window Size** 8, so we get $|E_i| = 30 * 8 = 240$. After calculating 5 time-domain features, the length of the vector is obtained which is $|T_i| = 7 * 8 = 56$ total features. Therefore, we get the final feature vector as $V_i = [E_i, T_i]$ which can be calculated as $|E_i| + |T_i| = 240 + 56 = 296$ features in total.

3.3 Classification

We used a feed-forward ANN with three layers; input, hidden and output layer. The number of input layers is equal to the number of features vectors, while the number of hidden layers is half of the number of input layer. The output layer consists of 16 elements in correspondence to the number of classes to be classified. Sigmoid function was used as a transfer function and to train the network, full batch gradient descent algorithm and cross-entropy cost function was used. The training dataset was used to train the network with 100 epochs and we utilized testing dataset to validate the accuracy of the network.

3.4 Post-processing

We get a label vector for each sEMG observation using the main window, with each label corresponding to the feature vector observed by the sub window. The threshold level τ is defined, and each observation in the main window is labeled by majority voting. If the amount in the vector of labels is higher than τ, it is assigned a label. If the sample signal level does not reach the TP threshold, it is classified as "no gesture." We found that $\tau = 70\%$ provided us with the most accurate recognition results, after trying various threshold levels through the hit and trial method.

4 Performance of the Proposed Model

The experiment was performed using a set of 15 gestures on a dataset obtained from 9 healthy subjects. The proposed model was evaluated using pre-processed signal and a bag of functions. All the tests were carried out on a desktop computer with an AMD R7 4800U processor and 16 GB of RAM. The confusion matrix in Fig. 6, shows the average recognition accuracy of all the subjects and number of samples correctly classified against each class. Our proposed model attains an average accuracy of 94.9% and above 90% for all 15 gestures. The confusion matrix illustrates that gesture "A" attained the highest accuracy of 99.3% and gesture "B" with lowest at 90.4%.

Discussing precision of the proposed model, the gesture "One" had the best precision among all gestures with 98.1% and two gestures "Ten" and "E" with lowest precision at 91.6%. Considering the error rate of the gestures, we found that the gesture "E" had the maximum probability to be wrongly classified as gesture "B". Additionally, signal strength of some samples is too low to pass the threshold level of the accuracy, so they are classified in the category of "no gesture".

4.1 Real-Time Response

The recognition time needs to be less than 300 ms to lie in the category of real-time SLR model. Our model presented an average time response of 195.19 ms with an average recognition rate of 94.9%. The fastest time response was 163.92 ms and lowest was 243.54 ms (Fig. 7).

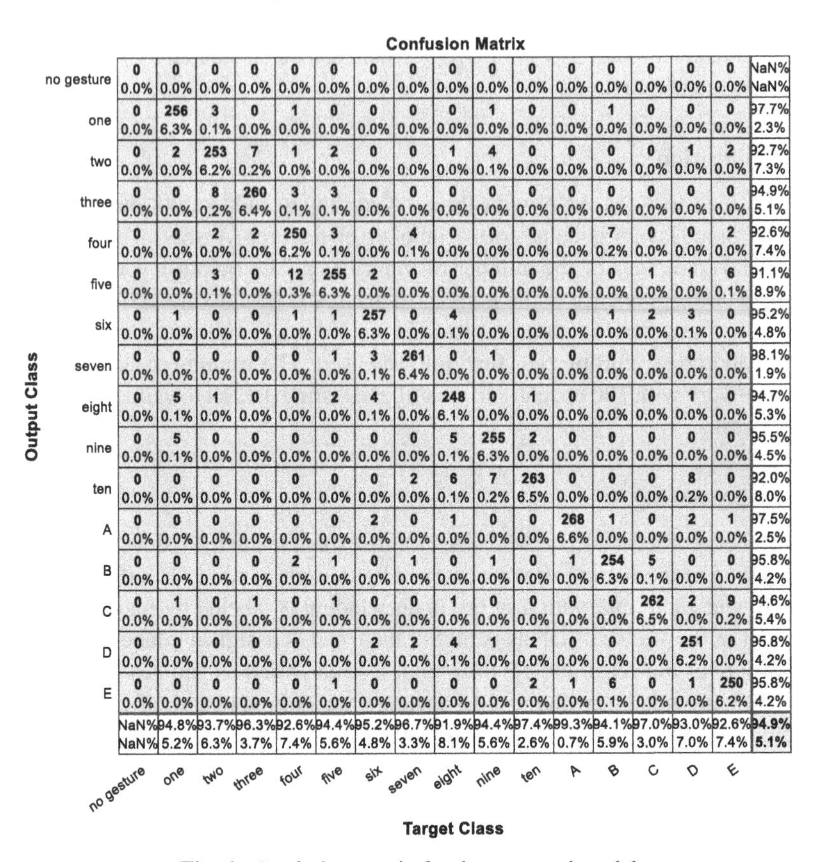

Fig. 6. Confusion matrix for the proposed model

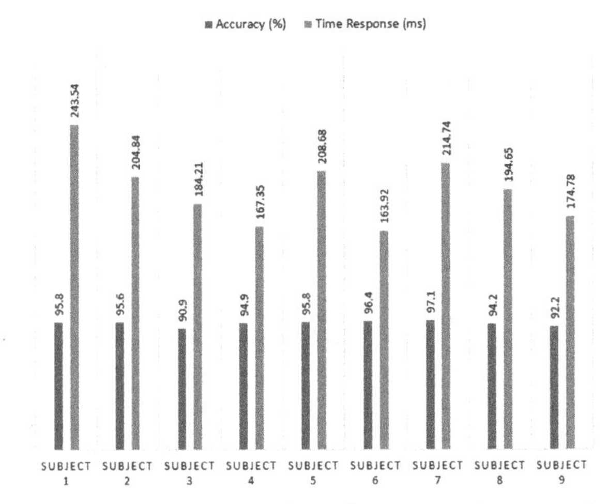

Fig. 7. Subject wise comparison of accuracy and response time

4.2 Comparison with Other Models

Table 2 shows that the proposed model, which uses both types of features (the preprocessed signal values and the results from the bag of functions), provided the best accuracy compared to the other models. The model that uses only the preprocessed signal values has quicker response and its recognition accuracy is higher than the model that uses only the results from the bag of functions. However, the model that uses only the bag of functions has the lowest training time because of less complex architecture. Table 1 also indicates that the proposed model has a time response of 195.19 ms that is lower than the real time limit (300 ms).

Table 2. Comparison of proposed model with existing models

Model	Accuracy (%)	Response (ms)
Evaluated models:		
Models using only the results from the bag of functions	88.4	193.48
Model using only the preprocessed signal values	91.8	178.54
Proposed model	94.9	195.19
Other models:		
Model using k-NN and DTW [4]	90.7	–
Model using ANN [7]	94.54	–
Model using discriminant analysis [15]	89.92	–
Model using random forest [15]	98.31	–
Model using deep learning [16]	86.0	–

Table 2 shows that the proposed model performs better in terms of accuracy than the existing models which used the DTW algorithm with K-NN classifier [4] and another model using discriminant analysis and random forest algorithms [15]. However, a HGR model using deep learning [16] outclasses all machine learning models with 98.31% accuracy.

5 Conclusion

In this study, we have proposed a real time SLR system which works on a set of 15 gestures. The raw s-EMG data was collected using the MYO armband on nine healthy subjects. We used a feed forward ANN to classify the signals and a test method that recognizes the gesture when the identified label times exceed the ANN classifier's activation times threshold mark. This model is 94.9% accurate and reacts in around 195.19 ms (real-time).

Future prospects include working on increasing capacity of proposed model to add more gesture classes other than the gestures used in this experiment. Furthermore, different feature extraction techniques and other parametric and non-parametric classification models can be studied further.

References

1. Gao, W., Fang, G., Zhao, D., Chen, Y.: A Chinese sign language recognition system based on SOFM/SRN/HMM. J. Pattern Recognit. **37**, 2389–2402 (2004)
2. Liu, K., Kehtarnavaz, N.: Real-time robust vision-based hand gesture recognition using stereo images. J. Real-Time Image Proc. **11**(1), 201–209 (2013). https://doi.org/10.1007/s11554-013-0333-6
3. Benalcazar, M.E., et al.: Real-time hand gesture recognition using the Myo armband and muscle activity detection. In: Proceedings of the 2017 IEEE 2nd Ecuador Technical Chapters Meeting, pp. 1--6. Salinas, Ecuador (2017)
4. Benalcazar, M.E., Jaramillo, A.G., Zea, A., Paez, A., Andaluz, V.H.: Hand gesture recognition using machine learning and the Myo armband. In: Proceedings of the European Signal Processing Conference, pp. 1075--1079. Kos, Greece (2017)
5. Adib, F., Hsu, C., Mao, H., Katabi, D., Durand, F.: Capturing the human figure through a wall. J. ACM Trans. Graph. **34**, 1–13 (2015)
6. Rossi, M., Benatti, S., Farella, E., Benini, L.: Hybrid EMG classifier based on HMM and SVM for hand gesture recognition in prosthetics. In: 2015 IEEE International Conference on Industrial Technology (ICIT), pp. 1700–1705. Seville, Spain (2015)
7. Motoche, C., Benalcázar, M.E.: Real-time hand gesture recognition based on electromyographic signals and artificial neural networks. In: Proceedings of the International Conference on Artificial Neural Networks, pp.4–7, Rhodes, Greece (2018)
8. Joshi, A., Monnier, C., Betke, M., Sclaroff, S.: Comparing random forest approaches to segmenting and classifying gestures. Image Vis. J. Comput. **58**, 86–95 (2017)
9. Asif, A.R., et al.: Performance evaluation of convolutional neural network for hand gesture recognition using EMG. J. Sens. **20**, 1642 (2020)
10. Mizuno, H., Tsujiuchi, N., Koizumi, T.: Forearm motion discrimination technique using real-time emg signals. In: Engineering in Medicine and Biology Society. EMBC, Annual International Conference of the IEEE, pp. 4435–4438. IEEE (2011)
11. Zhang, Z., Yang, K., Qian, J., Zhang, L.: Real-time surface EMG pattern recognition for hand gestures based on an artificial neural network. J. Sens. (Basel). **19**(14), 3170 (2019)
12. Zhang, Z., Su, Z., Yang, G.: Real-time Chinese sign language recognition based on artificial neural networks. In: 2019 IEEE International Conference on Robotics and Biomimetics (ROBIO), pp. 1413–1417, Dali, China, (2019)
13. Benalcázar, M.E., Anchundia, C.E., Zea, J.A., Zambrano, P., Jaramillo, A.G., Segura, M.: Real-time hand gesture recognition based on artificial feed-forward neural networks and EMG. In: 2018 26th European Signal Processing Conference (EUSIPCO), pp. 1492–1496, Rome, Italy (2018)
14. Kundu, A.S., Mazumder, O., Lenka, P.K., Bhaumik, S.: Hand gesture recognition based omnidirectional wheelchair control using IMU and EMG sensors. J. raml. Robot. Syst. **3**, 1–13 (2017)
15. Wahid, M.F., Tafreshi, R., Al-Sowaidi, M., Langari, R.: Subject-independent hand gesture recognition using normalization and machine learning algorithms. J. Comput. SCI-NETH **27**, 69–76 (2018)
16. Coteallard, U., et al.: Deep learning for electromyographic hand gesture signal classification using transfer learning. J. IEEE Trans. Neural Syst. Rehabil. Eng. **27**(4), 760–771 (2019)

Melanoma Classification Method Based on Ensemble Convolutional Neural Network

Xin Shen[1(✉)], Lisheng Wei[2], Huacai Lu[2], and Xu Sheng[3]

[1] Anhui Polytechnic University, Wuhu 241000, Anhui, China
[2] Anhui Key Laboratory of Electric Drive and Control, Wuhu 241002, Anhui, China
[3] Nantong Vocational University, Nantong 226007, Jiangsu, China

Abstract. Melanoma is the most serious form of skin cancer, with more than 123,000 new cases worldwide each year. Reliable automatic melanoma screening system will be a great help for clinicians to detect malignant skin lesions as soon as possible. To address this problem, a computer-assisted approach to detecting melanoma using convolutional neural networks will enable effective and faster treatment of the disease. Many attempts have been made to use convolutional neural networks to solve this problem, but so far the performance in this aspect is not good. In this project, Xception and VGG-16 models will be fine-tuned on the IEEE International Symposium on Biomedical Imaging (ISBI) Official Skin Data Set 2016 and combined into an integrated framework to predict whether skin disease images are benign or malignant. Our experimental results show that the ensemble model can achieve better classification results and the proposed method is competitive in this field.

Keywords: Melanoma · Convolutional neural network · Ensemble · Deep learning

1 Introduction

Skin cancer, also known as malignant melanoma, is one of the deadliest cancers if not caught in time. More than 100,000 new cases are involved in the United States each year, with more than 9,000 deaths [1]. Although China is a country with a low incidence of melanoma compared with European and American countries, in recent years, there are about 20,000 new cases of melanoma disease in China every year, and melanoma has become one of the diseases harmful to the health of Chinese people [2].

Early detection and diagnosis are the most effective ways to treat skin cancer. Schaefer et al. used automatic boundary detection method [3] to segment the lesion area, and then combined the extracted features such as shape, texture and color for melanoma recognition [4]. Due to the large intra-class differences and small inter-class differences between melanoma and non-melanoma, the effect of manual feature extraction is not ideal. In recent years, with the rapid development of deep learning, major breakthroughs beyond human capabilities have been made in many fields such as image classification [5, 6] and speech recognition [7]. The advantage of CNN is that it can automatically

M. Fei et al. (Eds.): LSMS 2021/ICSEE 2021, CCIS 1467, pp. 123–133, 2021.
https://doi.org/10.1007/978-981-16-7207-1_13

learn the feature representation of classification tasks through data sets [8], and it has good performance in many applications [9]. Kawahara et al. proposed a complete CNN based on AlexNet [5] to extract the representative features of melanoma [10]. Y U et al. proposed a new method based on deep CNN to deal with the challenge of automatic melanoma recognition in dermoscopic images, which includes two steps: segmentation and classification [11]. It is necessary to apply deep learning technology to medical images, which can help doctors diagnose diseases and improve the efficiency of doctors' diagnosis.

The structure of this paper is as follows: The first part is the introduction, which briefly introduces the research background, development status and research significance of this paper; The second part introduces the research method of this paper; The third part is to carry on the experiment simulation to the research method of this paper, and analyze the result; The fifth part gives the conclusion of this paper.

2 Methodology

2.1 Data Preprocessing

In order to adapt to the network training requirements, ensure the training can be carried out normally and get the best results, it is necessary to preprocess the data. The image is preprocessed as follows:

Adjust the image size to (224,224,3). Since the final classification layer in the network model used in this paper is the "softmax" layer, the parameter "class mode" is set to "categorical". Since the depth of the image is 3, the "color mode" parameter is set to "RGB". When the model is trained, the parameter "batch size" is set to 32, with 32 images in each group. The images are input into the CNN model in batches, which can improve the training time of the model.

2.2 Data Augmentation

In order to suppress model overfitting, obtain a better model training effect and increase the number of training data, data enhancement technology is applied to the training set. The test set and validation set are not enhanced because the data is only used to validate or test the model. The augmented techniques applied on the training set are as follows:

Firstly, normalize the data and set the parameter "rescale" to 1/255, which will rescale all pixel values of the image between 1 and 255, so as to standardize the image input into the network; Then set the parameter "horizontal flip" to True and flip the image horizontally. Finally, set the parameter "vertical flip" to True to flip the image vertically.

2.3 Loss Function

Since the data in the ISBI2016 dataset has obvious category imbalance, the Focal Loss loss function was used in this paper. Focal Loss is mainly to solve the problem that the proportion of positive and negative samples is seriously unbalanced in one-stage target detection. Focal Loss is modified on the basis of cross entropy loss function. First, review the cross loss of binary classification:

$$L = -y \log y' - (1 - y) \log(1 - y') = \begin{cases} -\log y', & y = 1 \\ -\log(1 - y'), & y = 0 \end{cases} \quad (1)$$

Focal Loss first adds a factor on the basis of the original one, in which $\gamma > 0$ is used to reduce the loss of easily classified samples and pay more attention to the difficult and misclassified samples. In addition, the balance factor alpha is added to balance the uneven proportion of positive and negative samples themselves. Here's the expression for the Focal Loss loss function:

$$L_{fl} = \begin{cases} -\alpha(1-y')^{\gamma} \log y' \quad , & y=1 \\ -(1-\alpha) y'^{\gamma} \log(1-y'), & y=0 \end{cases} . \quad (2)$$

2.4 Train Method

Firstly, this paper aims to obtain two basic models with the best performance through independent training and fine-tuning Xception and VGG-16 models. Then the two base models are combined by an average integration strategy, and the data is tested and evaluated. Figure 1 below is an ensemble framework for the classification of melanoma in this paper.

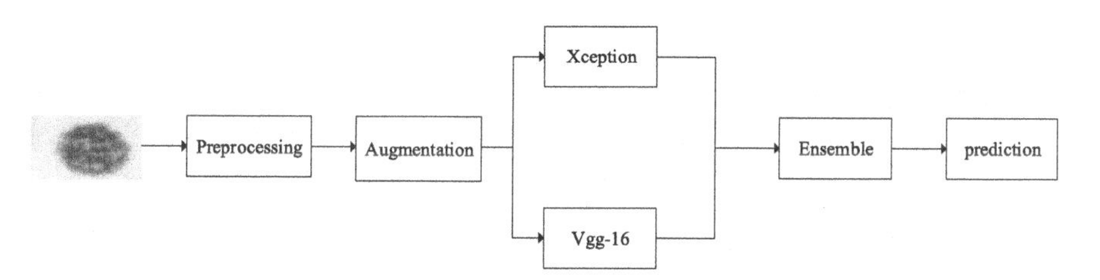

Fig. 1. Ensemble framework for the classification of melanoma

1) The Xception Architecture

Xception Model is a pre-trained convolutional neural network model trained on ImageNet datasets. It has 36 convolution layers, which can be visualized as convolution layer, pooling layer, regularization layer, activation layer and other blocks. For this project, Xception models have been fine-tuned in two stages on the ISBI 2016 official skin dataset.

The first phase begins with the download of Xception pre-training models and ImageNet weights to remove the classification layer from the model. Then set the parameter "trainable" to "False" so that the Xception model cannot train itself to update the weight of its neurons. Next, create a sequence model, using the Xception model as the first layer. Since the classification layer of the Xception model was removed, the output of the sequence model was a 3D feature map, which was then flattened into a 1D tensor via Keras's Flatten layer. The next step is to add two fully connected layers, 1024 and 512 neurons each, to the sequential model and use the ReLU activation function through the dense layer of Keras. Then through the dense layer of Keras, add 2 neurons and the "softmax" activation function to the sequential model as the final classification layer. The final classification layer is the output layer, the output skin disease image is "malignant" melanoma or "benign" melanoma. These two neurons represent two classes, and the probability that the image belongs to a particular class will be the output of the neuron representing that class. The sequential model is then compiled by setting the loss function, optimizer, learning rate, and "metrics" parameters. Finally, the fit() function was used to input the training set data into the model for training, and the Epoch was set as 25, and the model was verified with the validation set data. Figure 2 shows the architecture of Xception model after fine tuning.

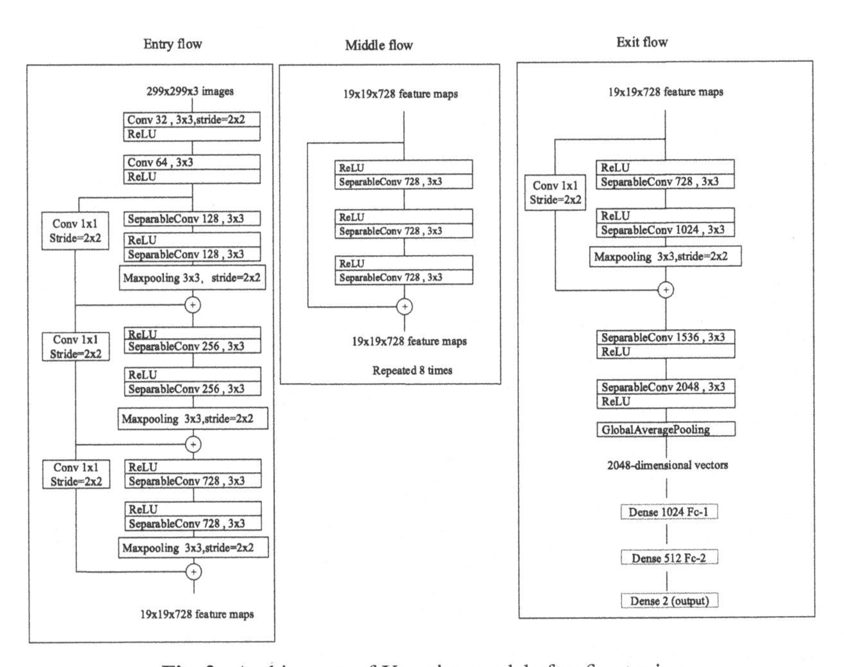

Fig. 2. Architecture of Xception model after fine tuning

The second Phase formally fine-tuned the model. Before the "Block2 SepConv2" layer, all layers of the Xception model remain frozen and unable to participate in training. If the Xception model layer is a "Block2 SepConv2" layer, practice by setting the trainable parameter to "True". Layers in the training Xception model fine-tune their weights according to the characteristics of the ISBI2016 data set to fit the current data set. The sequential model is then compiled by setting the loss function, optimizer, learning rate, and parameter "metrics". Finally, the fit() function was used to input the training set data into the model for training, and the Epoch was set as 50, and the model was verified with the validation set data. This phase also uses model checkpoints to get the best performing model.

After the training process is complete, the model is loaded with the best weights recorded in the model checkpoint and evaluated against the test data to obtain performance metrics.

2) **The Vgg-16 Architecture**

VGG-16 model is a pre-training model trained on ImageNet dataset. In this model, there are two convolution layer blocks before the classification layer, followed by a maximum pooling layer. Then there are three convolutional layer blocks, followed by the maximum pooling layer and the fully connected layer. In this project, the VGG-16 model has undergone two phases of fine-tuning on the ISBI2016 dataset.

In the first phase, the VGG-16 pre-trained model is downloaded along with the ImageNet weights, and the classification layer is removed from the model by setting the parameter "include top" to "False". Then set the parameter "trainable" to "False" so that the VGG-16 model cannot train itself to update the weight of its neurons. The next step is to build a sequential model, using the VGG-16 model as the first layer. Since the classification layer of the VGG-16 model has been removed, the output of the sequential model is a 3D feature map. It is then flattened into a 1D tensor through the Keras Flatten layer. Next, a full connection layer with 512 neurons and "ReLU" activation and a batch normalization layer were added to the sequential model through the dense layer and the batch normalization layer of Keras, respectively. Then through the dense layer of Keras, add 2 neurons and the "softmax" activation function to the sequential model as the final classification layer. The final classification layer is the output layer, the output skin disease image is "malignant" melanoma or "non-malignant" melanoma. These two neurons represent two classes, and the probability that the image belongs to a particular class will be the output of the neuron representing that class. The sequential model is then compiled by setting the loss function, optimizer, learning rate, and parameter "metrics". Finally, the fit() function was used to input the training set data into the model for training, and the Epoch was set as 25, and the model was verified with the validation set data. Figure 3 shows the architecture of Vgg-16 model after fine tuning.

The second phase formally fine-tuned the model. Before the "Block3 Conv2" layer, all layers of the VGG-16 model remain frozen and cannot participate in training. If the VGG-16 model layer is "Block3 Conv2" layer, the training is performed by setting the trainable parameter to "True". The layers in the VGG-16 model that participated in the training fine-tune their weights according to the characteristics of the ISBI2016 data set to adapt to the current data set. The sequential model is then compiled by setting the loss function, optimizer, learning rate, and parameter "metrics". Finally, the fit() function is used to input the training set data into the model for training, and the Epoch is set as 50, and the model is verified with the validation set data. This phase also uses model checkpoints to get the best performing model.

After the training process is complete, the model is loaded with the best weights recorded in the model checkpoint and evaluated against the test data to obtain performance metrics.

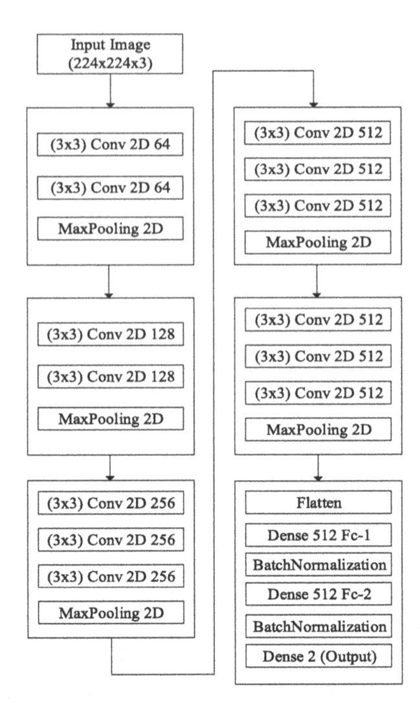

Fig. 3. Architecture of Vgg-16 model after fine tuning

3) **Ensemble Model**

The ensemble model structure is a combination of Xception model and VGG-16 model. By training the two models on the same training data, the image category with the highest average output probability is determined. Here's how to do it:

Firstly, Xception and VGG-16 models are loaded with the weight of the fine tuning training for the best performance. Then the input to the ensemble model is to input skin disease images into Xception and VGG-16 models with fine-tuned weights and set the shapes and sizes of the images to (224,224,3) to obtain the output layers of the two base models. The output layer of the ensemble model is the average of the output layers of the two underlying models. Finally, the model is compiled by setting loss function, optimizer and learning rate, and evaluated by test set data.

3 Experiments and Results

Experiment 1 and Experiment 2 are carried out in the corresponding stage of training the basic model, and Experiment 2 is improved on the basis of Experiment 1 to get the best basic model. Finally, the average integration strategy is used to combine the two basic models to test and evaluate the data.

3.1 The Xception Architecture

In Experiment 1, the loss function and optimizer of the model in the whole training stage are set as cross entropy and SGD, and the parameter "metrics" are both "accuracy". The learning rate of the model in the first stage is 0.0001, and the learning rate of the model in the second stage is 1e−20. Then, the epoch of the first stage is set as 25, and the epoch of the second stage is set as 50, then input ISBI2016 data set to train and test the model; The difference between Experiment 2 and Experiment 1 is that Experiment 2 sets the loss function and optimizer of the whole training stage of the model as Focal Loss and Adam, and other parameters are the same, finally, the ISBI2016 data set is input to train and test the model. Figure 4 and Fig. 5 show the accuracy and loss of Xception model in Experiment 1 and Experiment 2.

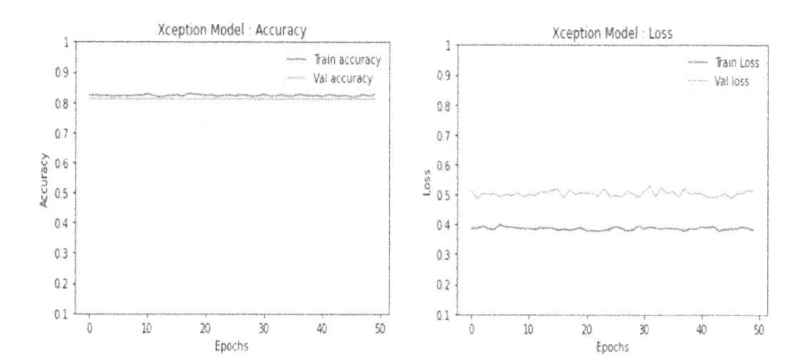

Fig. 4. Accuracy and loss of Xception model in Experiment 1

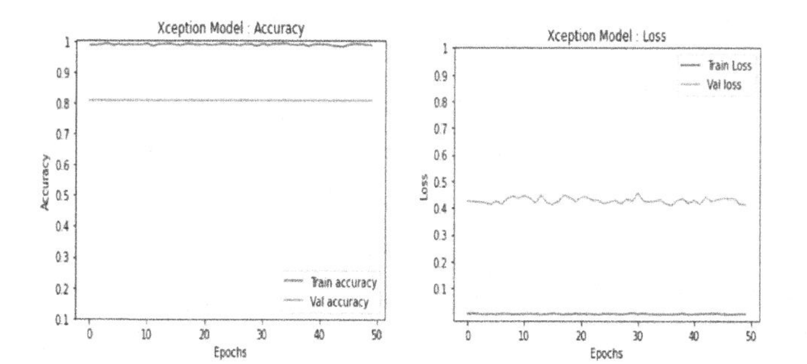

Fig. 5. Accuracy and loss of Xception model in Experiment 2

3.2 The Vgg-16 Architecture

In Experiment 1, the loss function and optimizer of the model in the whole training stage are set as cross entropy and SGD, and the parameter "metrics" are both "accuracy". The learning rate of the model in the first stage is 0.0001, and the learning rate of the model in the second stage is 1e−10. Then, the epoch of the first stage is set as 25, and the epoch of the second stage is set as 50, then input ISBI2016 data set to train and test the model; The difference between Experiment 2 and Experiment 1 is that Experiment 2 sets the loss function and optimizer of the whole training stage of the model as Focal Loss and Adam, and other parameters are the same, finally, the ISBI2016 data set is input to train and test the model.

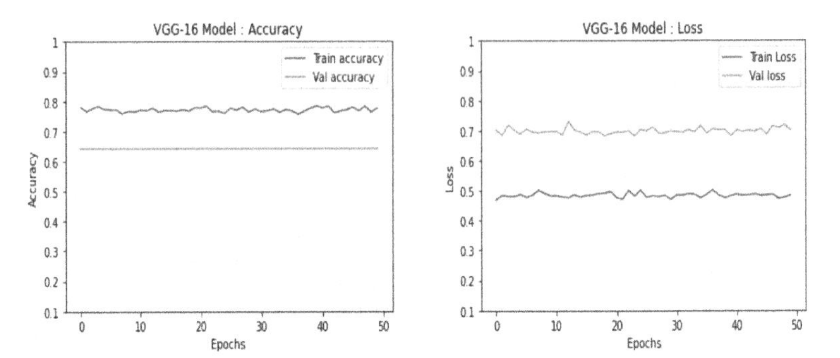

Fig. 6. Accuracy and loss of Vgg-16 model in Experiment 1

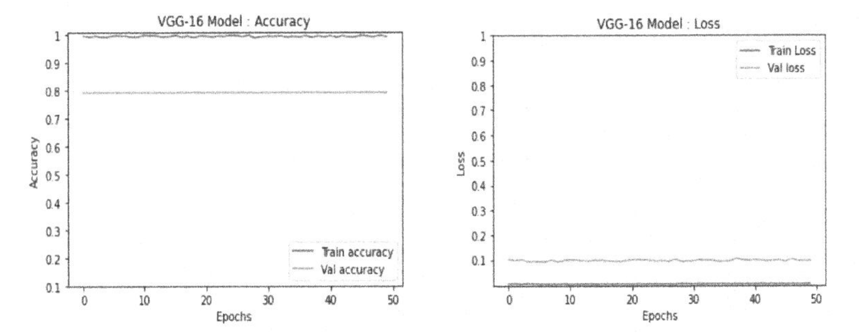

Fig. 7. Accuracy and loss of Vgg-16 model in Experiment 2

As can be seen from the comparison of Fig. 4 and Fig. 5, as well as Fig. 6 and Fig. 7, in Experiment 1, Xception model and VGG-16 model have high training loss and validation loss, and low accuracy in the validation set and test set. In particular, the performance of VGG-16 model is poor, which may be caused by the model overfitting due to the small data set and the imbalance of the categories of data sets. Experiment 2 improves Experiment 1 by using Focal Loss function and Adam optimizer to optimize the model. Table 1 shows the comparison of Xception model performance in Experiment 1 and Experiment 2, and Table 2 shows the comparison of VGG-16 model performance in Experiment 1 and Experiment 2. As can be seen from Table 1 and Table 2, the training and validation loss of Xception and VGG-16 models are significantly reduced, and the accuracy is also improved.

Table 1. Comparison of Xception model performance in Experiment 1 and Experiment 2

Parameters		Experiment 1	Experiment 2
Accuracy	Train	82.45%	**97.64%**
	Validation	81.01%	**81.01%**
	Test	80.21	**80.47%**
Loss	Train	0.3888	**0.0038**
	Validation	0.5098	**0.4054**
	Test	0.5573	**0.4692**

Table 2. Comparison of Vgg-16 model performance in Experiment 1 and Experiment 2

Parameters		Experiment 1	Experiment 2
Accuracy	Train	76.70%	**99.38%**
	Validation	64.25%	**79.33%**
	Test	64.38%	**76.78%**
Loss	Train	0.4829	**0.0035**
	Validation	0.6917	**0.1005**
	Test	0.6968	**0.1035**

3.3 Ensemble Model

Through the fine-tuning and training of Experiment 2, Xception and VGG-16, the best basic models, are obtained. Then the model is compiled by setting the loss function of the ensemble model as Focal Loss, the optimizer as Adam, and the learning rate as 0.0001. Finally, the model is evaluated through the test set. Table 3 shows the performance of the ensemble model compared with the performance of the basic models in Experiment 2.

Table 3. Performance comparison between the ensemble model and the basic models in Experiment 2

Model	Accuracy		Loss	
	Validation	Test	Validation	Test
Xception	81.01%	80.47%	0.4054	0.4692
Vgg-16	79.33%	76.78%	0.1005	0.1035
Ensemble model	**82.68%**	**81.00%**	**0.0987**	0.1069

As can be seen from Table 3, the accuracy of the ensemble model in the validation set and test set has been improved to a certain extent, and the comprehensive performance of the integrated model is better than that of the two basic models. Therefore, the melanoma detection method based on integrated convolutional neural network proposed in this paper can achieve a better classification effect and alleviate the adverse effects caused by small data sets and the imbalance of data sets to a certain extent.

4 Conclusion

In this paper, we propose a melanoma detection method based on ensemble convolutional neural network, and the effectiveness of the method is proved by experiments. The Xception and VGG-16 models were first fine-tuned and trained in the ISBI 2016 official skin dataset, using data preprocessing, data augmentation, different loss functions and optimizer methods to get the best performance of the base model. Then the two basic models are combined and the average strategy is used to predict the performance of the ensemble model. Experimental results show that the ensemble model in this paper is better than the basic model in classification accuracy, reaching 82.68% in the validation set and 81.00% in the test set.

Acknowledgments. This work was supported by the Natural Science Research Programme of Colleges and Universities of Anhui Province under grant KJ2020ZD39, and the Open Research Fund of Anhui Key Laboratory of Detection Technology and Energy Saving Devices under grant DTESD2020A02, the Scientific Research Project of "333 project" in Jiangsu Province under grant BRA2018218 and the Postdoctoral Research Foundation of Jiangsu Province under grant 2020Z389, and Qing Lan Project of colleges and universities in Jiangsu province.

References

1. Rebecca, L., Siegel, M.P.H., Kimberly, D., Miller, M.P.H., Jemal, D.V.M.A.: Cancer statistics. J. CA Cancer J. Clin. **65**(1), 457–480 (2015)
2. Si, L., Guo, J.: Interpretation of clinical practice guidelines for management of melanoma in China. J. Chin. Clin. Oncol. **17**(2), 172–173 (2012)
3. Emre, C.M., Hitoshi, I., Gerald, S., Stoecker, W.V.: Lesion border detection in dermoscopy images. J. Comput. Med. Imaging Graph. Official J. Comput. Med. Imaging Soc. **33**(2),148–153 (2009)
4. Schaefer, G., Krawczyk, B., Celebi, M.E., Iyatomi, H.: An ensemble classification approach for melanoma diagnosis. J. Memetic Comput. **6**(4), 233–240 (2014)
5. Krizhevsky, A., Sutskever, I., Hinton, G.E.: ImageNet classification with deep convolutional neural networks. J. Commun. ACM. **60**(6), 84–90 (2017)
6. Karen, S., Andrew, Z.: Very deep convolutional networks for large-scale image recognition. J. arXiv preprint arXiv, pp. 1409–1556 (2014)
7. Abdel-Hamid, O., Mohamed, A.-R., Jiang, H., Deng, L., Penn, G., Yu, D.: Convolutional neural networks for speech recognition. J. IEEE/ACM Trans. Audio Speech Lang. Process. (TASLP). **22**(10), 1533–1545 (2014)
8. LeCun, Y., Bengio, Y., Hinton, G.: Deep learning. Nature **521**(7553), 436–444 (2015)
9. Ronneberger, O., Fischer, P., Brox, T.: U-net: convolutional networks for biomedical image segmentation. In: Proceedings of the International Conference on Medical Image Computing, pp. 234–241. Springer, Comput-Assist, Intervent, Cham, Switzerland (2018). https://doi.org/10.1007/978-3-319-24574-4_28

10. Kawahara, J., BenTaieb, A., Hamarneh, G.: Deep features to classify skin lesions. In: 2016 IEEE 13th International Symposium on Biomedical Imaging (ISBI), pp. 1397–1400 (2016)
11. Yu, L.Q., Chen, H., Dou, Q., Qin, J., Pheng-Ann, H.: Automated melanoma recognition in dermoscopy images via very deep residual networks. J. IEEE Trans. Med. Imaging **36**(4), 994–1004 (2017)

A Study of Force Feedback Master-Slave Teleoperation System Based on Biological Tissue Interaction

Ze Cui[1], Peng Bao[1], Hao Xu[1], Mengyu Gong[1], Kui Li[1], Saishuai Huang[1], Zenghao Chen[1,3], and Hongbo Zhang[2](✉)

[1] School of Mechatronic Engineering and Automation, Shanghai University, Shanghai 200444, China
[2] Central State University, Wilberforce, USA
hzhang@centralstate.edu
[3] Shanghai Aerospace Control Technology Research Institute, Shanghai 200233, China

Abstract. The use of the teleoperated surgical system in surgery is a tremendous advancement in medicine, assisting surgeons with more complex treatments. However, there are still many details of the existing technology that need to be improved, and one of the key ones is the integration of force feedback into teleoperated surgical systems. In this paper, the interaction model of epithelial tissue, muscle fibers, and connective tissue were established through an in-depth study of biological tissue interaction models, selecting the Hunt-Crossley model, and using porcine pancreatic meat for relaxation testing. Based on the established soft tissue interaction model, a force-feedback master-slave teleoperated system without force sensors was built and the corresponding control strategy was developed. Finally, through trajectory tracking experiment, force feedback experiment, and lateral stretching experiment the good performance and practical value of the system.

Keywords: Teleoperation · Force feedback · Master-slave isomeric · Soft tissue interaction model

1 Introduction

Teleoperation surgery technology can assist surgeons in clinical operations to complete more complex treatments that are even impossible to achieve under normal circumstances, especially during the prevention and treatment of major infectious diseases such as the COVID-19. The lack of force feedback in the existing teleoperations will limit the normal level of the surgeon. The main purpose of force feedback is to enable the surgeon to gain a full perception of the mechanical properties of biological tissues, so as to obtain a better sense of presence during the operation [1].

A number of scholars have already contributed to this research; Wagner [2] investigated the role of different force feedback scales in a blunt instrument dissection task, and

M. Fei et al. (Eds.): LSMS 2021/ICSEE 2021, CCIS 1467, pp. 134–144, 2021.
https://doi.org/10.1007/978-981-16-7207-1_14

the number of errors committed by subjects during surgical procedures decreased significantly as the force feedback gain increased. The RAVEN teleoperated surgical robot, developed by the University of Washington [3], has small size and light weight. The University of Coimbra, Portugal [4] proposed a dual kinematic approach for robot-assisted remote ultrasound imaging that exploits the kinematic similarity of master and slave manipulators and proposes a teleoperated architecture based on joint space mapping.

In order to realize force feedback in teleoperation surgery, the following two methods are generally adopted: one is to directly measure force data for interactive modeling, the other is to estimate the force through appropriate mathematical modeling of the tool-organization interaction, and to establish a mathematical model of the interaction process by studying the interaction model between the tool and the "external environment". The output information is extracted and fed back to the master controller. The second method is widely accepted and applied, and there have been many related studies on the tool-organization interaction model. Yo [5] proposed a fractional Maxwell model with fractional Hilbert transformation term. The model and experimental results are highly correlated in terms of flexibility diagram, phase diagram and mechanical impedance. Rimantas [6] discussed the dissipative normal nonlinear Hertz contact problem widely discussed in the discrete element simulation. By studying some viscous damping models, they proved the contribution of these models to the force propagation in the contact particle chain. Fabien [7] proposed a log-linear approximation method for the Hunt-Crossley (HC) contact force model, using the recursive least square method to achieve fast and real-time identification.

This paper focuses on the realization of force feedback in teleoperation. It conducts force estimation through appropriate mathematical modeling of tool-tissue interaction, analyzes and compares existing models, and selects the HC model to construct interaction with biological soft tissues. The model parameters are obtained through experiments; based on this, a teleoperation experiment platform is built, and the feasibility and application prospects of the force feedback teleoperation are verified through experiments.

2 Teleoperation System Construction and Force Interaction Strategy

2.1 Teleoperation Platform Construction

Platform Architecture. As shown in Fig. 1, the master-slave teleoperation system built in this paper mainly includes: operator, master controller, PC, lower computer, slave controller, and environment (target). The teleoperation task process is as follows: the operator holds the master controller to move, collects the movement information of the master controller through the PC, and after processing, obtains the motion control instruction and sends it to the lower computer, and then the lower computer parses the instruction to control the slave controller. The hand moves accordingly to achieve the purpose of interacting with the environment, and then through the soft tissue interaction model, according to the displacement and speed information returned from the operator, the PC is processed, and the master controller provides real-time force feedback to the operator.

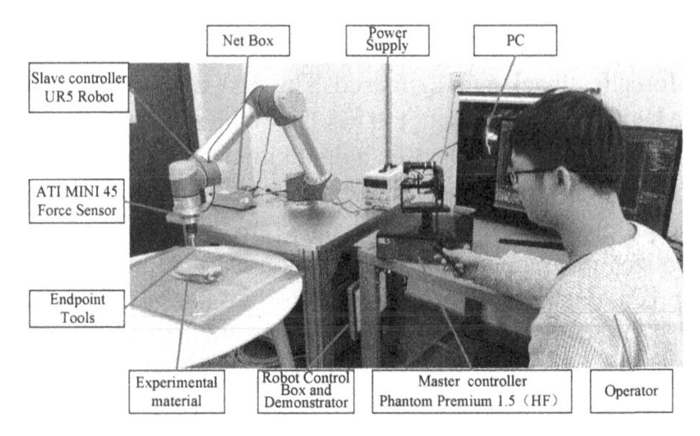

Fig. 1. Force feedback experiment system.

In order to provide force feedback to the operator during the interaction between the operator and the master controller, in the teleoperation system in this paper, the master controller uses Phantom Premium 1.5 (HF) force feedback equipment; the upper computer uses a desktop PC; the slave controller uses the UR5 collaborative robot; the lower computer is the control box equipped with the UR5 robot, and its structure is shown in Fig. 2. The entire software system is mainly based on the information interaction

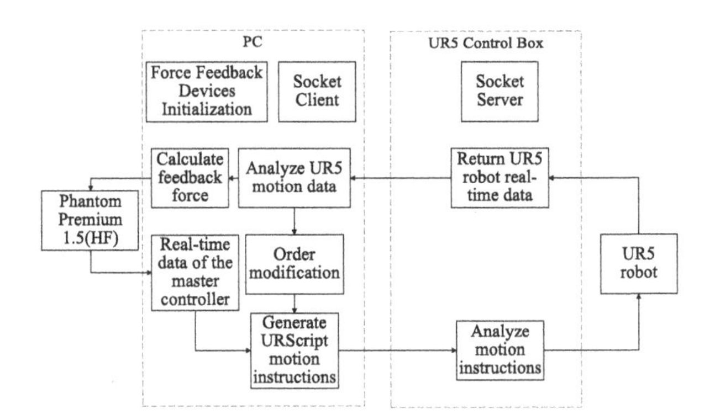

Fig. 2. Software platform structure.

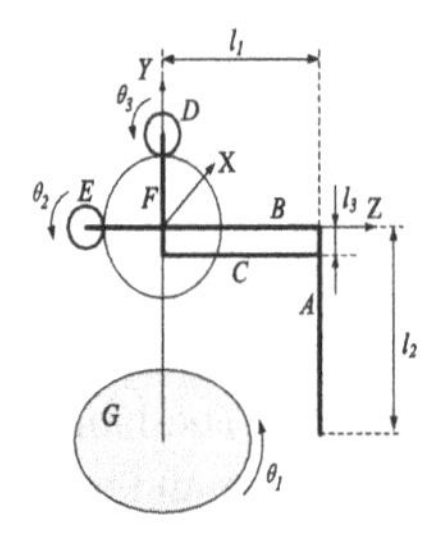

Fig. 3. Simplified diagram of Phantom Premium 1.5(HF) structure.

between the upper computer and the lower computer to realize the closed-loop control of the pose and the force feedback function.

Gravity Compensation. In order to get a better sense of presence, gravity compensation is also carried out for the master controller. The master controller Phantom Premium 1.5 (HF) wrist axis of the 3 revolving joints intersect at one point, which is simplified to the model shown in Fig. 3.

The complete kinetic model of Phantom Premium 1.5 (HF) was developed using the Lagrangian method in conjunction with the spin volume theorem and the exponential product equation.

$$M(q)\ddot{q} + C(q)\dot{q} + F(q, \dot{q}) + G(q) = \tau. \tag{1}$$

The gravity compensation of the master controller is achieved by solving the dynamics of the master controller and establishing a gravity model for each of its moving components in space to reduce or even eliminate the interference of the gravity of each moving component of the master controller on the operator's force feedback perception, and its gravity compensation model is:

$$\tau_G = \begin{bmatrix} 0 \\ \left(m_a l_1 + \frac{1}{2} m_c l_1 + m_{be} l_4\right) g \cos\theta_2 \\ \left(m_c l_3 + \frac{1}{2} m_a l_2 - m_{df} l_5\right) g \sin\theta_3 \end{bmatrix}. \tag{2}$$

2.2 Force Interaction Strategy and Control Strategy

Force Interaction Strategy. The combination of spring and damper is a common method to describe the viscoelastic behavior of contact force. This method includes linear model [8], quasi-linear model [9], nonlinear model [10, 11] and so on. Based on the viscoelasticity of biological tissues and the convenience of analysis and calculation, the Hunt–Crossley (HC) model is used as a biological soft tissue interaction model, which can more accurately describe the real behavior when in contact with soft tissues and choose pork belly pork as a soft tissue experimental sample for mechanical experiments.

Figure 4 shows the experimental platform used for the relaxation test. The experimental platform mainly includes the UR5 collaborative robot, the robot base, the stabilized power supply, the ATI MINI45 six-dimensional force sensor and its data collector, the end tool and its installation fixture and connectors, etc.

Using the above experimental platform, three groups of experiments were conducted to test the relaxation of epithelial tissue, muscle fibers and connective tissue of pork pancetta, ten times for each group, and the average value was taken to ensure the validity of the data obtained. The experimental procedure was as follows.

Control the UR5 robot to move down in the vertical direction until it touches the experimental material, and the initial force value read by the sensor is 0.5N to ensure that the tool and the material are actually in contact. Control the UR5 robot to move downwards in the vertical direction at a speed of 2 mm/s for 20 s, then stop the UR5 robot and leave it for 5 min. The six-dimensional force sensor records the force data of this process at a frequency of 1000 Hz.

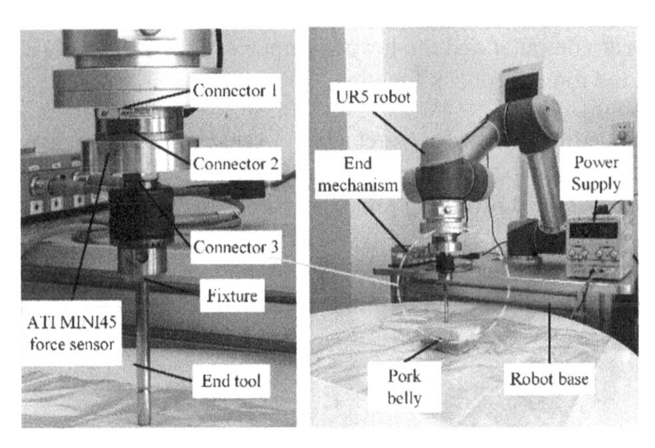

Fig. 4. Relaxation test platform.

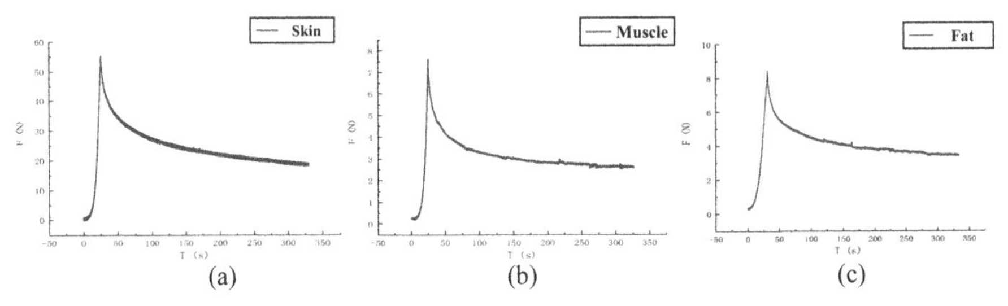

Fig. 5. Different tissue relaxation test force curves: a) Epithelial tissue relaxation curve, b) Muscle fiber relaxation curve, c) Connective tissue relaxation curve.

Table 1. HC model parameters.

	Epithelial tissue	Muscle fibers	Connective tissue
k	−0.499	−0.500	0.035
β	2.840	3.035	1.265
λ	0.250	0.250	0.018
R^2	0.997	0.990	0.982

The force curves measured in the three sets of experiments are shown in Fig. 5a), 5b), and 5c), respectively. The obtained displacement $x(t)$ and force $F(t)$ data were selected for the loading displacement time period, and the parameters k, β, and λ of the HC model were fitted by nonlinear least squares method using the Levenberg-Marquardt optimization algorithm, and the HC model parameters and the variance of different tissues were recorded in Table 1, respectively. And Fig. 6 shows the obtained HC model fitting curves with the original force curves and the residuals graphs.

The fitted coefficient of determination R^2 and the plot of the fitted curve against the original curve show that the HC model is well suited as a model to describe the interaction

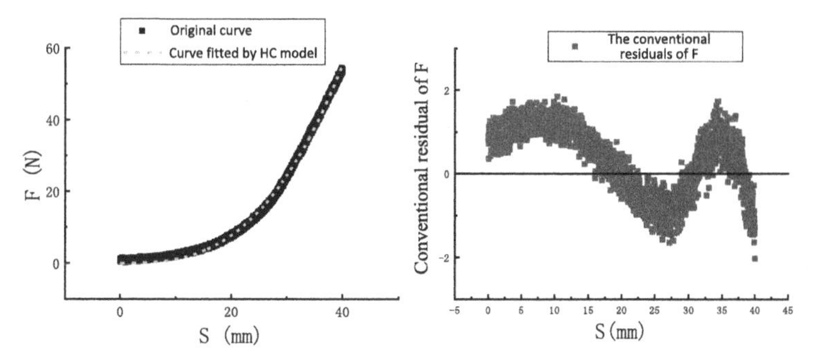

Fig. 6. HC model fitting curves and residual plots (with epithelial tissue as an example).

Table 2. Expressions for different parts of the HC model.

Tissue	Expressions
Epithelial tissue	$F_{Skin}(t) = -0.499x^{2.84}(t) + 0.25x^{2.84}(t)\dot{x}(t)$
Muscle fibers	$F_{Muscle}(t) = -0.5x^{3.035}(t) + 0.25x^{3.035}(t)\dot{x}(t)$
Connective tissue	$F_{Fat}(t) = 0.035x^{1.265}(t) + 0.018x^{1.265}(t)\dot{x}(t)$

with soft tissues. The HC model expressions corresponding to epithelial tissue, muscle fibers, and connective tissue were obtained as shown in Table 2.

Effectiveness Analysis of Control Strategy. The bilateral control system is mainly evaluated from three aspects of stability, tracking and transparency. The following is an analysis of the teleoperation system from these three aspects.

A. *Stability*

Time delay is the main factor that affects the stability of bilateral control. The stability of the system decreases as the time delay increases. It can ensure that the bilateral control system is stable when the time delay is less than a certain limit. The delay limit τ and related parameters are as follows shown.

$$\tau \leq \frac{(-a_1 - a_2 - d_1)[(a_1 + a_2)d_1 - b_2c_2]}{64[max(|a|, |a_2|, |b_2|, |d_1|)]^4}. \tag{3}$$

where M and B denote mass and damping, respectively, the subscripts m and s denote the master and slave controller, respectively, and B_c is the velocity gain. In the teleoperation system in this paper, $M_m = 8$ kg, $M_s = 18.4$ kg, $B_m = B_s = 0.051$, $B_c = 0.1$.

Substituting the above parameters into Eq. (3), the time delay limit $\tau = 755$ ms from the slave controller is calculated. In this system, the robot controller is connected to the PC through the network cable, and the time delay of the system is detected to be about 50 ms through experiments, which is much smaller than the time delay limit of the slave controller, so the system is stable.

B. *Traceability*
Traceability refers to the ability of the slave controller to track the movement of the master controller. It is also an important indicator for judging the performance of the bilateral system. The tracking performance formula of the system is proposed for the random movement position tracking of the system, as shown in Eq. (4). During exercise, the ratio Δ between the difference between the expected pose and the actual pose of the operator and the expected pose is used as an indicator of tracking performance.

$$\Delta = \left| \frac{g(P_m(t - t_0)) - P_s(t)}{g(P_m(t - t_0))} \right|. \tag{4}$$

where g denotes the spatial mapping relationship between the master controller and the slave controller, t_0 denotes the sampling time, and $P_s(t)$ denotes the actual pose change of the slave controller.

This paper adopts the strategy based on the incremental mapping method in the control strategy, multiplying the posture change of the master controller by a certain proportional coefficient k as the posture change of the slave hand. In this paper, the value of k is 1. And the position control method based on the move command is used for motion control, the maximum value of Δ is 0.082, the tracking performance of the system is good.

C. *Transparency*
The transparency of the system characterizes the operability of the system. When the transparency of the system is good, the operator can truly feel the force from the operator. In this study, the ratio λ between the feedback force of the master controller and the interaction force of the external environment is used as an indicator of system transparency, as shown in Eq. (5). The closer λ is to 1, the better the transparency of the system. When $\lambda = 1$, the system is completely transparent, the feedback force is equal to the external force, and the force feedback effect is the best.

$$\lambda = \frac{F_b}{F_e}. \tag{5}$$

According to the analysis of the results of the force feedback experiment, the maximum error of the force feedback by the master controller relative to the force collected by the force sensor occurs when the connective tissue is squeezed, and the maximum error value is $1.2\,N$, at this time $\lambda = 1.072$, which is close to 1, which fully proves that the teleoperating system in this paper performs well in transparency.

By analyzing the teleoperation system of this paper from the three aspects of stability, tracking and transparency, it can be concluded that the force feedback teleoperation system proposed in this paper is a stable, tracking, and a bilateral teleoperation system with good transparency, and has considerable application prospects in the field of teleoperation surgical robots.

3 Experimental Validation of Teleoperation System

3.1 Trajectory Tracking Experiment

In order to evaluate the trackability and stability of the system, the trajectory tracking experiment is conducted. The master controller was used to control the slave controller for random motion, and the motion trajectory was compared by taking two of the ten random motions. The comparison of the motion trajectory of the master and slave controller is shown in Fig. 7, We can see the maximum tracking error is 8.7%, and the effect is good.

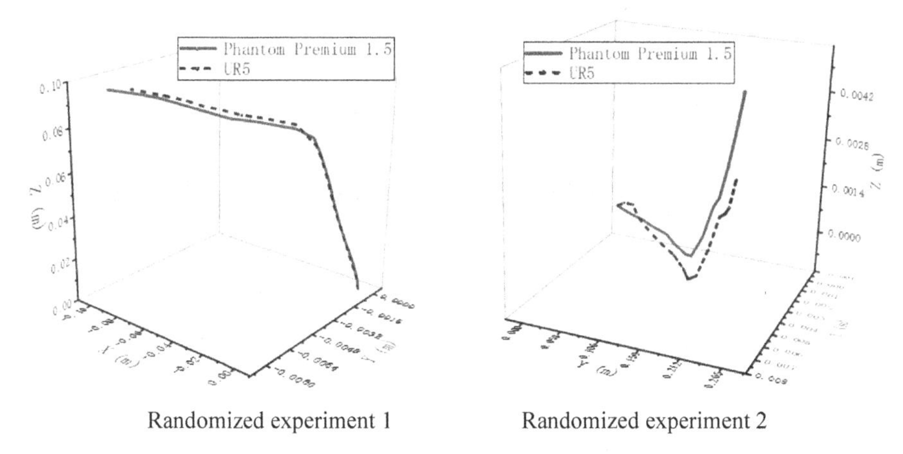

Randomized experiment 1 Randomized experiment 2

Fig. 7. Random trajectory tracking experiment.

3.2 Force Feedback Experiment

After completing the trajectory tracking experiment and the master-slave trajectory tracking effect is good, a force feedback experiment is also needed. As shown in Fig. 1, the operator holds the handle of the master controller, and the master controller is connected to the PC through the EPP interface. The robot control box is connected through the Ethernet port. The end of the UR5 robot is connected to the force sensor through the connector designed by the author. The force sensor is also connected to the fixture and the end tool through the connector designed by the author. The final tool is in contact with the experimental material. The operator interacts with the "environment" through the teleoperation system.

Through the above experimental system, force feedback experiments were conducted on different tissues of the experimental material (epithelial tissue, muscle fiber, connective tissue), and the force sensing data collected during the experiments and the feedback force data from the master controller to the operator were imported into Origin software for processing, and the force curves collected through the force sensors and the feedback force curves from the master controller were obtained as shown in Fig. 8.

From the comparison curve of the force collected by the force sensor and the feedback force of the master controller, it can be seen that based on the force feedback strategy of the soft tissue interaction model, the force feedback by the master controller to the operator has the same trend as the force collected by the force sensor, and the error is smaller, indicating that this force feedback method can more realistically restore the force during the interaction between the tool and the soft tissue. Moreover, it can be seen

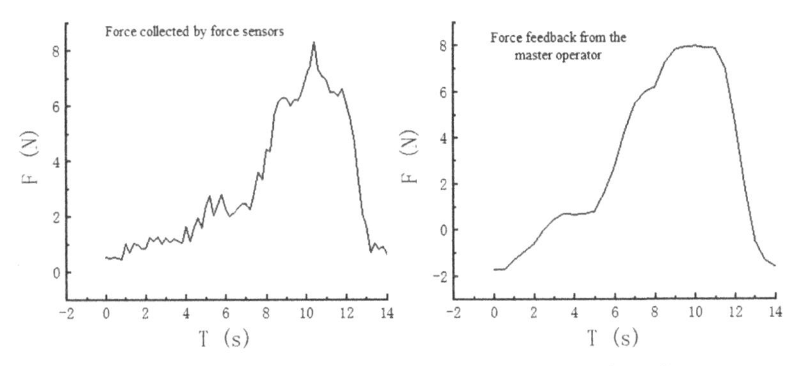

Fig. 8. Force collected by the force sensor versus the feedback force from the master controller (squeezing muscle fibers as an example).

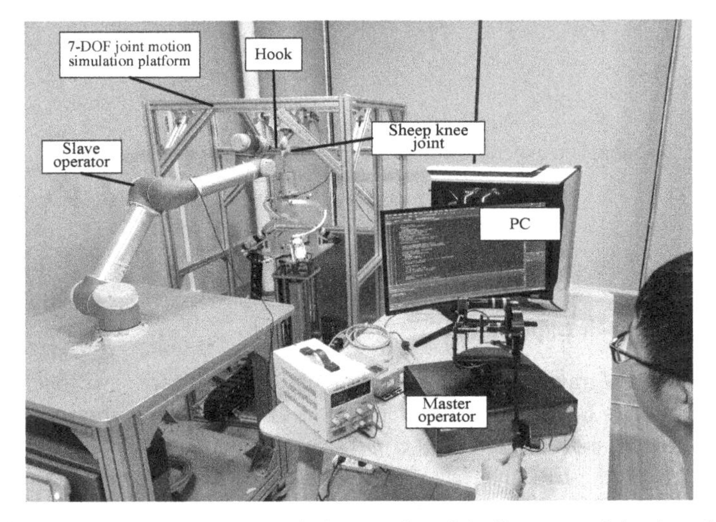

Fig. 9. Measurement of the mechanical properties of the ligaments of the sheep knee.

from Fig. 8 that the force fed back to the operator by the master controller is more stable than the force collected by the sensor, which reduces a lot of abnormal jitter. During the experiment, through the handle of the main operating hand, there is indeed a sense of touch with the soft tissues, and a better sense of presence.

3.3 Practical Application Experiment of Force Feedback Teleoperation Platform

In order to further explore the practical significance of the teleoperation system in the field of teleoperation surgical robots, combined with previous work [12], the force feedback teleoperation robot was used to measure the lateral tension of the ligament when the knee joint of the sheep was in the extended position (as shown in Fig. 9).

The HC model was established by performing a relaxation experiment on the ligaments of the sheep's knee joint. In the experimental scenario shown in Fig. 9, the 7-DOF

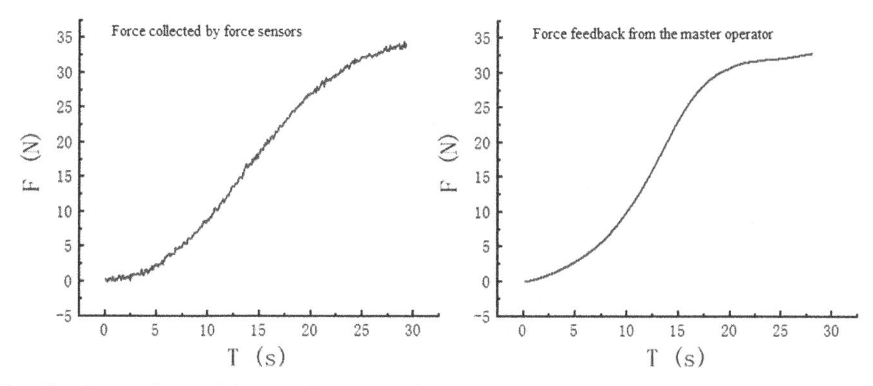

Fig. 10. Comparison of force collected by force sensor and feedback force from the master controller.

joint motion simulation platform was used to control the movement of the sheep's knee joint to a specified angle, and then the slave controller was controlled by the master controller. Then make the probing hook at the end of the slave controller hook the ligament of the sheep's knee joint and stretch it outwards. In the process, the data measured by the force sensor and the feedback force curve of the master controller are shown in Fig. 10.

It can be seen from the experimental results that in practical applications, the feedback force of the master controller is basically the same as the force collected by the force sensor, and the force feedback teleoperation system performs well in practical applications.

4 Conclusion

This paper takes the teleoperation system with force feedback as the research object, and conducts in-depth research on the soft tissue interaction model, the software and hardware of the master-slave heterogeneous teleoperation system, the control structure and control strategy of the teleoperation system, and the force feedback strategy. A force-feedback teleoperation system with no force sensor was developed, and it was verified through experiments in teleoperation trajectory tracking, force feedback, and practical applications. It has achieved good results and has a very good application prospect.

References

1. Picod, G., Jambon, A., Vinatier, D., Dubois, P.J.S.E.: Techniques O.I What can the operator actually feel when performing a laparoscopy. J. Surg. Endosc. **19**, 95–100 (2005). https://doi.org/10.1007/s00464-003-9330-3
2. Wagner, C.R., Stylopoulos, N., Jackson, P.G.: The benefit of force feedback in surgery: examination of blunt dissection. J. Presence-Teleop. Virt. **16**(3), 252–262 (2007)
3. Hannaford, B., et al.: Raven-II: an open platform for surgical robotics research. J. IEEE Trans. Bio-Med. Eng. **60**, 954–959 (2013)
4. Santos, L., Cortesão, R., Quint, J.: Twin kinematics approach for robotic-assisted tele-echography. In: 2019 IEEE/RSJ International Conference on Intelligent Robots and Systems (IROS), pp. 1339–1346. IEEE Press, Macau (2019)

5. Kobayashi, Y., Okamura, N., Tsukune, M., et al.: Non-minimum phase viscoelastic properties of soft biological tissues. J. Mol. Biol. **110**, 103795 (2018)
6. Kačianauskas, R., Kruggel, E.H., Zdancevičius, E., et al.: Comparative evaluation of normal viscoelastic contact force models in low velocity impact situations. J. Adv. Powder Technol. **27**(4), 1367–1379 (2016)
7. Courreges, F., Laribi, M.A., Arsicault, M., et al.: In vivo and in vitro comparative assessment of the log-linearized HuntCrossley model for impact-contact modeling in physical human–robot interactions. **233**(10), 1376–1391 (2019)
8. Takács, Á., Rudas, I.J., Haidegger, T.: Surface deformation and reaction force estimation of liver tissue based on a novel nonlinear mass–spring–damper viscoelastic model. Med. Biol. Eng. Comput. **54**(10), 1553–1562 (2015). https://doi.org/10.1007/s11517-015-1434-0
9. Moreira, P., Liu, C., Zemiti, N., et al.: Soft tissue force control using active observers and viscoelastic interaction model. In: 2012 IEEE International Conference on Robotics and Automation, pp. 4660--4666. IEEE Press, Saint Paul (2012)
10. Jolaei, M., Hooshiar, A., Dargahi, J.: Displacement-based model for estimation of contact force between RFA catheter and atrial tissue with ex-vivo validation. In: 2019 IEEE International Symposium on Robotic and Sensors Environments (ROSE), pp.1–7. IEEE Press, Ottawa (2019)
11. Jolaei, M., Hooshiar, A., Sayadi, A., et al.: Sensor-free force control of tendon-driven ablation catheters through position control and contact modeling. In: 2020 42nd Annual International Conference of the IEEE Engineering in Medicine & Biology Society (EMBC), pp. 5248–5251. IEEE Press, Montreal (2020)
12. Cui, Z., Huang, S.S., Chen, Z.H., et al.: Design, simulation and verification of a 7-DOF joint motion simulation platform. In: 2020 International Conference on Machine Learning and Intelligent Systems, vol. 332, pp. 331–336. MLIS 2020, Saint Paul (2020)

Image Enhancement and Corrosion Detection for UAV Visual Inspection of Pressure Vessels

Zixiang Fei, Erfu Yang(✉), Beiya Yang, and Leijian Yu

Department of Design, Manufacturing and Engineering Management, University of Strathclyde, Glasgow G1 1XJ, UK
{zixiang.fei,erfu.yang,beiya.yang,leijian.yu}@strath.ac.uk

Abstract. The condition of a pressure vessel is normally checked by human operators, which have health and safety risks as well as low working efficiency and high inspection cost. Visual inspection for pressure vessels can be done by an Unmanned Aerial Vehicle (UAV) with the sensing module. Image enhancement techniques and image processing techniques are vital in the UAV inspection of pressure vessels. However, there are several issues to be overcome in the UAV visual inspection of pressure vessels. The images captured by the UAV are of low quality under the cluttered environment due to poor lighting, noises and vibrations caused by the UAV. In this research, a system is developed for UAV visual inspection of pressure vessels using image processing and image enhancement techniques. In the developed system, the input image is captured by the UAV first. Next, efficient image enhancement techniques are applied to the images in order to enhance image qualities. After that, the corrosion part is detected and the percentage of the corrosion area in the entire image is measured. The proposed system has the potential to be implemented for the autonomous correction detection with the image enhancement techniques in UAV visual inspection for the pressure vessels.

Keywords: Image enhancement · Corrosion detection · UAV · Pressure vessel

1 Introduction

Nowadays, human operators are always needed to check the conditions of pressure vessels. However, when carrying out this work, human operators may face health and safety risks in addition to human-errors, low working efficiency and high inspection cost etc. To solve this problem, the visual inspection of pressure vessels can be done in a safer and efficient manner by utilizing a UAV with a high-quality sensing module.

In the UAV visual inspection of pressure vessels, image enhancement techniques and image processing techniques play important roles. However, there are several issues with the visual inspection of pressure vessels using UAV. The first issue is about the inspection of the internal coating, linings for signs of breakdown, crack detection of connecting welds to shell with high accuracy and good performance under the confined space for UAV. There is also an issue about the part in the inner pressure vessels where the access is limited and lighting condition is poor. The images captured by UAV are

M. Fei et al. (Eds.): LSMS 2021/ICSEE 2021, CCIS 1467, pp. 145–154, 2021.
https://doi.org/10.1007/978-981-16-7207-1_15

of low quality under the challenging environment due to the poor lighting, noises and vibrations caused by UAV [1–3]. As a result, it is difficult to detect the corrosion in these low-quality images. Therefore, the images quality needs to be enhanced to increase the detection accuracy.

The illumination is important for the sensing module of UAV, which has been discussed in many research works [4–6]. To provide a required illumination level, the lighting device and its installation need to be considered carefully. When the lighting device and its installation are fixed, it requires the developed algorithm to be robust in the harsh environment especially for its lighting conditions. For example, when the light intensity level is too high (over 1000 lx), it will result in the glare and contrast issues. On the other hand, if the light intensity level is too low, the visual inspection will be difficult. In image enhancement, there are both traditional techniques and the state-of-the-art techniques such as deep learning [7, 8]. As the deep learning based techniques need lots of computing resources, it requires a powerful UAV onboard platform. In this paper, therefore, the traditional techniques for image enhancement are focused on, such as histogram equalization and morphological operators [9, 10]. The morphological operators can be used to correct the non-uniform illumination. The histogram equalization can be used to enhance the image contrast for both grayscale images and RGB images [9]. After the images are enhanced with the image contrast, the corrosion part can be detected using the colour information in the image with higher accuracy. On the other hand, as the images are captured by a UAV when it is flying, the blurred images are caused by the vibrations of the UAV. A blurred image will affect the performance in detecting the accurate corrosion area. As a result, some techniques need to be applied to deblur these images. Wiener deconvolution can be useful to deblur the images when the point-spread function (PSF) and noise level are either known or estimated [9, 11, 12]. The PSF is used to describe the response of an imaging system to a point source or point object. Due to the relative motion between the object and the video camera, the obtained image is blurred in the image acquisition process. In order to deblur the motion-blurred images, the PSF needs to be estimated. After that, the Wiener deconvolution can be used to deblur the images for achieving better performance. In summary, by applying the image enhancement techniques, the image quality and the performance of corrosion detection will be improved.

In this paper, a system is developed for UAV visual inspection of pressure vessels using low computational image enhancement techniques. Within the developed system, the input is the image captured by the UAV first. Next, the image enhancement techniques are applied to these images. After that, the corrosion part is detected and the percentage of the corrosion in the entire image is measured. The system is designed for UAV's visual inspection without ideal lighting condition.

This paper is organized as follows. Section 1 has a general overview of the paper. Subsequently, the procedure of the system is introduced in Sect. 2. Section 3 to Sect. 5 describes the image enhancement techniques, contour detection techniques and corrosion detection techniques, respectively. Finally, a brief and key conclusion is made in Sect. 6.

2 Overview of the System

In the research, there are several steps for image enhancement and corrosion detection as shown in Fig. 1. The techniques have been combined and applied for images and videos of a real pressure vessel with the industrial scale.

To begin with, the input is the image captured by the UAV first. Next, the image enhancement techniques are applied to the obtained images. These image enhancement techniques are used to enhance the image contrast, correct the non-uniform illumination and de-blur the images which are caused by the UAV vibrations.

After that, the corrosion part is marked with yellow colour using colour based corrosion detection techniques shown in Fig. 1. Also, the percentage of the corrosion area in the entire image is measured. Finally, the contour of the corrosion part is detected after the corrosion detection is performed. The image is finally converted into a gray scale image for contour detection.

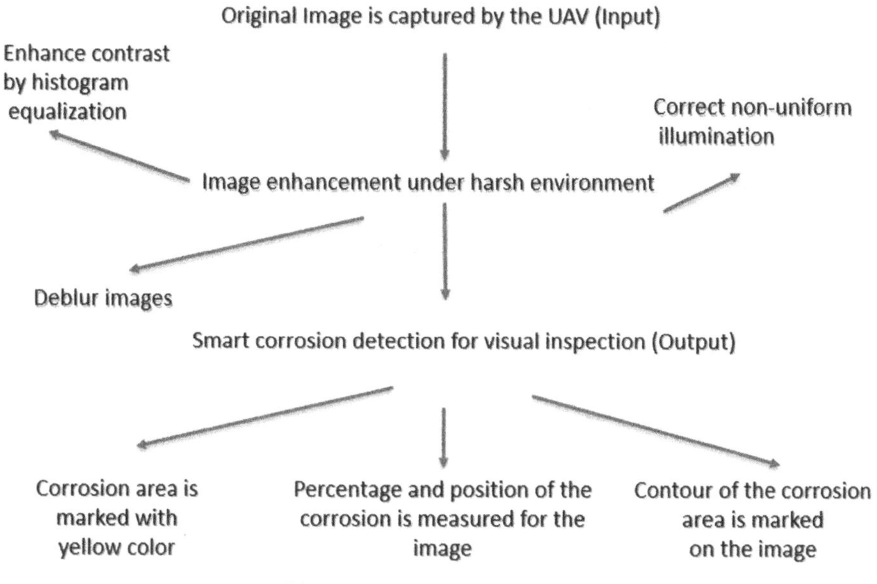

Fig. 1. Overview for the system

3 Image Enhancement

Image enhancement techniques are used to enhance the image quality and contrast before further image processing is conducted. Currently, the image enhancement techniques with Histogram equalization (HE) and Contrast Limited Adaptive Histogram Equalization (CLAHE) are adopted due to their low computational requirements. They can be applied to the RGB images for image enhancement [13–15]. The major idea of the HE is discussed in the following.

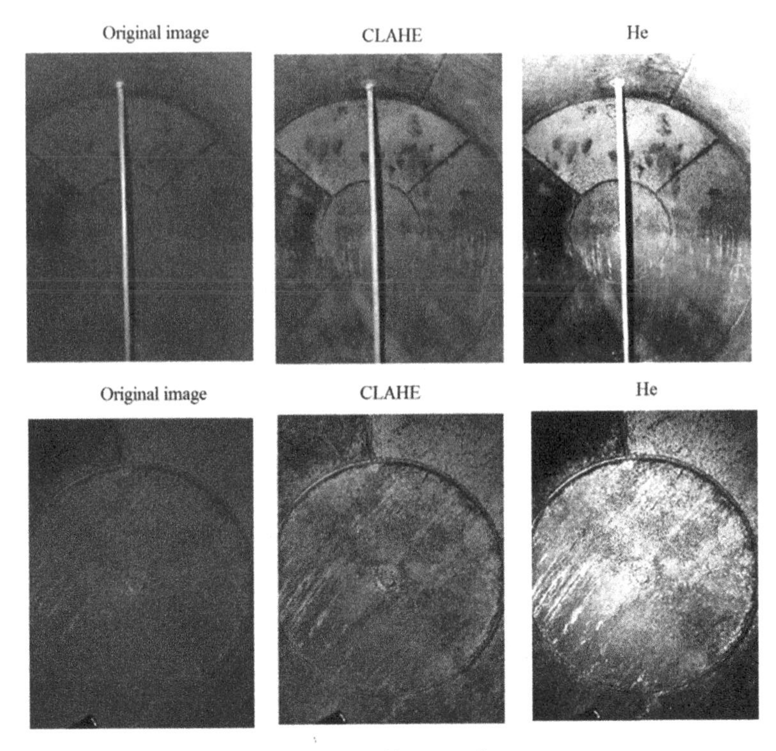

Fig. 2. Samples of image enhancement

In a discrete grayscale image$\{x\}$, n_s is the number of occurrences of gray level I [16]. The probability of an occurrence of a pixel for I can be represented by $p_x(s)$:

$$p_x(s) = p(x = s) = \frac{n_s}{n}, 0 \ll s < L. \tag{1}$$

In (1), L is the total number of gray levels, n is the total number of pixels. $p_x(s)$ is the image's histogram.

Let $\mathrm{cd}f_x(s)$ be the cumulative distribution function (CDF) to p_x:

$$\mathrm{cd}f_x(s) = \sum_{t=0}^{s} p_x(x = t). \tag{2}$$

For a new image$\{y\}$ which has a flat histogram, it has a linearized CDF:

$$\mathrm{cd}f_y(s) = sK. \tag{3}$$

Using the properties of the CDF, a transformation can be done:

$$\mathrm{cd}f_y\left(y'\right) = cdf_y(T(k)) = cdf_x(k). \tag{4}$$

Also, the following transformation needs to be applied to the result:

$$y' = y \cdot (\max\{x\} - \min\{x\}) + \min\{x\}. \tag{5}$$

The experiments were carried by using different internal pressure vessel images. In the current stage, the UAV is not ready for capturing the images of pressure vessels and the images captured by humans are used to carry out the experiments and evaluate the feasibility of the developed image enhancement and corrosion detection system. The experiment results are shown in Fig. 2. The original image is shown on the left. The enhanced images using the CLAHE technique and HE technique are shown in the middle and on the right, respectively. It demonstrates that the image using the CLAHE has improved the visual texture of the original image.

4 Contour Detection

Contour detection is a process that finds a curve joining all the continuous points along with the boundary that has the same colour or intensity. The contours are very useful for shape analysis, object detection and recognition. There are many contours detection techniques such as the Combination of Receptive Fields model (CORF), Gabor function model, multiple-cue inhibition operator [17–19]. The performance of various computational models in contour detection tasks has not been quantified and they have not been compared in complex visual tasks.

The Gabor filter is a linear filter for texture analysis in image processing [20]. Its mechanism in image analysis is similar to the perception in the human visual system. Given the point or region of analysis in an image, it is used to essentially analyze whether there is any specific frequency content in specific directions in the localized image region.

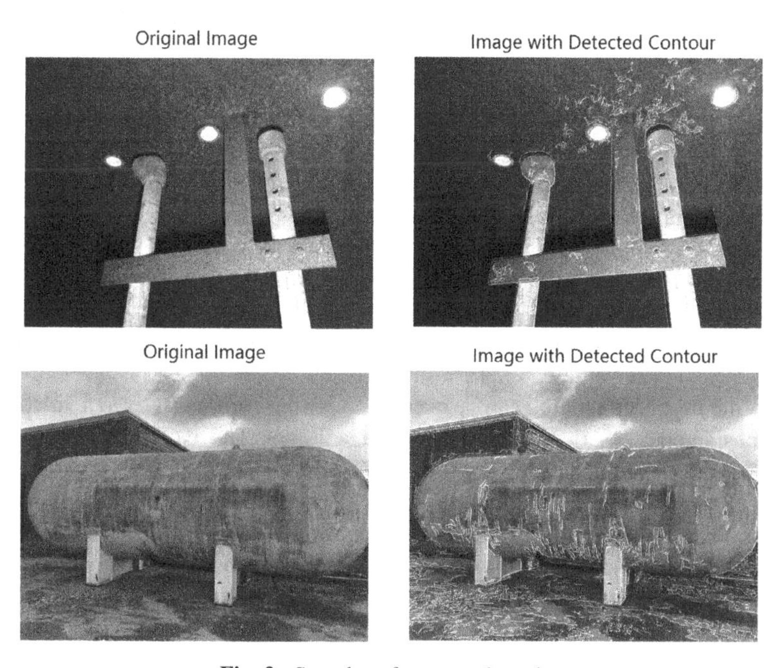

Fig. 3. Samples of contour detection

The Gabor filter is particularly suitable for the purpose of texture representation and discrimination.

The details for the CORF model can be referred to [19]. In the CORF, the response R is defined as the weighted geometric mean of all the sub unit responses which belong to the specifications determined by the set D:

$$R_D(x, y) = \left(\prod_{i=1}^{\lceil D \rceil} \left(D_{\delta_i,\sigma_i,\rho_i,\varphi_i}(x, y)\right)^{\omega_i}\right) 1 / \sum_{i=1}^{\lceil D \rceil} \omega_i. \tag{6}$$

$$\omega_i = exp^{-\frac{\rho_i^2}{2\alpha'^2}}. \tag{7}$$

$$\alpha' = \frac{1}{3} \max_{i \in \{1 \ldots \lceil D \rceil\}} \{\rho_i\}. \tag{8}$$

In (6), $\delta_i, \sigma_i, \rho_i, \varphi_i$ stand for the polarity of the sub unit, the scale parameter, the radius and the polar angle, respectively.

The experiment results with the CORF are given in Fig. 3. In the research, the CORF technique is used to detect the contours of the images taken from both the external and internal pressure vessel. The original image was converted to a gray scale image first and the parameters were carefully tuned to improve the performance. In Fig. 3, the original images are on the left and the image with the contour detected are on the right. It is shown that most of the contours have been detected successfully.

5 Corrosion Detection Under Cluttered Environment

Corrosion detection is an important part of visual inspection for pressure vessels. The proposed algorithm needs to detect the corrosion with high accuracy, short time and low cost under the cluttered environment such as poor lighting situations. In this part, the experiment is carried out to test the performance of the corrosion detection for the images of internal pressure vessels in the cluttered environment.

In the experiment, the RGB value is used to detect the corrosion in the image. From the training images, the range of RGB value of the corrosion area can be found. The range of RGB value is used to find the corrosion part in the testing images. The experiment result is shown in Fig. 4. The original images are shown above and the images with the detected corrosion in yellow points are shown below.

Also, the proposed method can be used to detect the corrosions from a video which can be considered as a series of continuous image frames. The experiment results on the corrosion detection from a video are shown in Fig. 5 and Table 1. The experiment was carried out on several continuous video frames. The interval between the two frames is 0.33 s. In the figure, a1 to a10 represent the original images of the video frame and b1 to b10 represent the images that show the detected corrosion area in yellow colour. In Table 1, the first column gives the order of the video frames and the second column shows the percentage of the corrosion area detected in the entire image in each video frame.

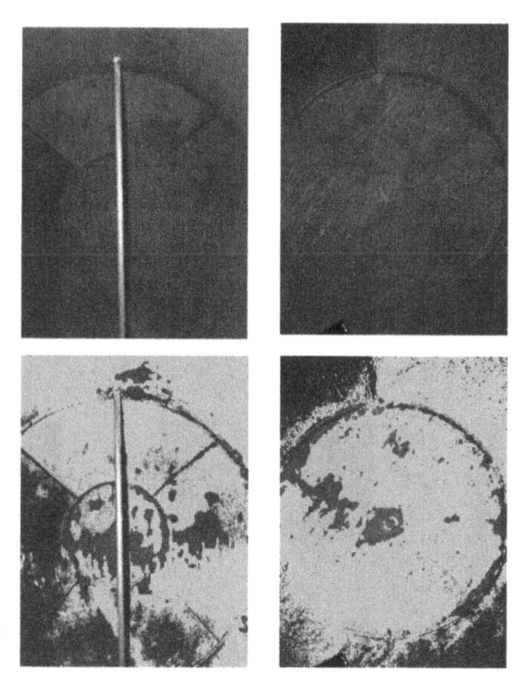

Fig. 4. Samples of corrosion detection under cluttered environment for pressure vessel

Table 1. Area of corrosion detected in a video

– Order	– Percentage of corrosion area detected (%)
– 1	– 69.7
– 2	– 72.8
– 3	– 78.9
– 4	– 79.0
– 5	– 81.5
– 6	– 83.7
– 7	– 82.6
– 8	– 82.2
– 9	– 75.2
*10	– 68.7

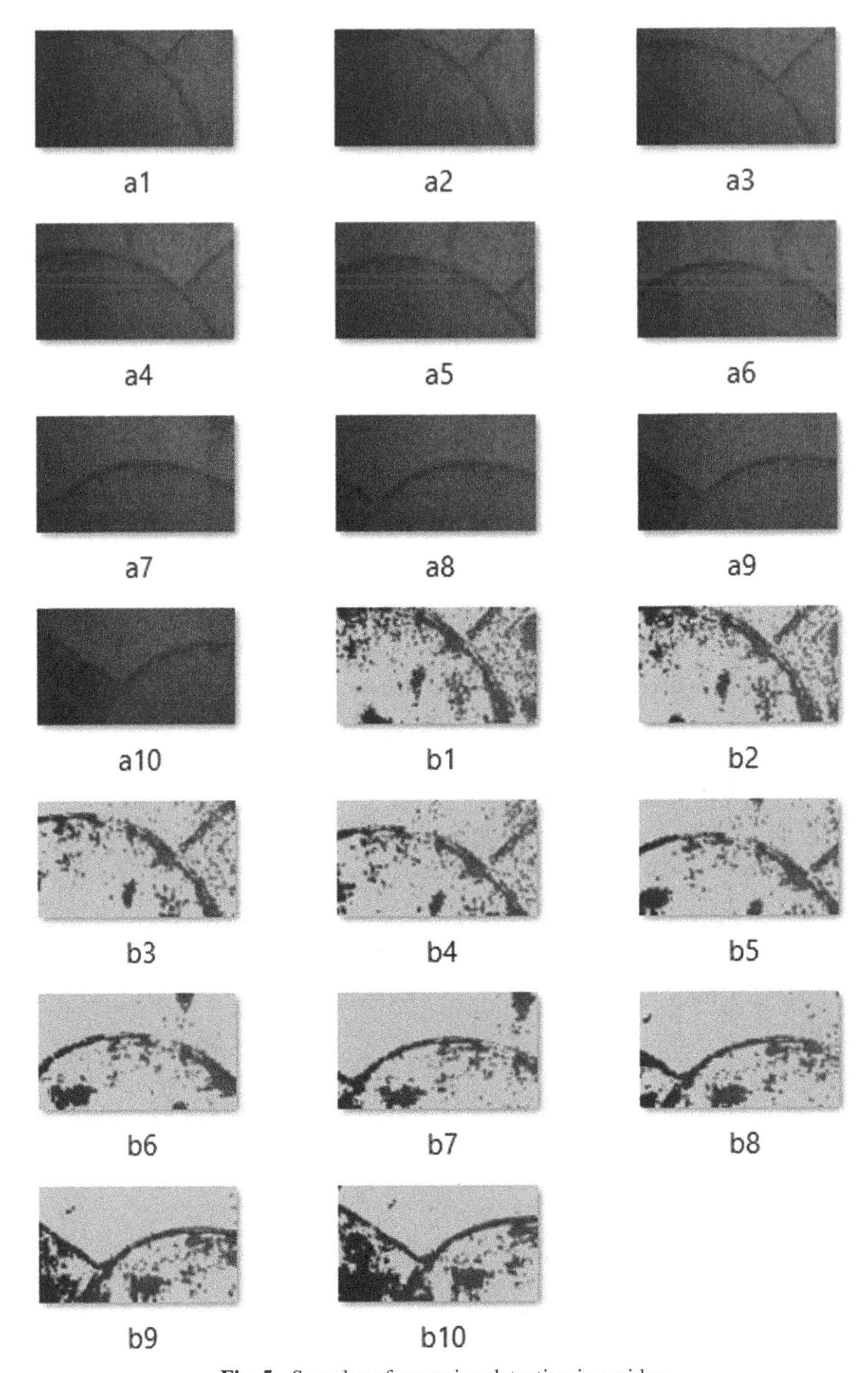

Fig. 5. Samples of corrosion detection in a video

6 Conclusion

This paper has presented a system that was developed for the UAV visual inspection of pressure vessels. It has been mainly focused on image enhancement and corrosion detection to address the challenging issues in the cluttered environment.

Within the developed system, the UAV is used to capture the videos from pressure vessels. The Histogram Equalization techniques are mainly applied to the images to enhance the contrast of the image under poor lighting condition in the internal pressure vessel. After that, the corrosion part is detected by using the image's colour information and the percentage of the corrosion in the entire image is measured. The experiments were carried out to evaluate the performance and demonstrate the feasibility of the developed system.

Though a good performance was demonstrated in the lab-based experiments, there are some limitations and future work. First of all, to complete the sensing and image processing module, a lighting system is required to install on the UAV to provide suitable illumination, which will be solved in future work. Also, the current experiments only tested the images captured manually. In future work, the image processing module needs to be integrated and tested with the images taken by the flying UAV in the internal pressure vessels. In addition, the CLAHE technique was used only in this paper as the image enhancement technique. Therefore, the other more efficient image enhancement techniques need to be tested and evaluated in future as well to identify the most suitable image enhancement techniques for the UAV visual inspection of pressure vessels in engineering practice.

Acknowledgements. This work is supported by the UK Oil and Gas Technology Centre (OGTC) under the LOCUST research project (2019–2021, Grant No.: AI-P-028). The authors thank all the supports from the OGTC and Strathclyde robotics project teams.

References

1. Li, C., Tan, F.: Effect of UAV vibration on imaging quality of binary optical elements. In: Proceedings of 2018 IEEE International Conference on Mechatronics and Automation, ICMA 2018, pp. 1693–1698. Institute of Electrical and Electronics Engineers Inc. (2018)
2. Jena, K.K., Mishra, S., Mishra, S., Bhoi, S.K.: Unmanned aerial vehicle assisted bridge crack severity inspection using edge detection methods. In: Proceedings of the 3rd International Conference on I-SMAC IoT in Social, Mobile, Analytics and Cloud, I-SMAC 2019, pp. 284–289. Institute of Electrical and Electronics Engineers Inc. (2019)
3. Liu, D., Sun, J., Jiang, P., Zhou, X.: Research on image stabilization control algorithm of UAV Gm-APD lidar PID. J. Presented at the December 2 (2020)
4. Teixeira, C.L.G., Nanni, M.R., Furlanetto, R.H., Cezar, E., Silva, G.F.C.: Reflectance calibration of UAV-based visible and near-infrared digital images acquired under variant altitude and illumination conditions. J. Remote Sens. Appl. Soc. Environ. **18**, 100312 (2020)
5. Wang, T., Huang, T.: Passive radar based UAV detection using opportunistic RF illumination. In: Proceedings - IEEE International Conference on Industrial Internet Cloud, ICII 2019, pp. 6–11. Institute of Electrical and Electronics Engineers Inc. (2019)

6. Gerard, F., et al.: Deriving hyper spectral reflectance spectra from UAV data collected in changeable illumination conditions to assess vegetation condition. In: International Geoscience and Remote Sensing Symposium (IGARSS), pp. 8837–8840. Institute of Electrical and Electronics Engineers Inc. (2018)
7. Yu, C.-Y., Lin, H.-Y., Lin, C.-J.: Fuzzy theory using in image contrast enhancement technology. Int. J. Fuzzy Syst. **19**(6), 1750–1758 (2017). https://doi.org/10.1007/s40815-017-0351-9
8. Kinoshita, Y., Kiya, H.: Hue-correction scheme considering CIEDE2000 for color-image enhancement including deep-learning-based algorithms. J. APSIPA Trans. Sig. Inf. Process. **9** (2020)
9. Dos Santos, J.C.M., Carrijo, G.A., De Fátima dos Santos Cardoso, C., Ferreira, J.C., Sousa, P.M., Patrocínio, A.C.: Fundus image quality enhancement for blood vessel detection via a neural network using CLAHE and Wiener filter. J. Res. Biomed. Eng. **36**, 107–119 (2020)
10. Nigam, M., Bhateja, V., Arya, A., Bhadauria, A.S.: An evaluation of contrast enhancement of brain MR images using morphological filters. In: Bhateja, V., Satapathy, S.C., Satori, H. (eds.) Embedded Systems and Artificial Intelligence. AISC, vol. 1076, pp. 571–577. Springer, Singapore (2020). https://doi.org/10.1007/978-981-15-0947-6_54
11. Luo, T., Fan, R., Chen, Z., Wang, X., Chen, D.: Deblurring streak image of streak tube imaging lidar using Wiener deconvolution filter. J. Optics Express. **27**, 37541 (2019)
12. Hu, W., Wang, W., Ji, J., Si, L.: The spatial resolution enhancement deconvolution technique of the optimized wiener filter in Terahertz band. In: Proceedings of IEEE 9th UK-Europe-China Workshop on Millimetre Waves and Terahertz Technologies, UCMMT 2016, pp. 96–99. Institute of Electrical and Electronics Engineers Inc. (2017)
13. Yao, Z., Lai, Z., Wang, C., Xia, W.: Brightness preserving and contrast limited bi-histogram equalization for image enhancement. In: 2016 3rd International Conference on Systems and Informatics, ICSAI 2016, pp. 866–870. Institute of Electrical and Electronics Engineers Inc. (2017)
14. Du, Y., Ma, Z., Yu, Y., Qin, H., Tan, W., Zhou, H.: Self-adaptive histogram equalization image enhancement based on canny operator. J. Presented at the October 24 (2017)
15. Liu, H., Yan, B., Lv, M., Wang, J., Wang, X., Wang, W.: Adaptive CLAHE image enhancement using imaging environment self-perception. In: Lecture Notes in Electrical Engineering, pp. 343–350. Springer Verlag (2018). https://doi.org/10.1007/978-981-10-6232-2_40
16. Bassiou, N., Kotropoulos, C.: Color image histogram equalization by absolute discounting back-off. J. Comput. Vis. Image Understand. **107**, 108–122 (2007)
17. Kang, X., Kong, Q., Zeng, Y., Xu, B.: A fast contour detection model inspired by biological mechanisms in primary vision system. J. Front. Comput. Neurosci. **12**, 28 (2018)
18. Lyakhov, P.A., Abdulsalyamova, A.S., Kiladze, M.R., Kaplun, D.I., Voznesensky, A.S.: Method of oriented contour detection on image using Lorentz function. In: 2020 9th Mediterranean Conference on Embedded Computing, MECO 2020. Institute of Electrical and Electronics Engineers Inc. (2020)
19. Azzopardi, G., Petkov, N.: A CORF computational model of a simple cell that relies on LGN input outperforms the Gabor function model. J. Biol. Cybern. **106**, 177–189 (2012)
20. Rad, G.A.R., Samiee, K.: Fast and modified image segmentation method based on active contours and gabor filter. In: 2008 3rd International Conference on Information and Communication Technologies: From Theory to Applications, ICTTA (2008)

Computational Method in Taxonomy Study and Neural Dynamics

PF-OLSR Routing Algorithm for Self-organizing Networks of Pigeon Flocks

Jianxi Song[1], Zhenlong Wang[2], Zijun Liu[1], and Weiyan Hou[1](✉)

[1] School of Information Engineering, Zhengzhou University, Zhengzhou 450001, China
[2] School of Life Sciences, Zhengzhou University, Zhengzhou 450001, China

Abstract. As the research vehicle for biobots, pigeon have better concealment and flexibility in the reconnaissance process. The optimal link-state routing protocol is a common routing protocol for classical UAV clusters. The greedy algorithm of this protocol cannot obtain the set with the minimum number of nodes. The protocol is difficult to guarantee the stability of the population network topology in a high-speed moving pigeon flocks' network. The flight of pigeons is influenced by various factors. Considering two biological perspectives, energy and speed of pigeon flocks, this paper proposes an ant colony algorithm based on the flight characteristics of pigeon flocks. The algorithm is applied to the pigeon flock network, and a new pigeon flock optimized link state routing protocol is proposed. The results of the three-layer modeling simulation by OPNET software show a significant reduction in the number of nodes in the set of multipoint relay nodes. The topology of the whole network changes less, the stability is improved by about 20%, and the throughput is improved by 9.545%. The new algorithm improves significantly over the greedy algorithm.

Keywords: OLSR protocol · Greedy algorithm · Ant colony algorithm · OPNET simulation · Flock control

1 Introduction

In nature, flying animals have their own unique characteristics. If artificial signals could be applied to the nervous system of animals to make them act and fly according to human's commands, then these biobots would be of significant military value [1].

Today, biology and communication networks are rapidly developing. The use of techniques such as electrical or optical stimulation to stimulate the nerves of birds or insects can interfere with their behavior. This allows for remote control of biobots [2, 3]. Biobots means implanted microchips into the young body of a living creature, and as the creature grows, the chip is embedded in the body with the accompanying growth and becomes part of the body. Then mounted the circuit board on the neck or back of the pigeon, which has a control module composed of sensors, GPS and wireless sensing devices [4], at which point it is possible to stimulate the nerves and muscles of the pigeon through wireless signals and circuit calls in order to control or interfere with its flight.

M. Fei et al. (Eds.): LSMS 2021/ICSEE 2021, CCIS 1467, pp. 157–166, 2021.
https://doi.org/10.1007/978-981-16-7207-1_16

As a flying animal, the pigeon has the advantages of small size, fast flight speed, and long flight distance [5]. It is disorienting because it can fly flexibly in the city and in the field.

Optimized Link State Routing (OLSR) protocol [6] is commonly used for communication during UAV swarm flight, which is one of the mobile ad hoc networks, but unlike UAV, there are many unique biological characteristics of pigeon flocks' flight:

1. Pigeons still have an individual tendency to fly while under control, and changes in position during the flight of a flock may cause serious changes in the network topology [7]. Simultaneously, the problem of rapid network topology changes is accentuated by the fact that pigeons fly very densely in flocks.
2. The pigeon's energy consumption during flight is affected by the length of its metatarsus and its own weight, and experiments have shown that the longer the attachment the better the flight endurance of the pigeon and the longer the flight duration of the pigeon with more weight.
3. Since the willingness of pigeons to fly is significantly reduced when they carry devices weighing more than 45 g [8], the battery weight needs to be strictly limited in order to reduce the weight of the wireless control devices carried by each individual.

2 Disadvantages Analysis of OLSR Protocol

The OLSR protocol is a routing protocol developed specifically for mobile ad hoc networks, which enables the detection of the entire network topology and the calculation of routes by exchanging information between nodes [9]. The MPR function in the OLSR protocol can effectively cut down the control messages [10]. MPR function is a simple and effective mechanism to control message flooding, which reduces the message redundancy among nodes in the network by reducing the number of nodes broadcasting messages, and is suitable for groups like pigeons that need to conserve energy and keep operating for long periods of time.

Conventional protocols select MPRs using a greedy algorithm, which is simple and fast to compute, but is not necessarily the optimal solution [11], because it does not take into account the intersection of the coverage area of each MPR node. The greedy algorithm is divided into the following steps.

Step 1: Define an empty set as the MPR set s_{MPR};

Step 2: Set the source node S. For all the nodes in the range of one-hop node set N_1, get the number of nodes they can cover in the two-hop node set N_2 respectively;

Step 3: For any one-hop neighbor node N_1^k of S, if there exists a two-hop node of S that can only be reached uniquely through that one-hop node, add node N_1^k to S;

Step 4: If there are still nodes in the set of two-hop node set N_2 that have not been reached by any node in the set of S, then for the nodes in the set of one-hop node set N_1 that have not been selected into the set of S, the node with the most connectivity in the set of two-hop node set N_2 is added to S;

Step **5:** Skip to step 4 until all nodes in the two-hop node set N_2 are covered.

According to the node distribution state in Fig. 1, the greedy algorithm yields the MPR node set of S nodes as s_{MPR}= {A,B,C,D,E,}, but the optimal set of nodes is s_{MPR}

= {A,B,C,E} or s_{MPR}= {A,B,D,E}, which can then be selected based on the flight speed of the pigeons on the one-hop nodes C, D and the remaining energy of the battery.

Pigeon colony networks need to further reduce network overhead and save node energy consumption. In this paper, an ant colony algorithm based on OLSR protocol optimization is proposed to address the drawbacks.

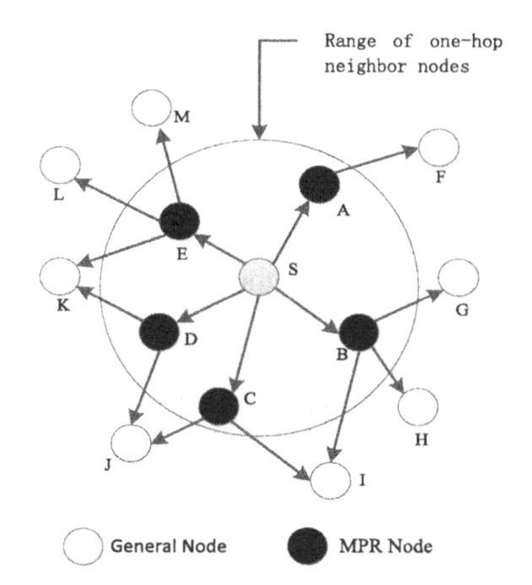

Fig. 1. Schematic diagram of greedy algorithm

3 Improved PF-OLSR Protocol for Pigeon Flocks' Characteristics

Ant Colony Optimization (ACO) algorithm [12] is a population intelligence algorithm manifested by a group of individuals interacting with each other [13]. The advantage of the Ant Colony algorithm is the randomness of its search. Each experiment does not get the same result. The pheromones make the algorithm a positive feedback process [14].

The ant colony algorithm is based on the probabilistic selection of the nodes for the next hop path and the pheromone update after each iteration [15]. The probabilistic selection is determined by the pheromone concentration on the neighboring nodes of a hop according to a set probabilistic selection formula. A round of iterations ends when the set of selected nodes can cover all the second-hop nodes. After each round of iteration, the pheromone is updated based on the current round of iteration. Finally, iterative optimization of the pheromone is performed.

a) **Probabilistic selection process**

In the ant colony algorithm of PF-OLSR protocol, the probabilistic selection formula is not only based on the pheromone concentration of adequate neighbor nodes, but also takes into account the biological nature of the pigeons.

In the ant colony algorithm, the source node S selects the MPR node from the one-hop neighbor nodes based on the path probability. The selection of the next node shifted in the probabilistic selection formula is influenced by several factors. In addition to the size of the pheromone, the number of second-hop nodes connected to the first-hop node is also taken into account. The more second-hop nodes, the more pigeons are clustered in that pigeon node. Then the node is more suitable to be an MPR node. Pigeons have different energy consumption due to attachment length, weight. Flight speed has an impact on future flight intentions. In order to keep the flock intact and the network topology stable, the weights of energy and speed are introduced. Therefore, the probability selection formula is shown in Eq. 1.

$$p_s^k = \frac{\mu(k)^a v(k)^b s(k)^c}{\sum_{k=1}^{|N_1|} \mu(k)^a v(k)^b s(k)^c}. \tag{1}$$

where $\mu(k)$ denotes the pheromone concentration on node N_1, the larger the pheromone, the higher the probability that the path will be selected, it will be updated with each iteration. $v(k)$ is the number of two-hop neighbor nodes, the more nodes are connected to the two-hop nodes, the higher the chance that the node will be selected as the MPR node, adding this factor makes the selection of MPR nodes iterate in the right direction. $s(k)$ is the energy and velocity weights formula, which is defined as Eq. 2.

$$s(k) = \alpha E_i + \beta E_r - \gamma V. \tag{2}$$

E_i represents the initial energy of the node, E_r is represented as the remaining energy of the node, and V represents the movement speed of the node. The initial energy gives the tendency of the ant to move at the beginning, which makes the result iterate in the desired direction and speeds up the initial iteration. α, β, and γ correspond to the weight factors of these three attributes of the nodes, respectively, and are set as shown in Table 1.

Table 1. Weight parameter setting table

Weighting parameters	Numerical value
α	0.3
β	0.5
γ	0.2
a	0.4
b	0.1
c	0.5

b) **Pheromone concentration update process**

One iteration is completed when all ants have constructed their paths. The pheromones on all paths are updated before the next round of iterations. Firstly, a certain amount of pheromone is volatilized between the source node and node i. The update rule is shown in Eq. 3, where ρ takes on the value range $\{\rho|0 \leq \rho \leq 1\}$.

$$\mu(i) \leftarrow (1-\rho)\mu(i). \tag{3}$$

If an ant passes through node i, then the corresponding pheromone has to be increased on that path. Each ant releases the same size of pheromone in one iteration, so the shorter the path constructed by the ant, the more pheromones it accumulates on the path it experiences, and the greater the probability that the ant will choose that path in the next time.

$$\mu(i) \leftarrow \mu(i) + \sum\nolimits_{k=1}^{m} \Delta\mu^k(i). \tag{4}$$

where $\Delta\mu^k(i)$ is defined by Eq. 5.

$$\Delta\mu^k(i) = \begin{cases} \frac{\mu_{inc}}{C^k} & \text{if exist} \\ 0 & \text{otherwise} \end{cases}. \tag{5}$$

C^k represents the length of the path travelled by ant k. μ_{inc} is the total amount of pheromones that released by a single ant. Because an ant releases the same total pheromone in one iteration, the shorter C^k is, the more pheromone is accumulated on the path that the ant travels through.

It is known that the pheromone update is divided into two parts, volatile and superimposed, but the difference is that volatile is a global node at the same time, while superimposed will only be performed at the nodes on the path through which this iteration passes. Combining the two parts and considering the pheromone concentration already present at the nodes, the pheromone update rule is Eq. 6, where the value of ρ is taken as 0.8 and the value of μ_{inc} is taken as 1.

$$\mu(i) \leftarrow \rho\mu(i) + \sum\nolimits_{k=1}^{m} \frac{\mu_{inc}}{C^k}. \tag{6}$$

If a node is always unselected, the pheromones on this path will decay rapidly with iteration, and conversely, the superposition of pheromones on the selected path will increase the probability of being selected and thus superimpose more pheromones. In this case, it is easy to fall into the dilemma of local optimum. In order to guarantee the obtained result belongs to the global optimal solution, it is necessary to add an iterative optimization process to the pheromone update link.

Iterative optimization is to reduce the pheromone concentration difference between nodes. The nodes with low pheromone concentration are compensated and the nodes with high pheromone concentration are reduced. In each round of pheromone update, this paper adds pheromone compensation rules to effectively solve the problem that a node is difficult to be selected due to its own low pheromone.

$$\mu(i) = \mu(i) + \begin{cases} \frac{\max_{j=1}^{|N_1|} \mu(j) - \mu(i)}{|N_1|} & \textit{if } \mu(i) = \min \mu(i) \\ 0 & \textit{otherwise} \end{cases} \tag{7}$$

In addition, considering the ant colony algorithm may have the problem that the pheromone concentration of nodes is too large, which affects other reference factors, the corresponding reduction mechanism is also introduced in the algorithm.

$$\mu(i) = \mu(i) - \begin{cases} \frac{\mu(i) - \min_{j=1}^{|N_1|} \mu(j)}{|N_1|} & \textit{if } \mu(i) = \max \mu(i) \\ 0 & \textit{otherwise} \end{cases} \tag{8}$$

In the process of pheromone update, the algorithm is set to volatilize during the iterative process, and it is not accumulated without loss, so the value of ρ is 0.8, and the release of pheromone is specified as 1.

c) **Minimum set iteration process**

To obtain the minimum MPR set, the pheromones of the nodes are updated at the end of each iteration. The result set of the first iteration is brought into an empty set *ANS*. The number of nodes in the set is recorded $COUNT(ANS)$. Then, record the set *CUR* of each iteration and the number of nodes within that set $COUNT(CUR)$. Compare the size of $COUNT(ANS)$ with $COUNT(CUR)$, and the rules as in Eq. 9. After the maximum number of iterations is reached, the iteration is stopped and the set *ANS* is the minimum number of MPR sets, and the number of nodes in the set is $COUNT(ANS)$.

$$ANS = \begin{cases} ANS \; COUNT(ANS) < COUNT(CUR) \\ CUR \; COUNT(ANS) > COUNT(CUR) \end{cases} \tag{9}$$

$$S_{MPR} = ANS. \tag{10}$$

$$COUNT(ANS) = \min \sum_{k=1}^{|N_1|} S_{MPR}. \tag{11}$$

4 Model Construction and Simulation

The simulation process is performed on OPNET software to verify the performance of the algorithm through three-layer modeling. This paper compared the performance of PF-OLSR protocol and OLSR protocol with the same number of nodes and network topology.

a) Model building

The entire network uses a random movement model, imitating the irregular distribution of pigeon flocks, placed randomly in each corner, as shown in Fig. 2.

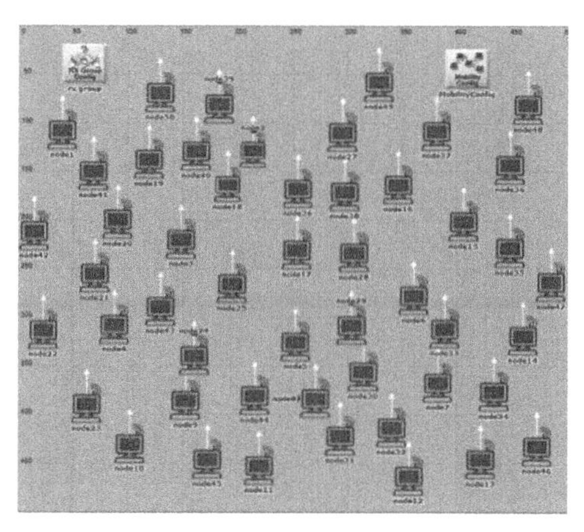

Fig. 2. PF-OLSR protocol network domain model

The number of nodes in the network is taken as 50 with reference to the common number of small-scale flocks, and other parameters are set as shown in Table 2.

Table 2. Link library table format

Parameters	Value
Number of nodes	50
Movement range	500 m * 500 m
Movement speed	10–15 m/s
Initial energy	0.1 J
Willingness	Willingness Default
Hello interval	2.0 s
TC interval	5.0 s

The node model of the network is shown in Fig. 3: where src is the service source module, which generates ON/OFF services, and the src source node is used to generate service packets; Sink is used to count the received data (delay, throughput, etc.); Router is the route calculation module, which is also the core of this paper, and its process domain model is shown in Fig. 5; Interface is the intermediate layer, which is the MAC and routing Intermediate layer, responsible for the data management and forwarding of upper and lower layers; Wireless_lan_mac is the MAC module, and 802.11 Mac protocol is used in this simulation.

In Fig. 4, the PF-OLSR routing protocol is initialized with the init state, and jumps to the wait state after obtaining the node attributes and parameter settings of the OLSR protocol. wait state connects up each condition as a wait state, and when the interrupt condition is triggered, the wait state will be transferred to the state corresponding to the condition.

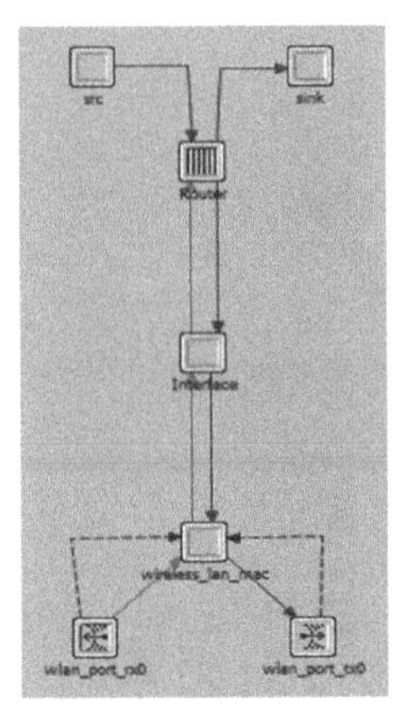

Fig. 3. Node model

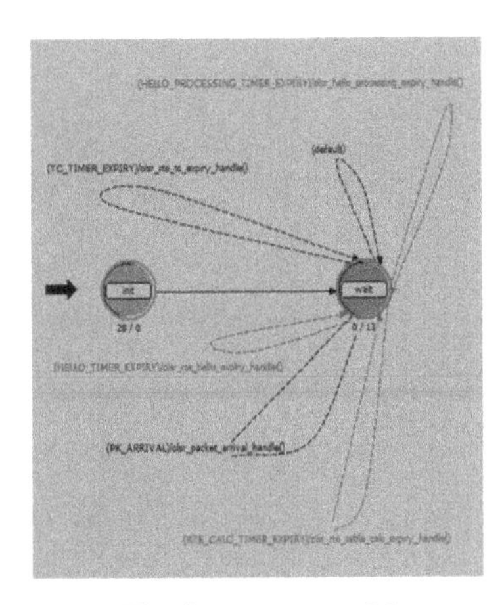

Fig. 4. Process model

b) Simulation results

The simulation results compare the performance of OLSR and PF-OLSR, and the results can be seen from Fig. 5.

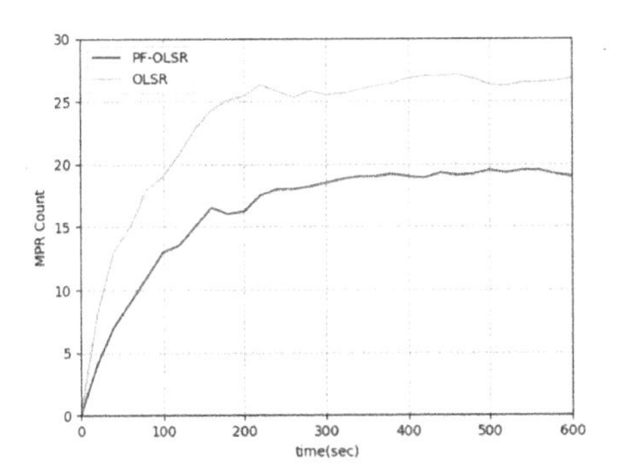

Fig. 5. Number of MPR nodes

The number of nodes in the MPR set iterated by PF-OLSR protocol is always less than that obtained by OLSR protocol. The average number of MPR nodes at any moment is reduced by 30.77%. This indicates that the PF-OLSR protocol message transmission has less redundancy and higher energy utilization than the OLSR protocol.

In a flock where individual flight is relatively random, the flight vector of pigeons cannot be set toward the UAV. However, due to the movement of network nodes in space, the network topology changes with it. A lot of information exchange is needed between nodes for new topology construction. Therefore, improving the stability of the network topology has a great improvement for the network performance. Figure 6 shows the network topology change diagram. The topology change scale of the optimized protocol is always smaller during the simulation. The stability of the pigeonhole network is improved by about 20% compared to the previous one.

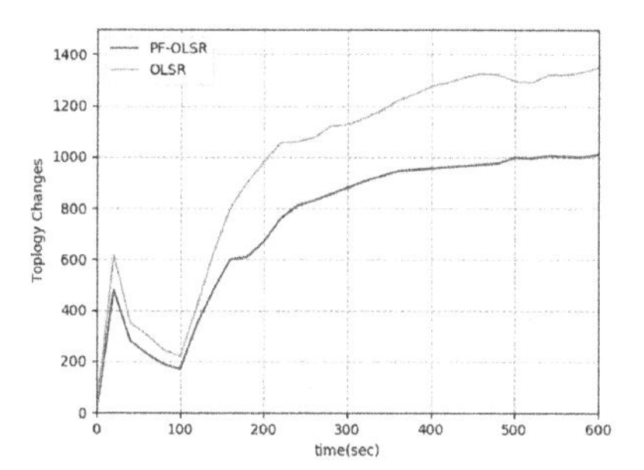

Fig. 6. Network topology change

Due to the reduction in the number of MPR nodes, the PF-OLSR routing protocol is able to exchange data more efficiently. It can be seen from Fig. 7 that the optimized protocol throughput is also higher than the OLSR protocol, with a 9.545% increase in the amount of information exchanged per unit time in the entire pigeonhole network.

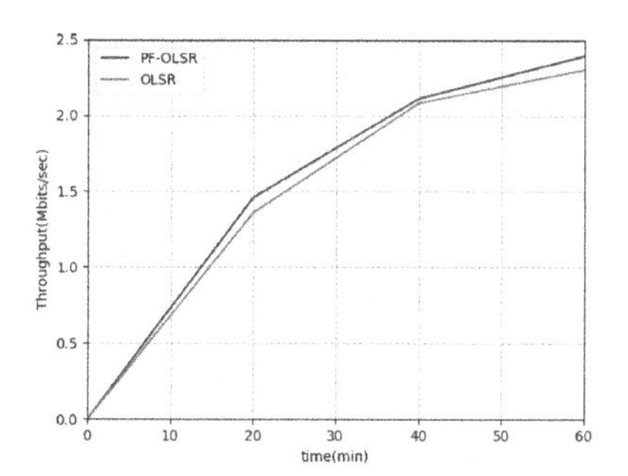

Fig. 7. Average throughput

5 Conclusions

OLSR is suitable for application to dense MANETs, and the key network technology is the MPR mechanism. The PF-OLSR protocol proposed in this paper. The protocol applies the ant colony optimization algorithm to the MPR mechanism. The new algorithm adds dynamic update of state information and iterative optimization. It makes up for the defects of the ant colony algorithm and improves the convergence speed of the algorithm. At the same time, the speed and energy factors are added to the probability selection of the ant colony algorithm. The regional missing node routing information caused by energy depletion is avoided.

The simulation experimental results show that the MPR election of PF-OLSR has good applicability in pigeon flocks' network. A pigeon flock is a dense and large population, which is difficult to maintain in a single OLSR network. Therefore, the next step is to investigate the weighted clustering algorithm based on the biometric characteristics of the pilot class within the pigeon flocks.

This topic is derived from the National Natural Science Foundation of China, Major Research Program (92067106). Thanks for the help and support!

References

1. Wang, Z.L., Shi, L., Liu, X.Y., et al.: Application of animal flight control technology in the military. J. National Defense Technol. **34**(06), 28–32 (2013)
2. Reinacher, A.R.: Deep brain stimulation electrodes may rotate after implantation-an animal study. J. Neurosurg. Rev. **62**(03), 113–115 (2020)
3. American Institute of Physics: Rat spinal cords control neural function in biobots. J. NewsRx Health Sci. **12**(02), 322–328 (2020)
4. Yang, J.Q.: Design and experimental study of an electronic system of neural electrical stimulation for robotic pigeons. D. Nanjing: Southeast Univ. **59**(06), 16–23 (2019)
5. Mann, R.P., Armstrong, C., Meade, J.: Landscape complexity influences route-memory formation in navigating pigeons. J. Biol. Lett. **10**(1), 531–538 (2013)
6. Martine, W., Patrick, S., Lucas, R.: Enhanced CBL clustering performance versus GRP, OLSR and AODV in vehicular Ad Hoc networks. J. Telecommun. Syst. **89**(08), 36–42 (2020)
7. Luo, Q.N., Duan, B.H., Fan, Y.M.: Analysis of stability and aggregation characteristics of pigeon flock movement model. J. Chin. Sci. Tech. Sci. **49**(06), 652–660 (2019)
8. Tian, X.M., Gong, Z.W., Liu, H.W., et al.: Effects of dorsal load on locomotor behavior of domestic pigeons C. In: The 13th National Symposium on Wildlife Ecology and Resource Conservation and the 6th Western China Zoological Symposium. Sichuan: Sichuan Zoological Society, vol. 55(09), pp. 783–690 (2017)
9. Yan, C.F.: Research on OLSR routing protocols for mobile self-assembled networks. J. Inf. Commun. **37**(02), 177–179 (2019)
10. Dong, S.Y., Zhang, H.: Optimal MPR set selection algorithm based on OLSR routing protocol C. In: Proceedings of 2016 4th International Conference on Electrical & Electronics Engineering and Computer Science (ICEEECS 2016). Ed, vol. 67(09), pp. 756–761. Atlantis Press (2016)
11. Mohamed, B., Zeyad, Q., Merahi, B.: Ad Hoc network lifetime enhancement by energy optimization. J. Ad Hoc Sensor Wireless Netw. **28**(12), 327–335 (2015)
12. Wei, G.: Modified ant colony optimization with improved tour construction and pheromone updating strategies for traveling salesman problem. J. Soft Comput. **35**(16), 26–37 (2020)
13. Zeynel, A.Ç., Suleyman, M., Faruk, S.: Robotic disassembly line balancing problem: a mathematical model and ant colony optimization approach. J. Appl. Math. Modell. **86**(05), 113–122 (2020)
14. Shi, Y.Y.: Research on improved ant colony optimized routing algorithm based on WSN network. D. Nanchang: Nanchang Aviation Univ. **32**(06), 78–86 (2018)
15. Lu, M., Xu, B., Jiang, Z., Sheng, A., Zhu, P., Shi, J.: Automated tracking approach with ant colonies for different cell population density distribution. Soft. Comput. **21**(14), 3977–3992 (2016). https://doi.org/10.1007/s00500-016-2048-7
16. Han, L.Y., Yu, L.T., Cong, Y., et al.: Simulation analysis of network modeling based on OPNET. J. Changchun Univ. Technol. (Nat. Sci. Edn.). **42**(01), 119–122+127(2019)

Design of Mechanical Property Verification System for Drug Dissolution Meter Based on Non-contact Measurement

Bin Li(✉), ZhengYu Zhang, Jie Chen, and Lin Jian

School of Mechatronic Engineering and Automation,
Shanghai University, Shanghai 200444, China
zzy09914@shu.edu.cn

Abstract. Drug dissolution meter is an instrument for measuring the degree of dissolution of drugs, which can effectively simulate the bioavailability of drugs in human body. In addition to the factors of the drug itself, the mechanical properties of the dissolution meter, such as the rotating speed, the swing of the drive shaft, the position of the drive shaft and so on will affect its dissolution rate, so the mechanical properties of the drug dissolution meter must be guaranteed. Some methods of mechanical verification of dissolution instrument are analyzed and a non-contact measurement method based on laser distance measuring was proposed to verify the key mechanical properties of the drug dissolution meter. Through data analysis, the swing value, the rotating speed and the coaxiality of the drug dissolution meter were detected, and a set of testing device and software were developed for it. The experiment results show that the developed measuring system has good measurement accuracy and repeatability.

Keywords: Drug dissolution meter · Mechanical property verification · Non-contact measurement · Laser distance measuring

1 Introduction

Dissolution refers to the degree and rate of dissolution of the active drug from tablets, capsules or granules in a specified environment. Especially for oral solid preparations, dissolution can directly give the actual reference value of drug dosage, which is a very important index to reflect the quality of drugs. Therefore, it is very necessary to use appropriate, stable and accurate dissolution measurement method to monitor drug dissolution curve, which could not only ensure the dissolution value of different drugs, but also reduce the risk of biological inequivalence caused by batch production of the same drug [1].

However, dissolution results are not only affected by the drug itself, but also closely related to sample quality, testing environment, instrument parameters and operating procedures. Apart from sample quality, other factors can be attributed to the mechanical properties of the instrument [2, 3]. Therefore, in order to obtain accurate and reliable test data, the mechanical properties of the dissolution meter should be checked regularly.

M. Fei et al. (Eds.): LSMS 2021/ICSEE 2021, CCIS 1467, pp. 167–177, 2021.
https://doi.org/10.1007/978-981-16-7207-1_17

According to the different drive shaft, dissolution meter can be divided into impeller method and basket method. For two different solutions of the dissolution meter, the following mechanical properties need to be calibrated: the swing value of the drive shaft, the deviation of the drive shaft at the center line of the dissolution cup (coaxiality), the rotating speed of the drive shaft, the deviation of verticality between impeller and drive shaft, the vertical deviation of the dissolution cup, the temperature and so on. Among them, the first three properties are produced during the operation of the dissolution meter, while the other properties can be regarded as the inherent properties of the instrument. At present, contact measurement is adopted for the verification of the above three properties. A dial gauge is used to measure the swing value and coaxiality of the drive shaft, while a reflective sleeve and tachometer are used to measure the rotating speed of the drive shaft.

Although the dial gauge and tachometer can directly measure the three key mechanical properties of the dissolution meter, this contact measurement method requires the meter to be directly in contact with the instrument, which inevitably affects the instrument in operation [4, 5]. Since the swing value, coaxiality deviation and rotating speed deviation of the drive shaft is not obvious, so the influence of this contact measurement on the instrument properties cannot be ignored [6]. Therefore, a non-contact measurement scheme is proposed to verify the three key mechanical properties of the dissolution meter based on laser ranging. The swing value can be measured directly by displacement laser sensor, and a special fixture is developed to detect the coaxiality and rotating speed.

2 Non-contact Measurement Method

The swing value of the drive shaft can clearly reflect the stability of the dissolution, and it can be measured by a dial gauge directly without other devices. As shown in Fig. 1, the swing value of the bottom of the basket is measured by the dial gauge. In this paper, the laser measuring displacement sensor is selected to measure the displacement of the point on the drive shaft. During a period of time when the dissolution meter is turned on, the swing value can be calculated by the difference between the maximum and minimum values.

Fig. 1. Dial gauge used to measure the swing value.

Except for controlling the swing momentum, the drive shaft should be kept in the same position with the axis of the dissolution cup when the dissolution meter is working, which is called the coaxiality of dissolution cup and drive shaft. For measuring the coaxiality, the USP Dissolution Mechanical Verification Toolkit recommended using a dial gauge or an inside caliper, as is shown in Fig. 2.

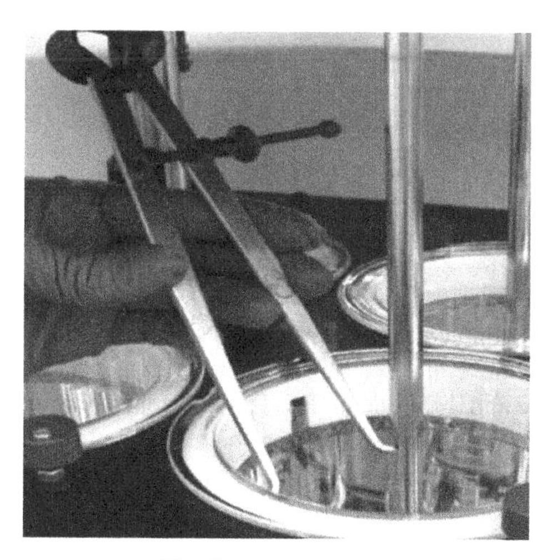

Fig. 2. Inside caliper.

On this basis, American Distek company has developed a dial meter, which is specially suitable for this measuring occasion as shown in Fig. 3.

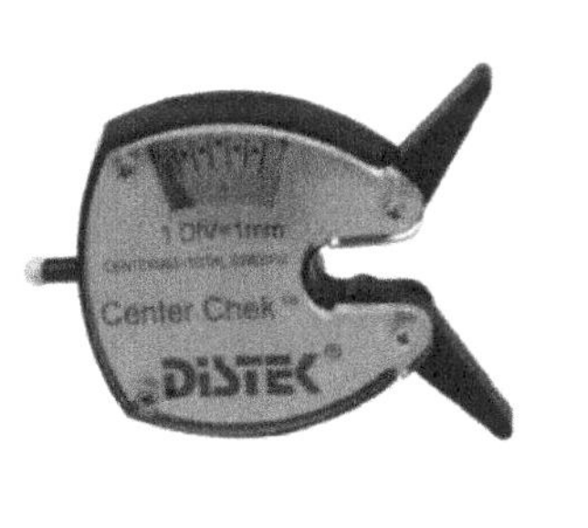

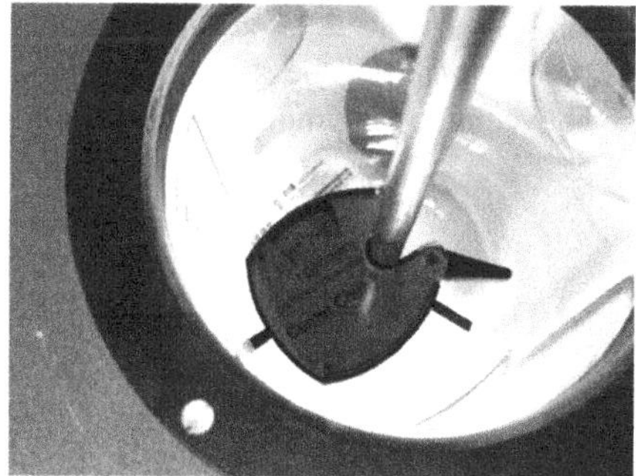

Fig. 3. Dial meter for coaxiality measuring.

Domestic Tianda Tifa has also developed a special tool to measure the coaxiality. It can be seen from Fig. 4 that the two fixed blocks are arranged above and below on the drive shaft and the dial gauge is fixed by the fixed block while the dial gauge measuring head is attached to the cup wall of the dissolution cup. However, the above devices still adopt the contact measurement method in principle.

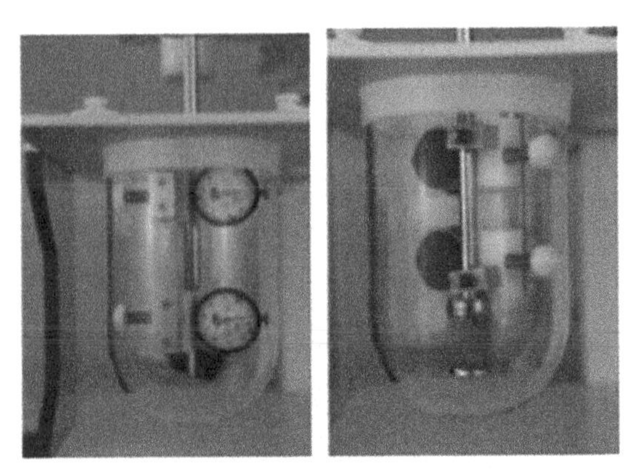

Fig. 4. Special tool for coaxiality measuring.

In this paper, two non-contact measurement schemes of coaxiality are proposed. The first one is to use several laser sensors to detect displacement at a certain point from the same distance and different angles. The difference of every laser sensor reflects the coaxiality deviation. The another is using for reference from the contact measuring thought of traditional dial. A special fixture was developed for Keyence IL-030 and the laser probe is fixed on the bottom of the drive shaft so that the laser is emitted to the cup wall. By recording the displacement data of the rotation of the drive shaft, the maximum displacement minus the average displacement is the eccentricity of the drive shaft relative to the dissolution cup axis canter. Considering the first scheme requires simultaneous use of multiple laser sensors which must be fixed around the drive shaft by a high-precision device to ensure that the probes emit laser converged at one point. The cost and technical difficulty are higher than the latter scheme. Therefore, the second special fixture was manufactured, as shown in Fig. 4. Once the laser probe is fixed on the fixture and the fixture is fixed on the bottom, the displacement between the laser probe and the cup wall will be recorded. After the drive shaft has rotated 1 round, the difference between the maximum and minimum values is the coaxiality (Figs. 5 and 6).

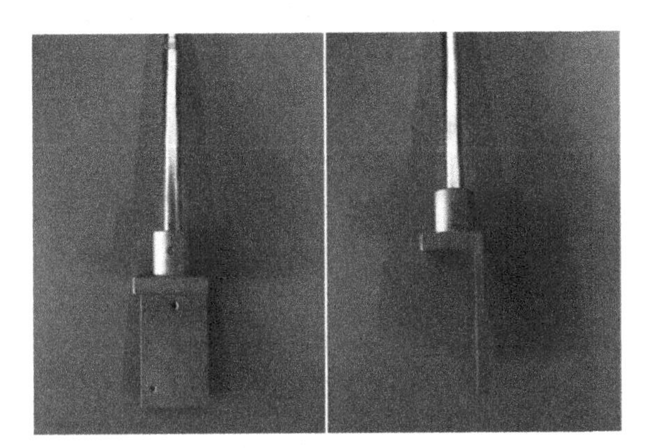

Fig. 5. Special fixture developed for Keyence IL-030.

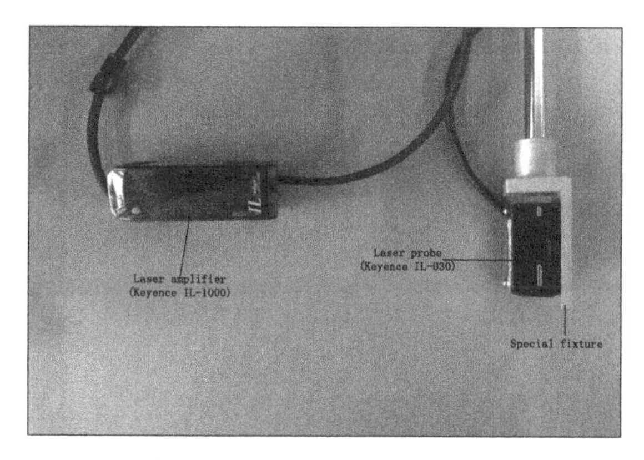

Fig. 6. Fixation of the laser sensor.

In the traditional detection scheme, the reflective bushing should be installed on the drive shaft and the special tachometer is used to realize the rotating speed measurement, which is really a complex job. As shown in Fig. 7, since the tester needs to hold the tachometer, it is possible to produce measurement error in the operation process. In this paper, the above mentioned fixture and laser sensor can finish rotating speed measurement without extra device. After finishing detecting the coaxiality, the laser probe will be removed from the special fixture and it is required to be aligned the center point of the upper edge of the fixture. When the dissolution meter is turned on, the displacement between the laser probe and the fixture is recorded. Since the measurement range of laser probe is narrow, the distance from the laser probe to the fixture may be beyond the range when the fixture rotates. Once the distance is out of the measuring range, the laser senor will output the maximum analog signal. It is found that there will be two mutation points in the output amplitude of the analog signal each rotation of the fixture. By calculating the cycle time of the mutation point, the rotating speed of the drive shaft is obtained.

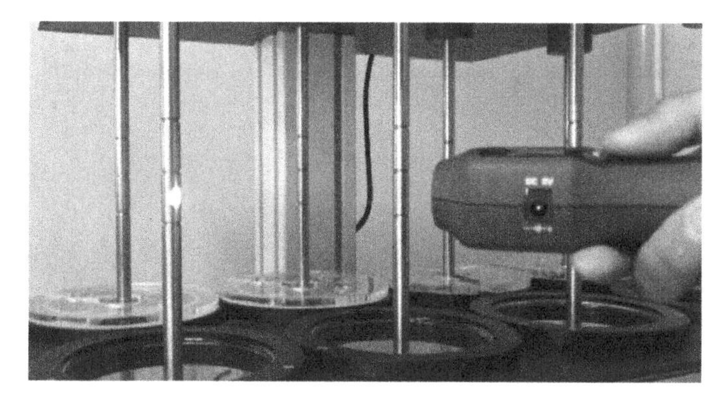

Fig. 7. Tachometer and reflective sleeve for measuring raotating speed.

3 Design of Measuring System

The overall detection system is composed of a dissolution instrument, a laser sensing unit and a data analysis unit, and the hardware block diagram is shown in Fig. 8.

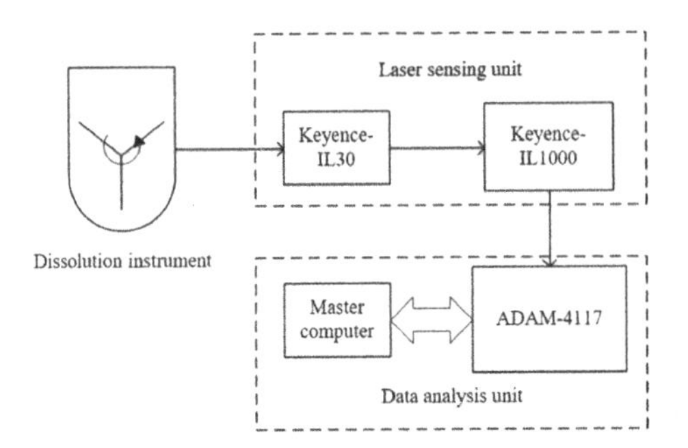

Fig. 8. Hardware block diagram.

In terms of laser sensing unit, the Keyence-IL30 sensor probe and the Keyence-IL1000 amplifier are selected. The reference range of the IL30 is 30 mm and the measurement range is 20–45 mm. The 655 nm wavelength red semiconductor laser is adopted and the minimum sampling period is 0.33 ms. The data analysis unit is for the AD conversion and the calculation of the data from laser sensing unit. The ADAM-4117 of Advantech is selected as the data acquisition unit, which is a 16-bit analog input module. After the AD conversion, the analog signal can be analyzed and calculated by the software program.

The software system of this paper is designed in LabWindows/CVI which is an interactive C language development platform launched by National Instruments Corporation. CVI provides rich callback functions and graphical controls, and a user-friendly software interface is designed.

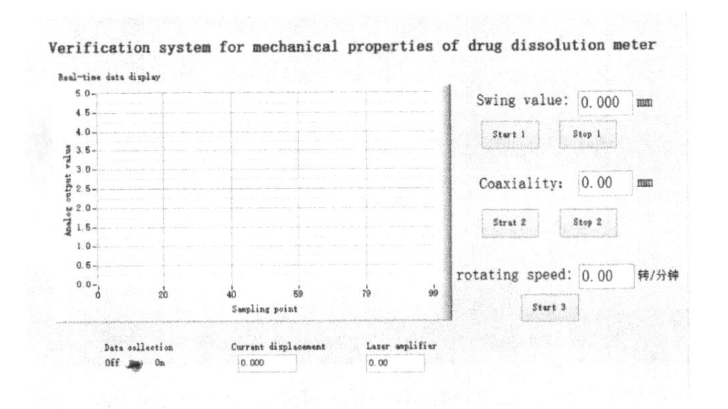

Fig. 9. Software interface.

As shown in the software interface in Fig. 9, the data collection switch can be regarded as the master switch. Once starting the data collection, the laser sensing unit will begin to transmit the data to the PC by the ADAM-4117. The data is presented as a curve in the real-time display in the form of 0 to 5 V voltage. The numerical values of the current displacement and the laser amplifier are converted and displayed in the interface so that the software user can observe the measuring condition in the process of measurement.

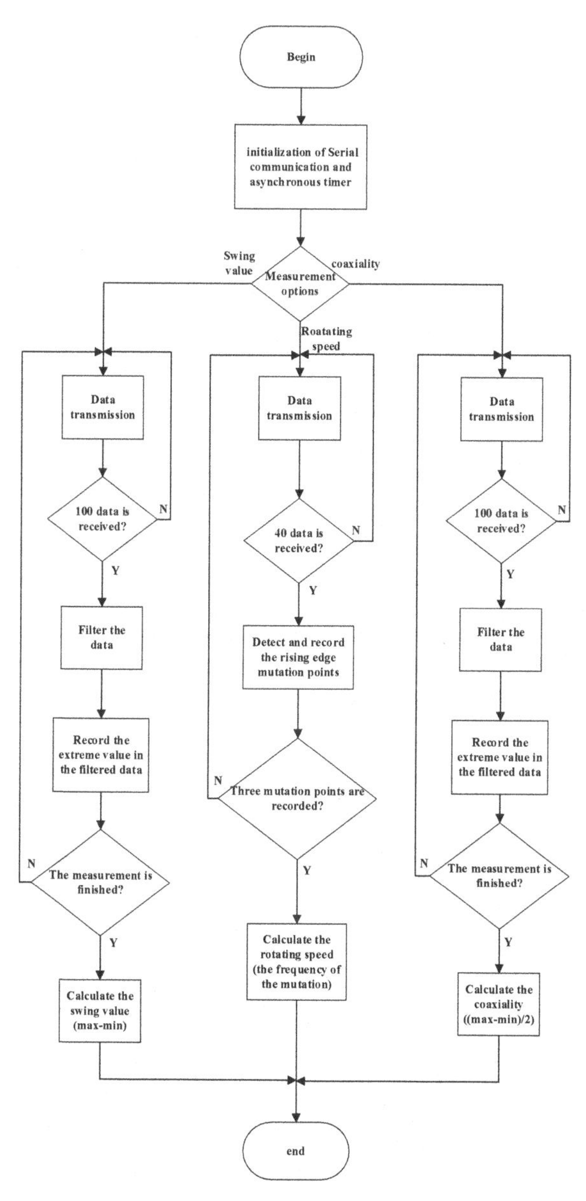

Fig. 10. Program chart of software system.

There are 3 measuring options in the software interface, the swing value, the coaxiality and the rotating speed. The measuring time of the first two items can be determined by the tester and the rotating speed item is real-time display. The program chart of the designed software is shown in the Fig. 10.

4 Experiment and Result

In order to verify the measuring accuracy of the mechanical property verification performance of the dissolution meter developed in this paper, this measuring system is applied to the RC806D dissolution meter made by Tianda Tifa to test the mechanical property. According to the Guiding Principles for Mechanical Verification of Drug Dissolution Instrument formulated by China Food and Drug Administration, the four dissolution cups in front of the dissolution meter were tested and the repeatability of the testing data was calculated by the following equations.

$$\overline{E} = \frac{1}{6}\sum_{n=1}^{6} E_{mn}. \tag{1}$$

$$(E_m)_r = [\frac{1}{5}\sum_{n=1}^{6} (E_{mn} - \overline{E}_m)^2]^{\frac{1}{2}}. \tag{2}$$

Where E_{mn} are the actual testing value of the nth (n = 1, 2, 3, 4, 5, 6) of the mth (m = 1, 2, 3, 4) dissolution cup, $\overline{E}_m$ are the average testing value of the mth (m = 1, 2, 3, 4) dissolution cup and $(E_m)_r$ are the repeatability of the testing data of the mth (m = 1, 2, 3, 4) dissolution cup. The specific measuring process is as follows.

First of all, the swing value of the drive draft was measured. The drive shaft swing is measured approximately 20 mm above the basket (impeller) which rotated at 50 revolutions per minute in 15 s, as shown in Fig. 11. Every dissolution cup was measured 6 times and the testing data are as shown in Table 1.

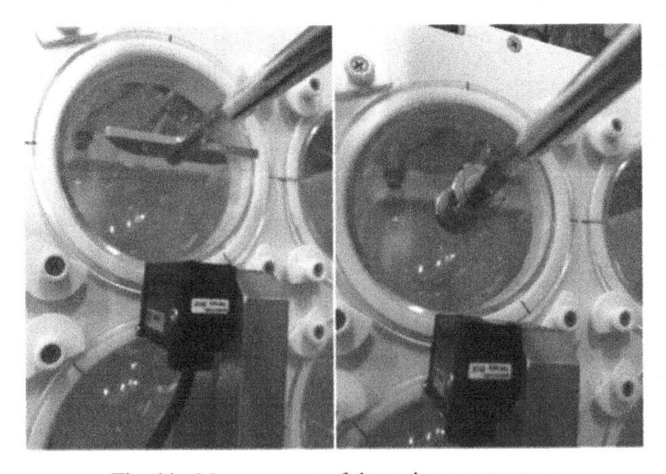

Fig. 11. Measurement of the swing momentum.

Then, the special fixture and sensor probe are fixed below the drive shaft, and the senor probe aimed at the inner wall of the dissolution cup, as shown in Fig. 12. The displacement from the senor probe to the inner wall is recorded completely in 1 round and the coaxiality is calculated, as shown in Table 2.

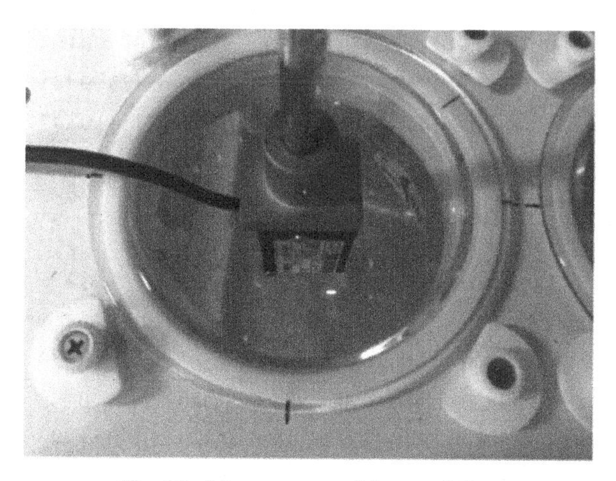

Fig. 12. Measurement of the coaxiality.

At last, the senor probe was removed from the fixture and aimed the center point of the upper edge of the fixture, as shown in Fig. 13. When the dissolution is turned on, the displacement between from the sensor probe to the fixture changes in the shape of a square wave and the rotating speed of the drive shaft could be calculated by the cycle time of the mutation points directly (Table 3).

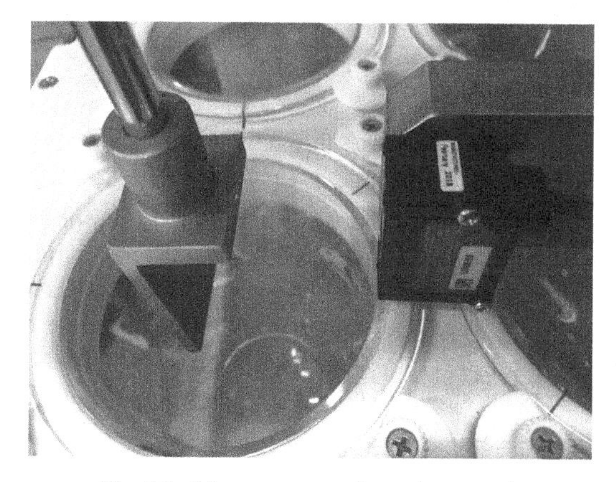

Fig. 13. Measurement of rotating speed.

Table 1. Measuring result of swing value.

	Swing value (mm)							
Dissolution cup number	1	2	3	4	5	6	Average	Repeatability
1	0.100	0.095	0.080	0.090	0.075	0.085	0.088	0.0088
2	0.095	0.102	0.083	0.092	0.085	0.095	0.092	0.0076
3	0.125	0.125	0.125	0.130	0.130	0.125	0.127	0.0026
4	0.130	0.145	0.160	0.145	0.150	0.150	0.147	0.0098

Table 2. Measuring result of coaxiality.

	Coaxiality (mm)							
Dissolution cup number	1	2	3	4	5	6	Average	Repeatability
1	0.750	0.760	0.730	0.740	0.780	0.760	0.753	0.058
2	0.420	0.460	0.480	0.460	0.440	0.410	0.445	0.027
3	1.420	1.410	1.480	1.440	1.440	1.470	1.443	0.027
4	0.800	0.740	0.660	0.810	0.840	0.790	0.773	0.064

Table 3. Measuring result of rotating speed.

Dissolution cup number	Set value (r/min)	Measuring value (r/min)	Allowable error (r/min)
1	100	100.84	4
2	100	100.84	4
3	50	50.21	2
4	50	50.42	2

5 Conclusion

In this paper, a set of mechanical property verification system of the drug dissolution meter based on laser ranging was constructed, and a special fixture for Keyence IL-030 was developed so that the coaxiality and rotating speed of the dissolution can be measured in a non-contact way. The interface and software were designed based on LabWindows/CVI, and the real-time verification of the RC806D dissolution was carried out. The experimental results shows that the measuring scheme and verification system have good measurement accuracy and repeatability.

References

1. Yang, X.Q., Ren, H.Y., Tian, A.L.: Study on the determination method of drug dissolution and its development direction. Chin. J. Pract. Med. **9**(19), 257–258 (2014)

2. Tang, S.F.: Study and discussion on the mechanical performance verification method of drug dissolution tester. Chin. J. New Drugs **23**(24), 2892–2896 (2014)
3. Li, G., Gu, H.F.: Discussion on verifying method of mechanical property of drug dissolution meter. J. Pharm. Eng. **18**(24), 41–42 (2011)
4. Martin, G.P., et al.: Overview of dissolution instrument qualification, including common pitfalls. Dissolution Technol. **18**(2), 6–10 (2011)
5. Martin, G.P., Gray, V.A.: Potential pitfalls when performing dissolution instrument qualification. J. GXP Compliance **15**, 6–12 (2011)
6. Zhang, J.R., Pang, Y.M., Zhu, J.Q., Zeng, X.Y.: Non-contact measurement of swing amplitude of drug dissolution meter. Meas. Meas. Technol. **45**(03), 48–49 (2018)

Classification of Coronary Artery Lesions Based on XGBoost

Rui Chen[1], Jianguo Wang[1], Junjie Pan[2], and Yuan Yao[3](✉)

[1] School of Mechatronical Engineering and Automation, Shanghai Key Lab of Power Station Automation Technology, Shanghai University, Shanghai 200072, China
jgwang@shu.edu.cn

[2] Department of Cardiology, Huashan Hospital, Fudan University, Wulumuqi Zhong Road, Shanghai 200040, China

[3] Department of Chemical Engineering, National Tsing-Hua University, Hsin-Chu 30013, Taiwan
yyao@mx.nthu.edu.tw

Abstract. XGBoost is an optimized distributed gradient enhancement library, which uses the gradient lifting method to build a strong learner after each iteration, and it is widely used in many classification and regression model scenarios. It is a scalable machine learning system of treeboosting, which can solve many data science problems quickly and accurately. Using retrospective analysis, 1552 patients were enrolled in the Department of Cardiology, Huashan Hospital affiliated to Fudan University from January 1, 2018 to December 31, 2020. A total of 1552 patients were treated and discharged from the hospital. According to the Gensini score, the patients were divided into four groups: low GSlow $<$ 24, middle GSmid $<$ 53, high GShigh $<$ 120 and critical recombination GSdanger $\geq$ 120. In this paper, the data set of patients' physiological indexes was analyzed by XGBoost algorithm, and the classification results of coronary artery lesions were predicted. The AUC values of the four groups were 0.88, 0.69, 0.87 and 0.96 respectively. The experimental results show that the algorithm has achieved good results in predicting the classification results of each group.

Keywords: Machine learning · XGBoost · Coronary artery disease · Gensini score

1 Background Introduction

1.1 Artificial Intelligence and Medical Big Data

With the emergence of big data and the improvement of people's ability to analyze big data, all fields of scientific and technological development are constantly innovating, among which the development of artificial intelligence is the most prominent, which also means that human society will usher in the era of intelligence more quickly. With the rapid development of artificial intelligence in recent years, various industries are constantly changing the development model to comply with the development of the

M. Fei et al. (Eds.): LSMS 2021/ICSEE 2021, CCIS 1467, pp. 178–187, 2021.
https://doi.org/10.1007/978-981-16-7207-1_18

times, of course, medicine is no exception. And artificial intelligence in the medical field has a wide range of application prospects and huge exploration space.

The promise of artificial intelligence and machine learning in cardiology is to provide a set of tools to augment and extend the effectiveness of the cardiologist [1]. According to the characteristics of artificial intelligence, in the medical field, everything that is "repetitive, has rules to follow and can be calculated by big data" can be replaced by artificial intelligence. So some repetitive work or labor in medical treatment will be replaced by artificial intelligence first. However, the introduction of artificial intelligence will not change the nature of the medical profession, let alone replace the diagnosis and treatment of doctors. Therefore, the goal of exploring the application of artificial intelligence in the medical field is to use artificial intelligence to better promote the development of medical disciplines, rather than relying on equipment and machines to replace medical specialties. In this paper, XGBoost algorithm is used to analyze the data set of patients' physiological indexes. The classification results of coronary artery lesions are predicted, and good results are obtained. To a certain extent, it effectively avoids doctors from making corresponding diagnosis to patients through invasive angiographic analysis, which has practical clinical significance.

1.2 Coronary Artery Disease

Coronary heart disease is the most common cause of death in the general population and in patients with ESRD [2]. Coronary artery disease mostly refers to coronary atherosclerotic disease, which is the most common stenotic coronary artery disease. Blood in the body enters the heart through two main coronary arteries and feeds the heart through a network of blood vessels on the surface of the heart muscle. Cholesterol and fat deposits form in the arteries, narrowing the channels, a condition known as atherosclerosis. Blood flowing in the arteries forms a thrombus that blocks the arteries. When under physical or psychological stress, the heart beats faster and needs more oxygen and nutrients, a condition that coronary arteries cannot cope with when they are severely narrowed or blocked. As a result, there is a lack of blood supply to the coronary artery, which can lead to angina or heartache. A sudden decrease in the amount of blood flowing to the heart muscle as a result of a thrombus blocking a coronary artery can lead to a heart attack, known as coronary atherosclerotic heart disease. The degree of coronary arteriosclerosis varies, mild lesions have no significant effect on the heart, and more severe lesions can cause lumen stenosis. When the workload of the heart increases, the blood supply to the myocardium is insufficient, resulting in angina pectoris, arrhythmia and heart failure.

1.3 Gensini Scoring

In the process of clinical treatment, angiographic analysis is needed to judge the degree of coronary artery disease. Gensini score [3] is a commonly used scoring system when evaluating the degree of coronary artery stenosis. As shown in Table 1, the calculation of the Gensini score is listed. The degree of stenosis $\leq 25\%$ is 1 point, $\geq 26\%$–$\leq 50\%$ is 2 points, $\geq 51\%$–$\leq 75\%$ is 4 points, $\geq 76\%$–$\leq 90\%$ is 8 points, $\geq 91\%$–$\leq 99\%$ is 16 points, and full closure is 32 points. The corresponding coefficients were determined according to different coronary artery branches [4], which were left trunk lesion $\times$ 5, proximal left anterior descending branch or circumflex branch × 2.5, middle left anterior descending branch × 1.5, distal left anterior descending branch × 1, middle and distal left circumflex branch × 1, right coronary artery × 1, small branch × 0.5. The basic score of each coronary artery stenosis is multiplied by the coefficient of the lesion site, that is, the score of the diseased vessel, and the sum of the score of each diseased vessel is the total score of the degree of coronary artery stenosis of the patient.

Table 1. Gensini scoring system.

Stenosis degree	Score	Lesion site	Score
1%–25%	1	Left trunk	5
26%–50%	2	Proximal left anterior descending or circumflex branch	2.5
51%–75%	4	Middle segment of left anterior descending branch	1.5
76%–90%	8	Distal segment of left anterior descending branch	1
91%–99%	16	Middle and distal segment of left circumflex branch	1
Completely close	32	Right coronary artery	1
		Small branch	0.5

2 Method

2.1 GBDT

Gradient Boosting Decision Tree [5] is a representative algorithm in boosting series of algorithms. It is an iterative decision tree algorithm, which is composed of multiple decision trees, and the conclusions of all trees add up as the final answer. We use the square error to represent the loss function, in which each regression tree learns the conclusions and residuals of all previous trees, and fits to get a current residual regression tree. The residual real value-predicted value, and the lifting tree is the accumulation of the regression tree generated by the whole iterative process.

In the process of GBDT iteration [6], suppose that the strong learner we got in the previous iteration is $f_{t-1}(x)$, loss function is $L(y, f_{t-1}(x))$, The goal of our iteration is to find a weak learner of the regression tree model $h_t(x)$, let the loss of this round $L(y, f_{t-1}(x)+h_t(x))$ to be smallest. In other words, the decision tree found in this iteration should make the loss function of the sample as small as possible.

For the binary classification problem, the commonly used loss function is:

$$L(y, f(x)) = \log\left(1 + e^{-yf(x)}\right).$$

The common loss functions for multi-classification problems are:

$$L(y, f(x)) = -\sum_{k=1}^{k} y_k \log(p_k(x)).$$

$$L(y, f(x)) = -\sum_{k=1}^{k} y_k \log(p_k(x)).$$

The expression of the probability of class k, $p_k(x)$ is as follows:

$$p_k(x) = \frac{\exp(f_k(x))}{\sum_{l=1}^{k} \exp(f_l(x))}.$$

Thus, the negative gradient error of the corresponding category of the i th sample of the t round can be calculated as follows:

$$r_{til} = -\left[\frac{\partial L(y_i, f(x_i))}{\partial f(x_i)}\right]_{f_k(x) = f_{t-1,l(x)}} = y_{il} - p_{t-1}, l(x_i).$$

The error here is the difference between the real probability of the corresponding category of sample I and the prediction probability of the t – 1 round. For the generated decision tree, the best residual fitting values for each leaf node are:

$$c_{tjl} = argmin \sum_{i=1}^{m} \sum_{k=1}^{k} L\left(y_k, f_{t-1,l(x)}\right) + \sum_{j=1}^{J} c_{jl}, I(x_i \epsilon R_{tj}).$$

As the above formula is difficult to optimize, approximate values can be used instead of:

$$c_{tjl} = \frac{K-1}{K} = \frac{\sum_{x_i \epsilon R_{tjl}} r_{til}}{\sum_{x_i \epsilon R_{til}} |r_{til}|(1 - |r_{til}|)}.$$

2.2 XGBoost Arithmetic

Although GBDT is quite perfect in both theoretical derivation and application scenario practice, there is a problem: the (approximate) residual of the nth tree needs to be used in the training of the nth tree. From this point of view, GBDT is more difficult to achieve distributed, XGBoost is to solve this problem. XGBoost [7] is an efficient implementation of GBDT. XGBoost is a tree integration model, which uses the sum of the predicted values of each tree of K (the total number of trees is K) as the prediction of the sample in the XGBoost system. The functions are defined as follows:

$$\hat{y_i} = \varnothing(x_i) = \sum_{k=1}^{k} f_k(x_i), f_k \in \mathcal{F}. \quad (1)$$

There are n samples and m features for the given data set, which are defined as:

$$D = \{(x_i, y_i)\}\left(|D| = n, x_i \epsilon R^m, y_i \epsilon R\right). \quad (2)$$

x_i Represents the i th sample, y_i represents the category label of the i th sample. The space of the CART tree is $\mathcal{F}$, as follows:

$$\mathcal{F} = \left\{f(x) = \omega_{q(x)}\right\}\left(q : R^m \to T, \omega \epsilon R^T\right). \quad (3)$$

Where Q represents the score that the structure of each tree maps each sample to the corresponding leaf node, that is, Q represents the model of the tree, input a sample, and map the sample to the leaf node output predicted score according to the model; $\omega_{q()}$ represents the set of scores of all leaf nodes of tree Q; T is the number of leaf nodes of tree Q. It can be seen from the formula (1) that the predicted value of XGBoost is the sum of the predicted values of each tree, that is, the sum of the scores of the corresponding leaf nodes of each tree.

The goal is to learn such a K-tree model $f(x)$. In order to learn the model $f(x)$, define the following objective function:

$$\mathcal{L}(\phi) = \sum\nolimits_i l(\hat{y_i}, y_i) + \sum\nolimits_k \Omega(f_k). \tag{4}$$

$$\Omega(f) = \gamma T + \frac{1}{2}\lambda||\omega||^2. \tag{5}$$

$\hat{y}_i$ represents the predicted value of the model, y_i represents the category label of i th sample, K represents the number of trees, f_k represents the k th tree model, T represents the number of leaf nodes per tree, ω represents the sets which consists of the scores of the leaf nodes of each tree, γ and λ represent coefficient, Parameters need to be adjusted in practical application.

Because the optimization parameter of the XGBoost model is that the model f (x), is not a specific value, we can not use the traditional optimization method to optimize in the Euclidean space, but use the additive training method to learn the model. Each time you leave the original model unchanged, add a new function f to the model, as follows:

$$\hat{y}_i^{(0)} = 0$$

$$\hat{y}_i^{(1)} = f_1(x_i) = \hat{y}_i^{(0)} + f_1(x_i)$$

$$\hat{y}_i^{(2)} = f_1(x_i) + f_2(x_i) = \hat{y}_i^{(1)} + f_2(x_i)$$

$$\ldots\ldots$$

$$\hat{y}_i^{(t)} = \sum\nolimits_{k=1}^{t} f_k(x_i) = \hat{y}_i^{(t-1)} + f_t(x_i)$$

The predicted value adds a new function f to each iteration in order to minimize the objective function as much as possible. Because the ultimate goal is to get the model $f(x)$, when minimizing $\mathcal{L}(\phi)$, but there is no parameter $f(x)$, in $\mathcal{L}(\phi)$, we substitute the last formula in the above figure into (4) to get the following formula:

$$\mathcal{L}^{(t)} = \sum\nolimits_{i=1}^{n} l\left(y_i, \hat{y_i}^{(t-1)} + f_t(x_i)\right) + \Omega(f_t). \tag{6}$$

Taylor expansion is generally used to define an approximate objective function to facilitate our further calculation. According to the following Taylor expansion, remove the higher-order infinitesimal term, and get:

$$f(x + \Delta x) \approx f(x) + f'(x)\Delta x + \frac{1}{2}f''(x)\Delta x^2.$$

$$g_i = \frac{\partial l(y_i, \hat{y}^{(t-1)})}{\partial \hat{y}^{(t-1)}}, h_i = \frac{\partial^2 l(y_i, \hat{y}^{(t-1)})}{\partial (\hat{y}^{(t-1)})^2}..$$

then formula (6) is equivalent to formula (7)

$$\mathcal{L}^{(t)} \approx \sum\nolimits_{i=1}^{n} \left[l\left(y_i, \hat{y_i}^{(t-1)}\right) + g_i f_t(x_i) + \frac{1}{2} h_i f_t^2(x_i) \right] + \Omega(f_t). \tag{7}$$

Since the goal is to find the $f(x)$ (variable of the model when $\mathcal{L}(\phi)$ is minimized), the minimum value of the model changes when the constant term is removed, but the variable with the minimum value remains the same. So, to simplify the calculation, we remove the constant term and get the following objective function:

$$\overset{\sim(t)}{\mathcal{L}} \approx \sum\nolimits_{i=1}^{n} \left[g_i f_t(x_i) + \frac{1}{2} h_i f_t^2(x_i) \right] + \Omega(f_t). \tag{8}$$

$I_j = \{i | q(x_i) = j\}$ is an example of a leaf node, (8) can be written as follows:

$$\begin{aligned} \overset{\sim(t)}{\mathcal{L}} &\approx \sum\nolimits_{i=1}^{n} [g_i f_t(x_i) + \frac{1}{2} h_i f_t^2(x_i)] + \Omega(f_t) \\ &= \sum_{i=1}^{n} \left[g_i \omega_{q(x_i)} + \frac{1}{2} h_i \omega_{q(x_i)}^2 \right] + \gamma T + \lambda \frac{1}{2} \sum_{j=1}^{T} \omega_j^2 \\ &= \sum_{j=1}^{n} \left[(\sum_{i \in I_j} g_i) \omega_j + \frac{1}{2} (\sum_{i \in I_j} h_i) \omega_j^2 \right] + \gamma T + \lambda \frac{1}{2} \sum_{j=1}^{T} \omega_j^2 \\ &= \sum_{j=1}^{n} \left[(\sum_{i \in I_j} g_i) \omega_j + \frac{1}{2} (\sum_{i \in I_j} h_i + \lambda) \omega_j^2 \right] + \gamma T \end{aligned}$$

At this time, the goal is to find the ω_j of the leaf node j of each tree, after finding ω_j, the ω_j of each tree is added to get the final predicted score. And in order to get the optimal value of ω_j, that is, to minimize our objective function, so the above formula takes the partial derivative of d and makes the partial derivative 0, and the ω_j^* at this time is:

$$\omega_j^* = -\frac{\sum_{i \in I_j} g_i}{\sum_{i \in I_j} h_i + \lambda}.$$

Definite: $G_j = \sum_{i \in I_j} g_i, H_j = \sum_{i \in I_j} h_i$:

$$\omega_j^* = -\frac{G_j}{H_j + \lambda}.$$

Replace ω_j^* into the original:

$$\overset{\sim(t)}{\mathcal{L}} = -\frac{1}{2} \sum\nolimits_{j=1}^{T} \frac{G_j^2}{H_j + \lambda} + \gamma T. \tag{9}$$

Equation (9) can be used as a score function to measure the quality of the tree structure Q, which is similar to evaluating the impurity score of the decision tree. According to the prediction results of the decision tree, the gradient data of each sample are obtained, and then the actual structure score is calculated. The smaller the score, the better the structure of the tree.

When looking for the optimal tree structure, we can constantly enumerate the structure of different trees, use this score function to find an optimal tree structure, add it to the model, and then repeat this operation. However, enumerating all tree structures is not feasible, here XGBoost uses the commonly used greedy algorithm or approximation algorithm, which will not be discussed here.

3 Case Study

3.1 Dataset Introduction

The data set used in this study included 1552 patients who were admitted to the Department of Cardiology, Huashan Hospital affiliated to Fudan University from January 1, 2018 to December 31, 2020. The corresponding Gensini score was calculated according to the angiographic results of each patient, and the results were divided into four categories according to the low score GSlow < 24, the middle score GSmid < 53, the high score GShigh < 120 and the critical group GSdanger ≥ 120. Each patient analyzed had 69 independent variables and one dependent variable.

3.2 Data Preprocessing

Due to the great differences in the numerical range of each feature in the data set, it is necessary to select an appropriate method to scale the data. In this paper, the method of feature standardization is adopted to make the value of each feature have zero mean and unit variance. The formula for this method is as follows:

$$x^{'} = \frac{x - \text{mean}(x)}{\text{std}(x)}.$$

The standard deviation is defined as:

$$std(x) = \sqrt{\frac{\sum (x - mean(x))^2}{n}}.$$

After the feature scaling, the missing values are filled by means of mode and mean respectively for the category variables and numerical variables in the data set. At the same time, the K-fold cross-verification method is adopted in the process of dividing the data set in this experiment. That is, it is divided for many times, and each time, training, testing and evaluation are carried out on different data sets, so as to get an evaluation result. As shown in Fig. 1, there is a 10-fold cross-validation, which means to divide the original data set ten times, each time for training and evaluation, and finally get the evaluation results after ten partitions. generally, we take the average of these evaluation results to get the final results.

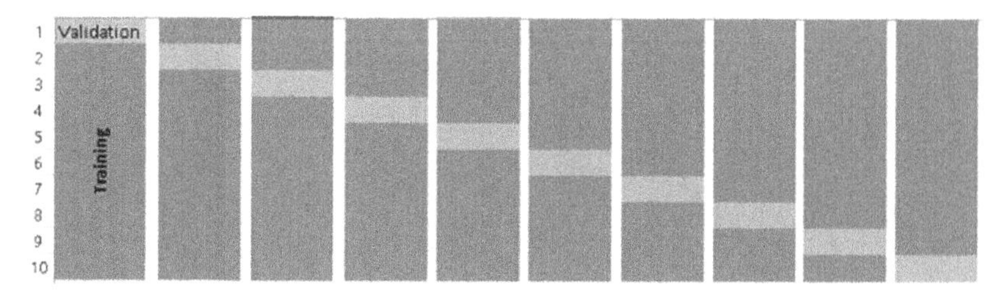

Fig. 1. Schematic diagram of cross-verification with 10% discount.

In addition, using the idea of hierarchical verification, the distribution of the four classification results in the original data set is calculated, and based on this, the proportion of each category in the training set and test set is determined. In this way, every 90% discount is better represented as a whole, resulting in better prediction results.

3.3 Experimental Analysis

In this paper, random forest, logical regression, naive Bayesian and support vector machine algorithm and XGboost algorithm are tested on the same data set, and the ROC [8] curves of each model are drawn and compared. For the XGboost algorithm, the grid search method is used to automatically adjust the parameters to get the optimal model parameters. As shown in Fig. 2 and Fig. 3, the ROC curves of each model are shown.

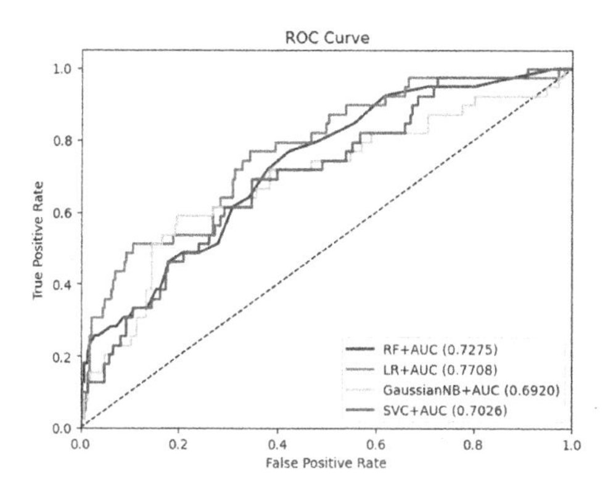

Fig. 2. Contrast model ROC curve.

Figure 3 contains the ROC graph of the classification results of the RF, LR, GaussianNB, and SVC algorithms in the same data set. It can be seen that the AUC [9] value of LR algorithm is higher, which is 0.7708. The ROC curve and the total ROC curve of the XGboost algorithm under the four classification results are drawn in Fig. 3 Compared with Fig. 2, it can be concluded that the overall classification effect of XGboost algorithm has obvious advantages.

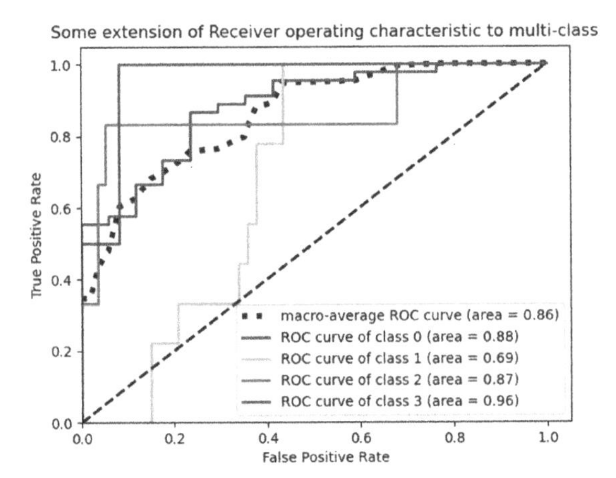

Fig. 3. ROC curve of XGboost model.

4 Conclusion

XGBoost is a novel tree-based Boost integration algorithm, which has been widely used in the field of ML. XGBoost outlines the creation of classification and regression tree. In this tree, by optimizing the customized objective function, the classification accuracy is improved one level at a time, the GBDT algorithm is implemented efficiently, and many improvements in algorithm and engineering are made. In this paper, XGboost algorithm is used to predict the results of four classifications based on Gensini score, and good results are obtained. In the data set used in this paper, all the data do not need to be obtained through radiography [10], so as to effectively avoid the possible physical harm to patients and reduce the treatment cost of patients. In the process of clinical treatment, it is helpful for doctors to judge the degree of coronary artery disease more quickly, so as to give a more reasonable treatment plan, which has high practical value.

References

1. Johnson, K.W., et al.: Artificial intelligence in cardiology. J. Am. Coll. Cardiol. **71**, 2668–2679 (2018). JACC focus seminar
2. Peter, A., Cullough, M.C.: Coronary artery disease. Clin. J. Am. Soc. Nephrol. **2**, 611–616 (2007)
3. Gensini, G.: A more meaningful scoring system for determining the severity of coronary heart disease. Am. J. Cardiol. **51**, 606 (1983)
4. Rampidis, G.P., Benetos, G., Benz, D.C., Giannopoulos, A.A., Buechel, R.R.: A guide for Gensini Score calculation. Atherosclerosis **287**, 181–183 (2019)
5. Zhang, Z.D., Jung, C.: Gradient boosted decision trees for multiple outputs. IEEE Trans. Neural Netw. Learn. Syst. **99**, 1–12 (2020)
6. Ogunleye, A., Wang, Q.G.: XGBoost model for chronic kidney disease diagnosis. IEEE/ACM Trans. Comput. Biol. Bioinform. **17**, 2131–2140 (2020)
7. Metz, C.E.: Basic principles of ROC analysis. Semin. Nucl. Med. **8**, 283–298 (1978)
8. Obuchowski, N.A., Bullen, J.A.: Receiver operating characteristic (ROC) curves, review of methods with applications in diagnostic medicine. J. Phys. Med. Biol. **63**(7), 1–28 (2018)

9. Hancock, J.T., Khoshgoftaar, T.M.: CatBoost for big data: an interdisciplinary review. J. Big Data **7**(1), 1–45 (2020). https://doi.org/10.1186/s40537-020-00369-8
10. Mattoon, J.S.: Digital radiography. Vet. Comp. Orthop. Traumatol. **19**, 123–132 (2006)

Application of a New Attribute Reduction Algorithm in Intracranial Aneurysm Prediction

Yang Li[1], Wenju Zhou[1], Yulin Xu[1](✉), Xiangrong Ni[1], and Yu Zhou[2]

[1] School of Mechanical and Electrical Engineering and Automation, Shanghai University, Shanghai 200444, China
xuyulin@shu.edu.cn

[2] Shanghai Tang Electronics Co., Shanghai 201401, China

Abstract. In order to solve the problem that the attribute reduction algorithm of neighborhood rough set only considers the influence of a single attribute on the decision attributes, but fails to consider the correlation between the attributes, this paper proposes an attribute reduction algorithm of neighborhood rough set based on chi-square test (ChiS-NRS). Firstly, the chi-square test is used to calculate the correlation, and the influence of related attributes is considered when selecting important attributes, which reduces the time complexity and improves the classification accuracy. Then, he improved algorithm uses the Gradient Boosting Decision Tree (GBDT) algorithm to build a classification model, and the model is verified on the UCI data set. Finally, the model is applied to predict the existence of intracranial aneurysms. The experimental results show that the proposed model can better predict the existence of intracranial aneurysms and assist doctors to make more accurate diagnosis.

Keywords: Rough set · Attribute reduction · Gradient boost · Decision tree

1 Introduction

An aneurysm is a cystic protrusion formed on the artery wall after a blood vessel wall is damaged or diseased. Cerebral aneurysm is the main cause of subarachnoid hemorrhage, the proportion reaches 85% [1]. The mortality rate of aneurysmal subarachnoid hemorrhage is 23%–51% [2], and 10%–20% of patients are permanently disabled [3, 4] and cannot live independently. Therefore, before the aneurysm ruptures, it is of important research significance to find the disease in time.

In recent years, some models and methods have been introduced at home and abroad to predict the occurrence of microvascular invasion to assist clinicians in making diagnostic decisions [5, 6]. Literature [7] used univariate analysis and logistic multivariate regression analysis methods, and studies proved that tumor pathological grade is the key predictor of recurrence and tumor-free survival of hepatocellular carcinoma without microvascular invasion. The literature [8] discussed the relationship between the occurrence of intracranial aneurysm and the characteristics of imaging omics from the perspective of imaging omics. After image segmentation, feature extraction, feature

M. Fei et al. (Eds.): LSMS 2021/ICSEE 2021, CCIS 1467, pp. 188–197, 2021.
https://doi.org/10.1007/978-981-16-7207-1_19

screening and classification and discrimination, the area under the receiver operating curve reached 0.76 by using the Support Vector Machine (SVM) model. Literature [9] considered the texture characteristics of arterial magnetic resonance T2-weighted imaging images and analyzed the texture characteristics statistically, and the area under the receiver operating curve reached 0.78. The above-mentioned methods focus on the study of the factors related to the occurrence of intracranial aneurysms and the use of traditional machine learning algorithms for modeling analysis. In terms of feature extraction, the redundancy and correlation between features are not fully considered, resulting in unsatisfactory results.

Rough Sets is a theoretical tool for data analysis. It is used to deal with fuzzy, incomplete and massive data. It can perform dimensionality reduction and feature extraction on data. It has a wide range of applications in various fields [10]. Therefore, this paper introduces the chi-square test to calculate the correlation method, and proposes a Neighborhood Rough Set attribute reduction algorithm on Chi-Square test (ChiS-NRS). It fully considers the correlation between attributes, screens out the most important attributes, and uses the gradient boosting tree model to establish an aneurysm prediction model to provide effective help for preoperative tumor diagnosis.

2 Related Theory

2.1 Neighborhood Rough Set

Rough set theory was proposed by Polish mathematician Pawlak in 1982. It is a mathematical tool for processing and analyzing uncertain knowledge and fuzzy data. The main idea is to approximate the inaccurate or uncertain knowledge with the knowledge that already exists in the knowledge base. The core content of rough set theory is attribute reduction, which is a process of eliminating redundant attributes.

The related definition of neighborhood rough set is as follows:

Definition 1. In a given M-dimensional real number space Ω, $\Delta = R^N \times R^N \to R$, then Δ is called a distance metric on R^N.

Definition 2. A non-empty finite set $U = \{x_1, x_2, \cdots, x_n\}$ existing in a given real number space Ω, the δ neighborhood of $\forall x_i$ is defined as:

$$\delta(x_i) = \{x | x \in U, \Delta(x_i, x_j) \le \delta\} \delta \ge 0. \tag{1}$$

Definition 3. Given a non-empty finite set $U = \{x_1, x_2, \cdots, x_n\}$ on a real number space Ω and the neighborhood relationship N on it, that is, the two-tuple $N = (U, N)$, $\forall x \in U$, Then the upper approximation and lower approximation of X in the neighborhood approximation space $N = (U, N)$ are:

$$\overline{N}X = \{x_i | \delta(x_i) \cap X \ne \emptyset, x_i \in U\}. \tag{2}$$

$$\underline{N}X = \{\delta(x_i) \subseteq X, x_i \in U\}. \tag{3}$$

Then the approximate boundary of X can be obtained as:

$$BN(X) = \overline{N}X - \underline{N}X. \quad (4)$$

Among them, the lower approximation $\underline{N}X$ of X is the positive domain, and the area completely unrelated to X is the negative domain, namely:

$$Pos(X) = \underline{N}X. \quad (5)$$

$$Neg(X) = U - \overline{N}X. \quad (6)$$

Definition 4. Given a neighborhood decision system $NDS = (U, C \cup D)$, the decision attribute D divides the universe U into N equivalence classes $(X_1, X_2, \cdots, X_n)$, $\forall B \subseteq C$, then the upper and lower approximations of decision attribute D with respect to subset B are:

$$\overline{N}_B D = \bigcup_{i=1}^{N} \overline{N}_B X_i. \quad (7)$$

$$\underline{N}_B D = \bigcup_{i=1}^{N} \underline{N}_B X_i. \quad (8)$$

Among them,

$$\overline{N}_B X = \{x_i | \delta_B(x_i) \cap X \neq \emptyset, x_i \in U\}. \quad (9)$$

$$\underline{N}_B X = \{x_i | \delta_B(x_i) \subseteq X, x_i \in U\}. \quad (10)$$

Similarly, the boundary of the decision-making system can be obtained as:

$$BN(D) = \overline{N}_B D - \underline{N}_B D. \quad (11)$$

The positive and negative domains of the neighborhood decision system are:

$$Pos_B(D) = \underline{N}_B D. \quad (12)$$

$$Neg_B(D) = U - \overline{N}_B D. \quad (13)$$

The dependence of decision attribute D on conditional attribute B is:

$$k_D = \gamma_B(D) = \frac{|Pos_B(D)|}{|U|}. \quad (14)$$

From Eq. (14), the dependence degree k_D is monotonic, if $B_1 \subseteq B_2 \subseteq \cdots A$, then, The importance of $\gamma_{B_1}(D) \leq \gamma_{B2}(D) \leq \cdots \leq \gamma_A(D)$ conditional attribute B relative to decision attribute D is:

$$sig_r(B, C, D) = \gamma_C(D) - \gamma_{C-B}(D). \quad (15)$$

2.2 Chi-Square Test and Gradient Boosting Decision Tree

Chi-square test is a commonly used mathematical tool to calculate the correlation between two variables. It mainly includes fitness test and independence test. In the independence test, statistics are the most commonly used. The commonly used method for binary classification problems is to use a 2×2 contingency table to calculate the correlation. In the prediction of aneurysm, there are only two types of aneurysm patients with aneurysm and no aneurysm.

Gradient Boosting Decision Tree (GBDT) is a decision tree classification algorithm based on a gradient boosting framework. Gradient boosting refers to the need to reduce the residual error in the previous iteration during each iteration, and build a new model in the direction of the gradient of the residual reduction. The decision tree divides the feature space into multiple regions according to a specific splitting principle, and each region returns a value as the decision value of the decision tree. Combine the idea of gradient boosting with the decision tree classification algorithm, that is, in each iteration, a new decision tree model is established in the direction of the gradient of the residual reduction of the model generated in the previous iteration. If the number of iterations is N, N decision tree models will be obtained. These N models are also called weak classifiers, which form a final GBDT classifier model by weighting or voting on the N weak classifiers.

3 Attribute Reduction Algorithm of Neighborhood Rough Set Based on Chi-Square Test

3.1 Description of the Main Idea of the Algorithm

From Eq. (15), we can see that the calculation formula of attribute importance is $sig_r(B, C, D) = \gamma_C(D) - \gamma_{C-B}(D)$. This means that the importance of a certain attribute B is equal to the degree of influence on the classification decision attribute after removing attribute B from conditional attribute C. When the importance value of a certain attribute is 0, it indicates that the attribute has no influence on the classification decision attribute and can be reduced and deleted.

From the above description, it can be seen that the neighborhood rough set attribute reduction algorithm only considers the direct impact of a single attribute on the decision attribute when calculating the importance of the dependency, and does not consider the interaction between multiple attributes. This may cause some important attributes to be deleted by mistake, leading to unsatisfactory reduction effect and affecting the final classification effect.

Literature [18] proposed an improved neighborhood rough set attribute reduction algorithm, which considers all other attributes when calculating the importance of an attribute, and the experimental results are relatively ideal. But the algorithm does not consider the size of the correlation between a certain attribute and other attributes, so this paper proposes a neighborhood rough set attribute reduction algorithm based on chi-square test (ChiS-NRS). Chi-square test is a hypothesis testing method that calculates the correlation between attributes. By combining with the neighborhood rough set attribute reduction algorithm, it not only considers the importance of a single conditional attribute, but also considers the attributes that are highly correlated with the attribute. So as to reduce the redundant attributes.

3.2 UCI Data Set Validation

In order to verify the effectiveness of the neighborhood rough set attribute reduction algorithm based on chi-square test proposed in this paper, the attribute reduction algorithm based on neighborhood rough set and the improved algorithm in this paper are used to perform attribute reduction experiments on the data set on UCI. And use the gradient boosting decision tree classifier model to calculate the classification accuracy of the two algorithms before and after the attribute reduction, and finally evaluate the effects of the two attribute reduction algorithms according to the classification accuracy. The attribute information of the UCI dataset selected in this paper is shown in Table 2.

Table 2. The attribute information of UCI data sets.

UCI data set	Number of samples	Number of attributes	Data category
Forest types	523	27	4
Dermatology	366	33	6
Parkinson	195	22	2
Sapfile1	131	21	3
Audiology	200	69	24
Biodeg	1055	41	2
Cleve	303	13	5
Credit	690	14	2
Heart	270	13	2
Sonar	208	60	2
Wdbc	569	30	2
Wpbc	198	33	2
Crx	690	15	2
Derm	366	34	6

Before performing algorithm verification, it is necessary to perform data cleaning operations on all data sets. For the processing of missing values, fill the column with fewer missing values in the data set with the average value of the column, and delete columns with more missing values. For the processing of noise data, the method of direct deletion is adopted. After the data set is cleaned, the improved neighborhood rough set attribute reduction algorithm based on chi-square test is applied to each data set for attribute reduction, and the matlab2014b software is used for simulation experiments. The reduction results are compared with the reduction results of other algorithms, and the comparison results between the algorithms are shown in Table 3. It can be seen from Table 3 that there is a certain gap in the attribute reduction results of the four algorithms. The number of attributes reduced by the PCA algorithm is the smallest compared to the other three algorithms. Except for the Sonar data set reduced to 13, the number of attributes reduced in all other data sets is less than 10. The LASSO algorithm has a large number of attributes after reduction, especially the Sonar data set has doubled the number of attributes after reduction compared to the PCA algorithm. Except for the two data sets Sapfile1 and Audiology, the number of attributes reduced by the NRS algorithm is between the above two algorithms. After the attribute reduction algorithm improved in this paper performs attribute reduction on the data set, the number of attributes obtained is more than other algorithms. This is because in the calculation process of attribute importance, the NRS algorithm does not consider the influence between related attributes, and the calculation process of the importance of a single attribute is independent. The improved algorithm in this paper emphasizes the influence of related attributes in the calculation of the attribute degree of a single attribute. Therefore, when performing attribute reduction, more important attributes and their related attributes will be considered, resulting in a relatively large number of attributes after reduction.

Table 3. Number of attributes after reduction by different attribute reduction algorithms.

UCI data set	Number of original attributes	PCA	LASSO	NRS	ChiS-NRS
Forest types	27	6	11	9	11
Dermatology	33	7	13	10	14
Parkinson	22	4	12	8	10
Sapfile1	21	8	8	7	9
Audiology	69	10	12	19	20
Biodeg	41	8	17	13	14
Cleve	13	4	7	9	10
Credit	14	5	10	13	11
Heart	13	5	9	7	8
Sonar	60	13	24	20	24
Wdbc	30	6	15	9	14
Wpbc	33	8	16	9	11
Crx	15	6	15	15	12
Derm	34	7	12	10	14

According to the attribute reduction result, we select the corresponding attributes and labels from the original data set to form a new data set. Then use the gradient boosting decision tree (GBDT) classifier model to classify and recognize the data set before and after the reduction, and obtain the recognition accuracy of several algorithms after attribute reduction as shown in Table 4.

Table 4. Recognition accuracy after reduction with several different attribute reduction algorithms.

UCI data set	Original data accuracy	PCA	LASSO	NRS	ChiS-NRS
Forest types	70.32%	71.41%	73.13%	72.37%	**76.92%**
Dermatology	98.28%	94.23%	96.82%	97.66%	**98.76%**
Parkinson	87.86%	90.72%	95.33%	94.54%	**98.33%**
Sapfile1	67.52%	65.27%	68.85%	70.39%	**78.62%**
Audiology	74.43%	77.68%	78.26%	79.22%	**84.00%**
Biodeg	84.57%	82.34%	80.65%	86.34%	**88.57%**
Cleve	66.83%	67.23%	69.94%	73.98%	**77.97%**
Credit	88.19%	87.97%	86.28%	89.11%	**90.10%**
Heart	87.20%	86.38%	83.39%	87.73%	**89.23%**
Sonar	86.01%	85.55%	86.64%	87.10%	**91.07%**
Wdbc	**97.11%**	94.22%	91.28%	93.29%	95.65%
Wpbc	78.51%	79.88%	83.44%	82.24%	**85.53%**
Crx	86.98%	83.45%	84.99%	86.98%	**87.76%**
Derm	**96.43%**	94.38%	95.77%	92.15%	94.05%

It can be seen from the table that compared with the other three algorithms, except for the Wdbc and Derm data sets, the ChiS-NRS algorithm proposed in this paper has the highest accuracy rate and the best performance on other data sets among them, the algorithm in this paper has the most obvious improvement effect on the four data sets of Sapfile1, Audiology, Biodeg and Cleve.

4 Prediction of Intracranial Aneurysm

4.1 Experimental Data Set and Evaluation Criteria

The data set selected in this paper is derived from the medical images of 206 patients who have undergone aneurysm surgery collected by radiologists. After image segmentation and feature extraction, a total of 64 image omics features are obtained for the prediction of aneurysm. And the presence or absence of each aneurysm has been accurately diagnosed.

There are various algorithms for attribute reduction. The improved neighborhood rough set attribute reduction algorithm (Chis-NRS reduction) in this paper is compared with unproduction, attribute reduction based on neighborhood rough set (NRS reduction) and attribute reduction based on chi-square test (ChiS reduction). Put the reduced attributes into the gradient boosting tree classification model to establish an aneurysm prediction model, and use 75% of the intracranial artery data as the training set to train the prediction model, and the remaining 25% as the test set to test the effect of the model. From the two aspects of the attributes after reduction and the classification accuracy after reduction, the results are shown in Fig. 5.

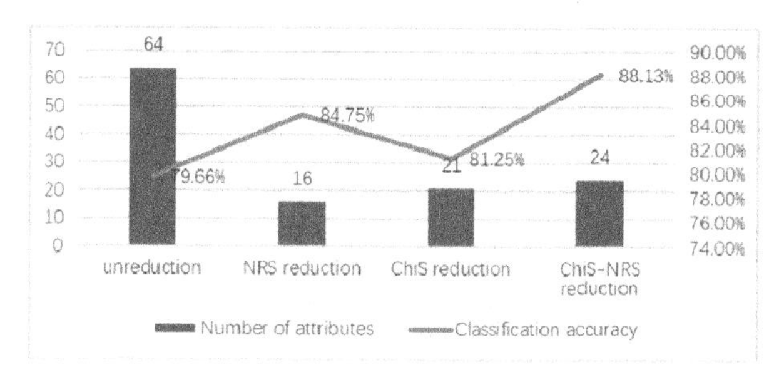

Fig. 5. Prediction model effect of aneurysm.

Through experiments, it is found that the number of attributes reduction in the neighborhood rough set is the least, but the final classification accuracy is higher than the attribute reduction algorithm based on chi-square test. The reason is that the attribute reduction algorithm based on the chi-square test needs to discretize the data when using the contingency table to calculate the attribute correlation, which may affect the classification accuracy. However, attribute reduction based on neighborhood rough set can directly process continuous data, and knowledge reduction is more accurate. The improved neighborhood rough set attribute reduction algorithm based on chi-square test in this paper has a few more attributes after reduction, but the accuracy rate is higher than other algorithms. This is because the algorithm in this paper takes into account the interaction between attributes, extracts more important attributes, and achieves the best results in the final comparison experiment.

4.2 Evaluation of the Effect of the Classifier Model

In this paper, the improved attribute reduction algorithm is combined with common classification models such as Convolutional Neural Network (CNN), Recurrent Neural Network (RNN), Support Vector Machine (SVM), Random Forest (RF) and other common classification models to establish intracranial aneurysm invasion prediction models. Compare with GBDT-based intracranial aneurysm prediction model, use confusion matrix to evaluate the classification accuracy, sensitivity and specificity of each model, and use Receiver Operating Curve (ROC) and Area Under Curve (AUC) to evaluate the model.

Substitute the values in the two-dimensional confusion matrix of this model and other models into the formula to calculate the sensitivity and specificity of each model on the test set, as shown in Table 6.

Table 6. Comparison of evaluation indexes among each prediction model.

	TP	FN	FP	TN	Accuracy	Sensitivity	Specificity
CNN	23	7	9	20	72.88%	76.67%	68.97%
RNN	21	9	5	24	76.27%	70.00%	82.76%
SVM	23	7	6	23	77.97%	76.67%	79.31%
RF	24	6	6	23	79.66%	80.00%	79.31%
GBDT	25	4	3	27	88.14%	86.21%	90.00%

Taking sensitivity as the vertical axis of the coordinate axis and 1-specificity as the horizontal axis, the ROC curve and the area under the curve (AUC) of each intracranial aneurysm prediction model on the test set are shown in Fig. 6.

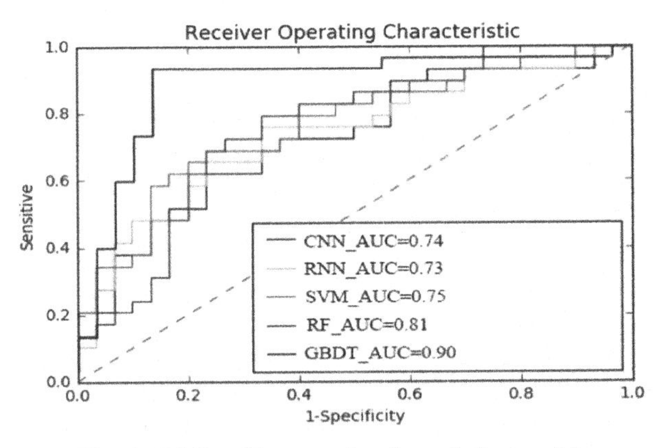

Fig. 6. ROC and its area of each prediction model.

It can be seen from Table 6 that the newer models CNN and RNN are not effective in processing medical data. Analyzing the reasons, it is concluded that the amount of data required by the deep learning model is huge, and the amount of data for the intracranial aneurysm in this article is only more than 200, and the deep model cannot learn more knowledge from less data. The two models commonly used in processing medical data, SVM and RF, are more effective than deep models. The accuracy, sensitivity, and specificity of the improved intracranial aneurysm prediction model in this paper on the test set reached 88.13%, 87.10% and 89.29%, respectively. Compared with other models, the effect is significantly improved. Figure 6 shows the ROC curve of each prediction model and the value of the area under the curve AUC. It can be seen that, compared with other classifier models, the AUC value of the area under the ROC curve has reached 0.90. It shows that this model has the highest accuracy and the best effect in predicting intracranial aneurysms. It can provide effective prediction and accurate diagnosis for the presence or absence of intracranial aneurysms in preoperative tumor patients.

5 Conclusion

The algorithm proposed in this paper carries out related research on the prediction of intracranial aneurysm. Comparing the effect of attribute reduction and the effect of the classifier model respectively, it verifies the effectiveness of the improved neighborhood rough set attribute reduction algorithm based on chi-square test in the feature reduction of tumor patients. And judging from the classification accuracy, sensitivity and specificity of the combination of the algorithm and the gradient boosting decision tree classification model, the prediction model has a good effect in predicting the presence or absence of intracranial aneurysms in tumor patients before surgery. It can play an active role in the diagnosis of medical intracranial aneurysms.

References

1. Gijn, J.V., Kerr, R.S., Rinkel, G.J.E.: Subarachnoid haemorrhage. Lancet **369**(9558), 306–318 (2007)
2. Ingall, T., Asplund, K., Mahonen, M., et al.: A multinational comparison of subarachnoid hemorrhage epidemiology in the WHO MONICA stroke study. Stroke **31**(5), 1054–1061 (2000)
3. Algra, A., Hop, W.C.J., Van, D.L.R., et al.: Case-fatality rates and functional outcome after subarachnoid hemorrhage: a systematic review. Stroke J. Cereb. Circ. **28**(3), 660 (1997)
4. Stember, J.N., et al.: Convolutional neural networks for the detection and measurement of cerebral aneurysms on magnetic resonance angiography. J. Digit. Imaging **32**(5), 808–815 (2018). https://doi.org/10.1007/s10278-018-0162-z
5. Yang, P., Si, A., Yang, J., et al.: A wide-margin liver resection im-proves long-term outcomes for patients with HBV-related hepatocellular carcinoma with microvascular invasion. Surgery **165**(4), 721–730 (2019)
6. Ma, H., Wang, Y., Yang, H.C., et al.: Clinical study on predicting microvascular invasion and early recurrence of hepatocellular carcinoma. Chin. J. Clin. Physicians (Electron. Ed.) **6**(20), 58–60 (2012)
7. Zhou, L., Rui, J., Zhou, W., et al.: Edmondson-Steiner grade: acrucial predictor of recurrence and survival in hepatocellular carcinoma without microvascular invasion. Pathol. Res. Pract. **213**(7), 824–830 (2017)
8. Liu, T.T., Dong, Y., Han, H., et al.: Prediction of microvascular invasion and tumor differentiation grade in hepatocellular carcinoma based on radiomics. Chin. J. Med. Comput. Imaging **24**(1), 83–87 (2018)
9. Wu, M.H., Tan, H.N., Wu, Q.X., et al.: Value of MRI T2-weighted image texture analysis in evaluating the microvascular invasion for hepatocellular carcinoma. Chin. J. Cancer **28**(3), 191–196 (2018)
10. Velayutham, C., Thangavel, K.: Detection and elimination of pectoral muscle in mammogram images using rough set theory. C. In: Proceedings of the 2012 IEEE International Conference on Advances in Engineering, Science and Management, pp. 48–54. IEEE (2012)

Constrained Adversarial Attacks Against Image-Based Malware Classification System

Hange Zhou(✉) and Hui Ma

College of Information Science and Engineering, Ocean University of China, Qingdao 266100, China
mahui@ouc.edu.cn

Abstract. In this paper, a multi-pixels attack is proposed to demonstrate the vulnerability of the image-based malware classifier. We adopt the self-adaptive differential evolutionary algorithm to reinforce the power of adversarial attack. The proposed scheme has the following advantages: 1) as a semi-black-box attack, it can fool the neural network without any knowledge about the gradients or model structure. 2) the transformation from binary files to images can train a classifier with higher performance. 3) the outside group is added while implementing the DE algorithm to avoid being trapped in the local optimal value. The experimental results show that our attack can make the classifier misjudge with high confidence rate and more false negatives under the restriction of not destroying the files' function. Besides, further defensive mechanisms are explored such as local median smoother, color depth reducing and adversarial training to show that the ℓ_0 norm attacks can be relieved.

Keywords: Convolutional neural network · Malware classification · Malicious attack · Adversarial training

1 Introduction

With deep learning gradually being the heart of artificial intelligence, more malware detection systems are built on the machine learning [1–5]. DNN-based approaches have gained a lot of popularity in many fields but it has been demonstrated to lack robustness while facing well-tuned adversarial perturbations [6–9]. Hence, great attention should be paid when applying DNN to the security field [10,11] such as malware detection for the reason that the intelligent system may be fooled and trigger a series of safety issues. It is sound to transfer the attacks methods from the well-studied computer vision field to security-critical field like malware classification task [12]. Notably, it has huge difference comparing with image perturbation which can be conveniently manipulated by changing pixels on a real and continuous scale and the constraint of image's modification is usually done by the norm for the aim of being undistinguished

M. Fei et al. (Eds.): LSMS 2021/ICSEE 2021, CCIS 1467, pp. 198–208, 2021.
https://doi.org/10.1007/978-981-16-7207-1_20

for a human observer. However, for the malware classification system based on machine learning, the requirement is that the modification should not jeopardize the application functionality. Rather than just operating on the binary files, what we do in this paper is to transform the files into images [13] which make it easier to refer various well-studied attack methods in the realm of computer vision. Similarly, we train the model using the data collected from *Kaggle* which include but not limited to Android malware portable execution files and other images in the Malimg dataset. When reshaping the images, the application's functionality should not be destroyed by just modifying the pixels in the extra added area instead of reducing any original pixels. The accuracy of our model is achieved up to an average rate of 98.92%, owning better performance than the model trained just using the binary files and we attribute this kind of success to the nature that images have higher entropy for the model to capture high-dimensional features. Considering that only part of the region can be modified, one-pixel attack method and its variation (multi-pixels attack) [14] may play a role, satisfying the requirement of little distance between original and adversarial samples. We choose to attack the redundant part that we replenish as a reshape, so it remained the application's functionality and has good generalization ability as well. Figure 1 shows an example of the main procedure of our experiment.

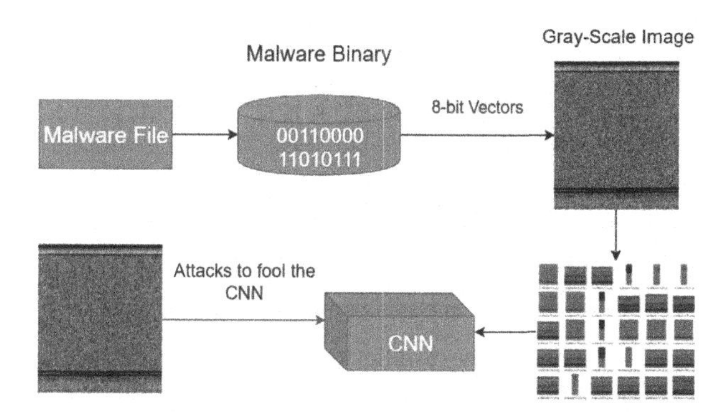

Fig. 1. The process of attacking the classifier based on file-to-image transformation

Contribution: In this paper, we have made some contributions as follows. Firstly, unlike the method that Kathrin Grosse et al. [12] have proposed which utilize the malware files to train a model, we consider on transforming the binary files into gray-scale images [13] for the classifier. This may be a new groundbreaking strategy to deliver well-studied adversarial attacks or defense in the computer vision (CV) filed to security-critical domains such as malicious programs detecting and classifying. Notably, our experiments have achieved pretty high model accuracy and a better trade-off among the accuracy, precision and recall rate.

Secondly, due to the restriction of preserving the functionality of malware files, we choose the extreme but suitable one-pixel attack to fool the model we have trained and we utilize the self-adaptive differential evolution algorithm as an optimal method to discover the object to modify. While applying the SaDE algorithm, we have added an extra-ethnic competition using random population after the operation of mutation and crossover. This trick helps a lot and make the objective function quickly converge to the global optimal value, further reducing the probability of falling into the local optimal value at the same time.

The rest of the paper is organized as followed. We introduced the related work and necessary background in the field of adversarial attack and corresponding defensive mechanisms in Sect. 2. Our method to confuse the classifier using adversarial attacks will be elaborated in Sect. 3. Finally, we discuss the experimental results as well as the future work in Sect. 4.

2 Problem Description

In this section, we firstly introduce the construction of the malware classifier. Secondly, we highlight the problem and difficulty of crafting adversarial perturbations under the strong restriction of not destroying the malware's function. Then, the ℓ_0 norm attack methods are proposed to solve the difficulty.

2.1 Preliminaries on the Model

Our object of research is a malware classifier of CNN structures trained with the images transformed from malware binary files, and we attempt to prove its vulnerability to adversarial attacks as well as exploring some corresponding defensive mechanism. In order to get the best trade-off among the accuracy, precision and recall rate, we adopt different kinds of CNN structures and train them using cross-entropy loss function: $H_{\hat{y}}(y) = -\sum_{i} \hat{y}_i log(y_i)$ where y_i represents the prediction and $\hat{y_i}$ is the ground truth. From the gradient of parameters: $\frac{1}{n}\sum_{x} x_j(\sigma(z) - y)$ where the activation function is sigmoid, we know that when the error $(\sigma(z) - y)$ between the prediction and ground truth is large, the gradient becomes large and the parameters $\mathbf{W}$ adjust faster meaning a much faster training process. The formulas of precision rate and recall rate are: $P = \frac{TP}{TP+FP}$ and $R = \frac{TP}{TP+FN}$ where TP represents true positive numbers and FN represents false negative numbers. Especially, we consider the recall rate as more important because false negative samples may cause much more devastating results.

2.2 Attack Strategy Against Model

To give a formal description of adversarial attack which can be regarded as an optimization problem with certain constraint such as the ℓ_p norm, we start with the **n**-dimensional inputs $\mathbf{X} = (x_1, x_2, \ldots, x_n)$ and the classifier $\mathbf{F}$ return a two dimensional vector $\mathbf{F}(X) = [\mathbf{F}(X)^0, \mathbf{F}(X)^1] \quad s.t. \quad \mathbf{F}(X)^0 + \mathbf{F}(X)^1 = 1$. Observing the output $\mathbf{F}(X)^0$ we can get the probability that the input is labeled as benign and correspondingly the $\mathbf{F}(X)^1$ encodes the belief on malware. Then, we define the classification label $\mathbf{y} = \arg\max_{i} \mathbf{F}(X)^i$ meaning that the index (label) of the higher probability one will be returned as the final result. While implementing the attack, our goal is to find an adversarial perturbation δ for the aim that $\hat{\mathbf{y}} = \mathbf{F}(X + \delta)$ is rather different from the original label i.e. $\hat{\mathbf{y}} \neq \mathbf{y}$.

The whole problem can be summarized as followed:

$$\min_{\delta^*} \quad \mathbf{F}(\mathbf{X_{malware}} + \delta)^1 \quad subject \quad to \quad \|\delta\|_0 \leq \mathbf{d},$$

and correspondingly to make the benign to change to malware:

$$\min_{\delta^*} \quad \mathbf{F}(\mathbf{X_{benign}} + \delta)^0 \quad subject \quad to \quad \|\delta\|_0 \leq \mathbf{d}.$$

In the above formulations, we attempt to reduce a sample's probability of being judged as its original label after adding perturbations. For example, a malware sample which is originally labelled 1 with a confidence probability of 0.7 and then it is reduced lower than 0.5 which means now the sample is misclassified as benign. The ℓ_0 norm is used to restrict the numbers of maximum modification and $\mathbf{d}$ is a constant that we set according to the pixels we want to modify and a special condition of one-pixel attack strategy is generated when $\mathbf{d} = 1$.

3 Attack Target and Method

In this section, we elaborate our approach to craft adversarial perturbation to fool the malware classifier. We start with describing the procedure of data processing and detailing how we train our own classifier with these data. Then, we describe how we conduct the one-pixel attack and further multiple-pixels attack strategies with definite restriction towards our neural network. There are 9,000 malware and corresponding benign images we get from the *Malimg*, which have been already transformed from the binary files and another 3,000 from DREBIN dataset [15], which we have conducted the file-to-image transformation. In view of remaining the functionality of the malware files, we replenish some extra pixels as a data augmentation when adjusting and reshaping the image size. We are allowed to only search and modify the point pixel in the restricted area, so we can not apply the attack methods using the gradient descent since they tend to search globally. That may explain why we adopt the one-pixel attack cause we do not need to know the inner structure, model parameters, or gradients.

3.1 Model Training

Since to the best of our knowledge there is no publicly available and mature malware classifying model whose structure is based on neural networks, we develop our own classifier. We utilize the already processed gray-scale images as input (each size: 512 × 512) to train our classifier based on CNN structure. Notably, all the features we extracted and trained with are static rather than some other dynamic-behavior based analysis since they are much more challenging so we leave the dynamic features training for future work. The Table 1 shows that the ResNet performs better than the other four models, therefore it is our target to implement the attacks.

Table 1. The results on each of the CNN-based model we have trained with 512×512 gray-scale images as inputs.

Model	Accuracy (%)	Precision (%)	Recall (%)
LeNet-5	97.85	97.00	97.70
VGG-16	96.43	97.41	95.40
AlexNet	97.14	97.96	96.28
ResNet	98.92	99.24	98.60
InceptionV_3	98.27	98.73	97.80

Table 2. The comparison of our best model with others on the accuracy (%), false positive rate (FPR %) and false negative rate (FNR %) under different malware ratio (MWR).

	MWR = 0.3			MWR = 0.4			MWR = 0.5		
	Accuracy	FPR	FNR	Accuracy	FPR	FNR	Accuracy	FPR	FNR
ResNet	98.41	8.18	1.97	98.29	7.02	3.11	98.78	6.32	3.29
M1 [12]	98.35	9.73	1.29	96.60	8.13	3.19	95.93	6.37	3.96
M2 [12]	97.85	17.30	1.71	97.35	10.14	2.29	95.65	6.01	4.25
M3 [12]	95.63	15.25	3.86	93.95	10.81	5.82	92.97	8.96	6.92

We have picked three different models M1[200,200], M2[20,200], M3[200,10] [12] as comparison and our best model ResNet only uses 3,000 DREBIN malware samples to train but gets better performance. The Table 2 shows that our model gets better results while applying 2,000 test dataset and the malware ratio represents malicious DREBIN samples' proportion.

3.2 Adversarial Samples Crafting

Among many attacks, it is common to craft adversarial samples with ℓ_2 or ℓ_∞-norm. As an extreme case of ℓ_0-norm attack, one-pixel attack only modifies one pixel during the whole perturbation-generating time. The differential evolution (DE) algorithm is divided into four parts: initialization, mutation, crossover and final selection. We have made further improvements by using the self-adaptive differential evolution (SaDE) algorithm because it updates the value of scaling factor and crossover probability for the aim of avoiding falling into the local

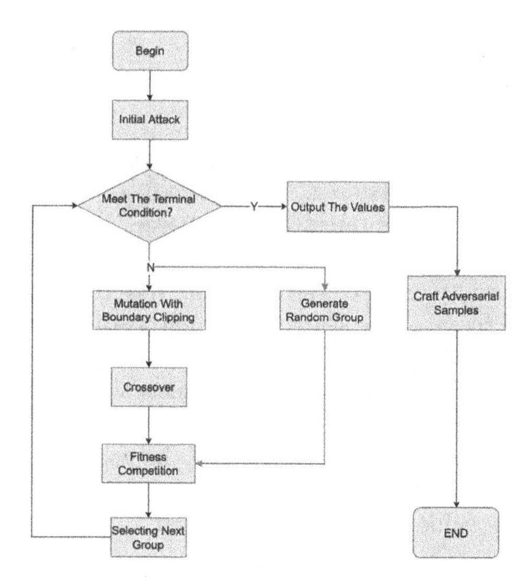

Fig. 2. The process of crafting adversarial samples using extra outside-group competition (red line) to reduce the probability of falling into the local optimal value. (Color figure online)

optimal value and speed up the search process. The Fig. 2 shows the whole process of SaDE, including initialization with NP numbers of group size (see Formula 1), mutation Formula 2 of which we adopt the DE/rand/1/bin strategy, crossover Formula 3, and final competition Formula 4 to pick more suitable individuals to compose the next generation. Also in order not to jump out of the local area while generating new individuals, we set two kinds of boundary processing(see formulation Formula 5 and Formula 6).

$$X_i^0 = (d_{i,1}^0, d_{i,2}^0, d_{i,k}^0 \dots d_{i,D}^0) \quad s.t. \quad i = 1, 2 \dots, NP \tag{1}$$

$$\text{DE/rand/1/bin:} \quad \upsilon_i = x_{r_1}^g + \mathbf{F}(x_{r_2}^g - x_{r_3}^g) \tag{2}$$

$$\mathbf{u}_i^g = \begin{cases} \upsilon_{i,j}^g, \; if \quad rand \leq \mathbf{CR} \quad or \quad randi(1, D) = j \\ x_{i,j}^g, \; if \quad rand > \mathbf{CR} \quad and \quad randi(1, D) \neq j \end{cases} \tag{3}$$

$$\mathbf{x}_i^{g+1} = \begin{cases} \upsilon_i^g, if \quad \mathbf{f}(\upsilon_i^g) \quad superior \quad to \quad \mathbf{f}(\mathbf{x}_i^g) \\ x_i^g, \qquad\qquad otherwise \end{cases} \tag{4}$$

$$\mathbf{x}_{i,j}^{g+1} = \begin{cases} \mathbf{x}_{min}, \; if \quad x_{i,j}^{g+1} < \mathbf{x}_{min} \\ \mathbf{x}_{max}, \; if \quad x_{i,j}^{g+1} > \mathbf{x}_{max} \\ x_{i,j}^{g+1}, \qquad otherwise \end{cases} \tag{5}$$

$$\mathbf{x}_{i,j}^{g+1} = \begin{cases} \mathbf{x}_{min} + rand().(\mathbf{x}_{max} - \mathbf{x}_{min}), if x_{i,j}^{g+1} < \mathbf{x}_{min} \quad or \quad x_{i,j}^{g+1} > \mathbf{x}_{max} \\ x_{i,j}^{g+1}, \quad otherwise \end{cases} \tag{6}$$

In our experiment, we set the attribute number **D** equal to 3 which represents the coordinates and pixel value e.g. (x, y, value), the scaling factor(mutation rate) **F** equal to 0.5 in the beginning, the group size NP equal to 50, the number of iteration equal to 100, the crossover probability equal to 0.4. The adaptive mutation rate $\hat{F}$ is used to replace the mutation rate **F** to increase the mutation rate and expand the scope of the search at the early stage of evolution. Similarly, if the fitness value obtained in the next generation is greater than the fitness value of the parent individual, the mutation rate $\hat{F}$ needs to be reduced accordingly. The Formula 7 shows how the mutation rate $\hat{F}$ conducts self-adjustment. The f_{max} means maximizing the fitness value and the f_{min} intends to minimize it. And the $f(v_i^g)$ represents the fitness value of current individual and the $f(x_{best}^g)$ represents the fitness value of the best individual.

$$\hat{F} = \begin{cases} 1 - \sqrt[3]{\frac{f(v_i^g)}{f(x_{best}^g)}}, if \quad f_{max} \\ 1 - \sqrt[3]{\frac{f(x_{best}^g)}{f(v_i^g)}}, if \quad f_{min} \end{cases} \tag{7}$$

4 Results and Evaluation

The results show that continually adding the pixels to be modified can not achieve much better misclassification rate (see Table 3), and we attribute this phenomenon to the reason that we can not discover the maximum value in the saliency map of gradients due to the restriction of local-area perturbation.

Table 3. The misclassification rate and the confidence rate on the fault label while increasing the attacked pixels.

Attacking-Numbers	MR (%)	Confidence (%)
1	74.09	64.72
3	83.25	69.54
5	86.53	72.79
7	86.47	76.48
9	87.20	81.92
11	87.47	88.40

However, the average confidence rates of the successfully misclassified images are gradually increasing, which we attribute to the larger difference between the input distribution. The Fig. 3 shows that we randomly sampled 500 malware and benign inputs (each of them has 250 samples) and the successful numbers which

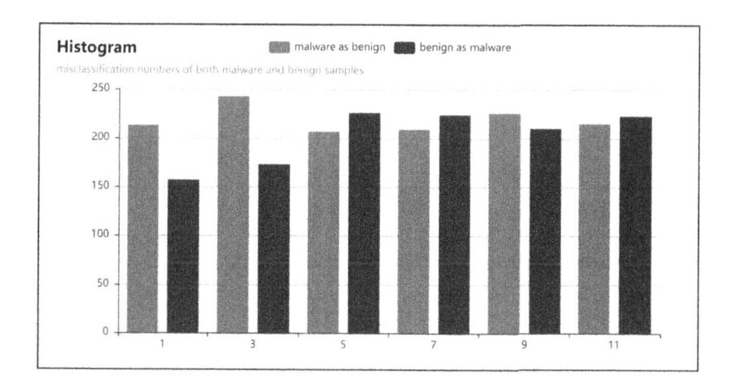

Fig. 3. The numbers of malware after modifying which are misclassified as benign and the numbers of benign as malware as the modified pixels increase, each of them has a number of 250 clean set.

have been judged an opposite label. We find that as the modified pixels numbers increase, the successful rate of each category gets closer. We further explored the defense mechanisms that can be deployed against the multi-pixels attack. Firstly, we adopt local median smoothing from the feature squeezing technology [16], which dedicates to reduce the difference among individual pixels. The local median smoothing runs a sliding window (e.g. a 3×3 window includes 9 pixels) on the image, where the center pixel is replaced by the median value of neighboring pixels in the sliding window.

Table 4. The classification results of 500 adversarial samples after increasing the size of sliding window.

$Size_{slidingwindow}$	Accuracy (%)	Precision (%)	Recall (%)
None	12.44	16.45	12.94
1×3	29.84	33.61	29.78
1×5	58.62	66.54	60.07
1×7	74.93	78.37	72.18
3×3	81.25	80.47	82.73
3×5	89.13	89.60	88.54

The pixels on the edges are filled with reflection padding. As can be seen in the Table 4, we have adopted 27-pixels attack to 500 clean samples. Then, we use median filter as a 'firewall' in front of the classifier and find that it indeed makes the classifier more robust. Figure 4 shows that the reduction of the color depth can effectively lower the success rate of multiple pixels attack. We attribute this to the reason that reducing the depth may restrict adversary's choices of attacking. Finally, we attempt to conduct adversarial training as well.

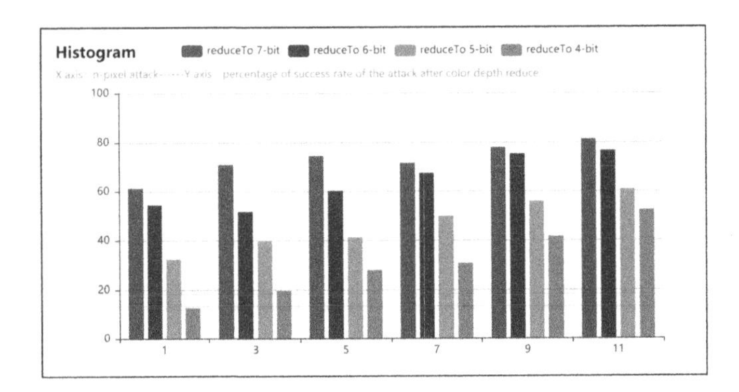

Fig. 4. The success rate of multiple-pixels attack after reducing different color bit depth.

Then we make a comparison of three defensive ability (we choose the settings of the best performance each e.g. 3 × 5-size window, 4-bit scaling) used in our experiment. The Table 5 shows that re-training has much more defensive power although the rest two mechanisms also achieve great performance. However, it is not difficult to see the adversarial training has severe limitations on defending other attack methods. So, we need to find more effective attacking strategies to fool the file-to-image based neural networks and utilize these strategies to conduct adversarial training.

Table 5. The success rate (%) of multiple-pixels attack after implementing different defensive strategies.

Attacked numbers	None defense	Median filter	Depth reduce	Re-training
1	74.09	11.87	12.43	5.14
3	83.25	24.36	19.56	6.03
5	86.53	30.14	27.87	5.47
7	86.47	34.59	32.04	6.96
9	87.20	36.93	41.49	5.91
11	87.47	40.14	52.38	6.32

4.1 Conclusions

In this paper, we have explored the potential adversarial attacks which can be crafted against the field of malware classifying, a totally different domain from computer vision due to more restrictions on adding adversarial perturbations. Hence, while designing effective attacks to fool the malware classifier to misjudge the malicious one as benign, attackers should make sure that the functionality of the malware is well retained. So we firstly add extra edge area on the images which are transformed from binary files, and then choose the multiple-pixels attack method to perturb the special area. That is how we attack the neural network while keeping the functionality well at the same time. The experimental results show that in spite of the much more excellent accuracy and performance of the classifier based on file-to-image transformation, the model still can be fooled by our attacks. We further prove that the defensive mechanisms such as median filter, color depth reducing and adversarial training make a positive difference against adversarial attacks, decreasing the misclassification rate a lot. However, there are some points needed to improve like exploring more latent attack methods and corresponding defensive strategies in this security-critical field and our adversarial training still has room for improvement. In this paper, all the features we have trained are static, but malicious features of the malware may be hidden firstly and vary along with time in reality, called dynamic features which we will leave for future work.

References

1. Li, J., Sun, L., Yan, Q., Li, Z., Srisa-an, W., Ye, H.: Significant permission identification for machine-learning-based android malware detection. IEEE Trans. Ind. Inf. **14**(7), 3216–3225 (2018)
2. Demontis, A., et al.: Yes, machine learning can be more secure! a case study on android malware detection. IEEE Trans. Dependable Secure Comput. **16**(4), 711–724 (2019)
3. Chen, X., et al.: Android HIV: a study of repackaging malware for evading machine-learning detection. IEEE Trans. Inf. Forensics Secur. **15**, 987–1001 (2020)
4. Li, J., Sun, L., Yan, Q., et al.: Significant permission identification for machine-learning-based android malware detection. IEEE Trans. Industr. Inf. **14**(7), 3216–3225 (2018)
5. Ahmadi, M., Ulyanov, D., Semenov, S., et al.: Novel feature extraction, selection and fusion for effective malware family classification. In: Proceedings of the Sixth ACM Conference on Data and Application Security and Privacy, pp. 183–194 (2016)
6. Darvish Rouani, B., Samragh, M., Javidi, T., Koushanfar, F.: Safe Machine Learning and Defeating Adversarial Attacks. IEEE Secur. Privacy **17**(2), 31–38 (2019)
7. Papernot, N., McDaniel, P., Jha, S., et al.: The limitations of deep learning in adversarial settings. IEEE European Symposium on Security and Privacy (EuroS&P), pp. 372–387 (2016)
8. Chen, S., Xue, M., Fan, L., et al.: Automated poisoning attacks and defenses in malware detection systems: an adversarial machine learning approach. Comput. Secur. **73**, 326–344 (2018)
9. Ilyas A, Engstrom L, Athalye A, et al.: Black-box adversarial attacks with limited queries and information. International Conference on Machine Learning(PMLR). 2137–2146 (2018)
10. Chen, C., Shao, Y., Bi, X.: Detection of anomalous crowd behavior based on the acceleration feature. IEEE Sensors J. **15**(12), 7252–7261 (2015)
11. Das, S., Liu, Y., Zhang, W., Chandramohan, M.: Semantics-based online malware detection: towards efficient real-time protection against malware. IEEE Trans. Inf. Forensics Secur. **11**(2), 289–302 (2016)

12. Grosse, K., Papernot, N., Manoharan, P., et al.: Adversarial perturbations against deep neural networks for malware classification. arXiv: 1606.04435 (2016)
13. Euh, S., Lee, H., Kim, D., Hwang, D.: Comparative analysis of low-dimensional features and tree-based ensembles for malware detection systems. IEEE Access. **8**, 76796–76808 (2020)
14. Su, J., Vargas, D.V., Sakurai, K.: One pixel attack for fooling deep neural networks. IEEE Trans. Evol. Comput. **23**(5), 828–841 (2019)
15. Arp, D., Spreitzenbarth, M., Hubner, M., et al.: Drebin: effective and explainable detection of android malware in your pocket. NDSS **14**, 23–26 (2014)
16. Xu, W., Evans, D., Qi, Y.: Feature squeezing: Detecting adversarial examples in deep neural networks. arXiv.1704.01155 (2017)

Intelligent Medical Apparatus, Clinical Applications and Intelligent Design of Biochips

Inelastic Modelling of Bone Damage Under Compressive Loading

Qiguo Rong(✉) and Qing Luo

Department of Mechanics and Engineering Science, College of Engineering, Peking University, Beijing 100871, China
qrong@pku.edu.cn

Abstract. The existing constitutive models of inelastic behavior of bone are all based on the assumed tissue behaviors, thus lacking direct physical meaning. To address the issue, this study proposed a semi-empirical constitutive model for describing the post-yield behavior of bone based on the experimental relationships of the elastic modulus, plastic strain, and viscoelastic responses of bone with the applied strain. Assuming that these decomposed properties are independent to each other, a simple but physically sound model was derived by combining all (elastic, plastic, and viscoelastic) behaviors in one governing equation. In this model, a decayed exponential function was applied to describe the modulus loss (damage accumulation) with increasing strain. In addition, a power law function was used to describe the increase of plastic deformation with damage accumulation. Finally, the linear viscoelastic model based on a convolution integral was proposed to describe the viscoelastic behavior of bone as the function of time and loading rate. To verify the model, all material constants required for the experimental relationships between the Young's modulus, plastic deformation, viscoelastic deformation and the applied deformation were obtained using a novel progressive loading protocol on human cortical bone. Then, the monotonic stress-strain curve obtained from bone specimens at the same anatomic location was curve-fitted with the proposed constitutive model and the material constants were predicted and compared with the experimental ones. The analysis indicated that the proposed constitutive model was reasonably accurate in assessing the post-yield behavior of bone.

Keywords: Bone · Constitutive model · Inelastic · Post-yield behavior

1 Introduction

Most of the energy absorbed by bone in the fracture process is dissipated in the post yield stage of bone. Therefore, the changes of bone mechanical properties in the post yield stage determine the toughening mechanism of bone. The post yield behavior of bone is a typical nonlinear behavior, including three distinct mechanical responses: the change of material stiffness, plastic deformation and viscoelastic behavior. It is well known that the accumulation of micro damage in bone tissue runs through the whole post yield stage and plays a very important role in the process of energy dissipation. The previous studies

M. Fei et al. (Eds.): LSMS 2021/ICSEE 2021, CCIS 1467, pp. 211–220, 2021.
https://doi.org/10.1007/978-981-16-7207-1_21

showed that the accumulation of micro damage is closely related to the elastic property, plastic deformation and viscoelastic response. But there is still no generally accepted constitutive model to describe them. For the constitutive relationship can quantitatively predict the changes of bone mechanical properties, it is very important to establish the constitutive relationship between micro damage and bone mechanical properties.

Mullins et al. [1] proposed an extended Drucker-Prager model for the post-yield behavior of cortical bone, accounting for pressure dependent yield in tension. Pawlikowski et al. [2] set up a HRK model, in which, anisotropy, non-homogeneity and bone remodeling were taken into consideration. Natali et al. [3] formulated a constitutive model of cortical bone considering an anisotropic configuration and post-elastic and time-dependent phenomena. However, these studies don't reflect the effect of damage accumulation on mechanical behavior of cortical bone, while the non-linear stress-strain behavior is mainly determined by damage accumulation [4]. Krajcinovic et al. [5] suggested the continuum damage model as a proper choice for the cortical bones. Fondrk et al. [4] developed a damage model with two internal state variables to model the nonlinear tensile behavior of cortical bone. However, the consensus is that the long bones are usually loaded in compression. In addition, the previous study has pointed that the mechanical behavior of cortical bone under compression is different from that under tension [6]. So is the constitutive equation. Therefore, it is necessary to develop a constitutive model for the nonlinear compressive behavior of cortical bone.

Based on the experimental results from previous studies [6, 7], we presented a physical sound constitutive relationship of cortical bone. Empirical determination of the correlation between the bulk and ultra/microstructural behavior of bone was utilized to derive the constitutive model, accounting for the effects of damage accumulation, plastic deformation and viscoelastic properties. We firstly develop the constitutive model and verify it by comparing the intrinsic material properties of bone predicted using the model with those determined from experiments. The accuracy of the new constitutive model demonstrate it can be used to numerically simulate the cortical bone behavior under compression, and some material properties can be determined from the model.

2 Experimental Observations

The experiment methods of progressive cyclic loading [6–9] was used to examine the non-linear stress-strain behavior of cortical bone in compression. The human bone specimens were divided into three orientation groups: longitudinal, radial, and circumferential groups. All specimens were tested in compression, according to the load-dwell-unload-dwell-reload protocol (Fig. 1). Mechanical properties of bone were obtained at each incremental strain level.

For each loading cycle, the instantaneous modulus (E_i) was determined as the tangent of the data between the end of the stress relaxation dwelling before unloading and the end of anelastic deformation dwelling after unloading. The relationship between the obtained instantaneous modulus and the applied strain was shown in Fig. 2.

A decayed exponential model was used to fit the experimental data, where E_0 is the initial modulus (unloaded bone), ε_i is the applied strain, and m is applied to describe the

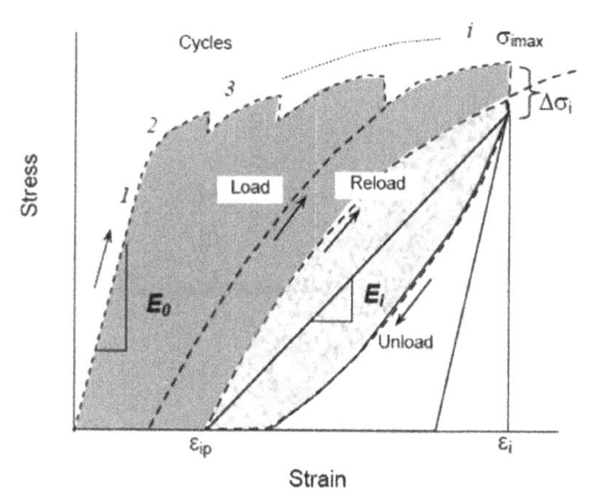

Fig. 1. The protocol of progressive cyclic loading including a load-dwell-unload-dwell-reload scheme was applied in the tests. E_i is the instantaneous modulus at each loading cycle which was determined as the tangent of the line between the ends of stress and the strain relaxation dwells. The plastic strain ε_{ip} is calculated at the end of the strain relaxation dwell. And the instantaneous strain ε_i is the maximum strain applied in each loading cycle.

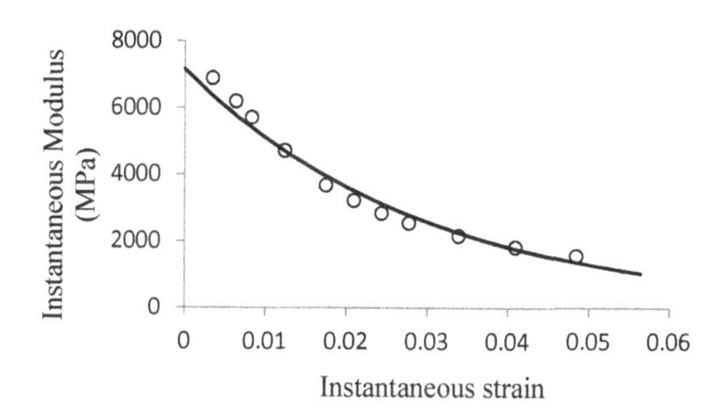

Fig. 2. A decayed exponential model can depict modulus loss with applied strain well.

sensitivity of cortical bone to damage accumulation:

$$E_i = E_0 e^{-m \cdot \varepsilon_i}. \tag{1}$$

The plastic strain ε_p (permanent strain) was defined as the residual strain at the end of anelastic deformation dwelling after unloading. As shown in Fig. 3a, very limited plastic strain appeared at small strain levels. If we represented damage as:

$$D = 1 - E_i / E_0. \tag{2}$$

The power law function was suitable for describing the relationship between plastic strain and damage accumulation (Fig. 3b).

$$\varepsilon_p = k \cdot D^n. \tag{3}$$

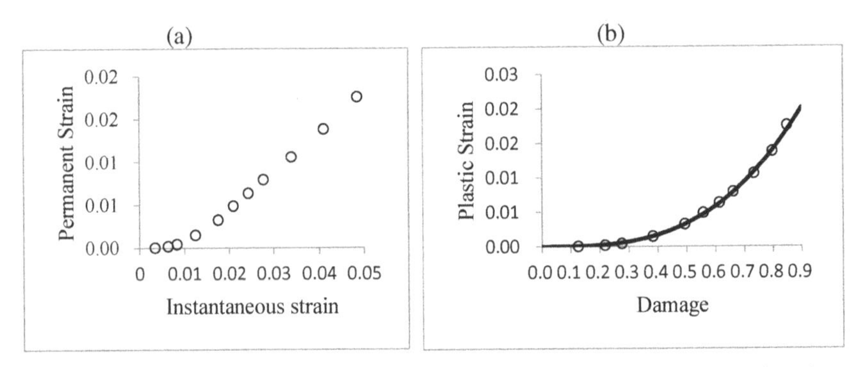

Fig. 3. Very limited plastic strain appeared at small strain level (a). A power law function was suitable for describing the relationship between plastic strain and damage accumulation (b).

$\Delta\sigma_i$ was defined as the relaxation stress, which was calculated by subtracting the maximum stress at each loading cycle and the stress at the end of the stress relaxation dwelling. $\Delta\sigma_i$ could be used to determine the viscoelastic behavior at different damage levels in bone. In the literature, a linear viscoelastic model has been usually applied to describe the viscoelastic response of mineralized biomaterials [10] and bones [3, 11]. In this study, we used the same model, in which $\Delta\sigma_i$ was defined by a convolution integral:

$$\Delta\sigma_i = \int_0^t Y_i(t-\tau)\frac{d\varepsilon(\tau)}{d\tau}d\tau. \tag{4}$$

where Y_i is the relaxation function of the tissue.

3 Constitutive Model Development

Assuming the instantaneous strain (ε_i) applied to bone is composed of elastic (ε_e) and plastic (ε_p) components when bone is the equilibrium state without any viscoelastic responses,

$$\varepsilon_i = \varepsilon_e + \varepsilon_p. \tag{5}$$

the elastic stress could be expressed as follows by incorporating Eq. (1), Eq. (2) and Eq. (3) into Eq. (5):

$$\sigma_{ei} = E_i\varepsilon_e = E_0 e^{-m\cdot\varepsilon_i}[\varepsilon_i - k\left(1 - e^{-m\cdot\varepsilon_i}\right)^n]. \tag{6}$$

Taking into account the viscoelastic response, the constitutive equation of the stress-strain relationship of bone would take the following form:

$$\sigma_i = \sigma_{ei} + \Delta\sigma_i = E_0 e^{-m\cdot\varepsilon_i}\left[\varepsilon_i - k\left(1 - e^{-m\cdot\varepsilon_i}\right)^n\right] + \int_0^t Y(t-\tau)\frac{d\varepsilon(\tau)}{d\tau}d\tau. \tag{7}$$

In fact, a rheological damage model can be used to describe the aforementioned constitutive relationship (Fig. 4). The upper part includes an elastic spring series connected with a plastic pad, while the lower part is a Maxwell body. Two parts were then

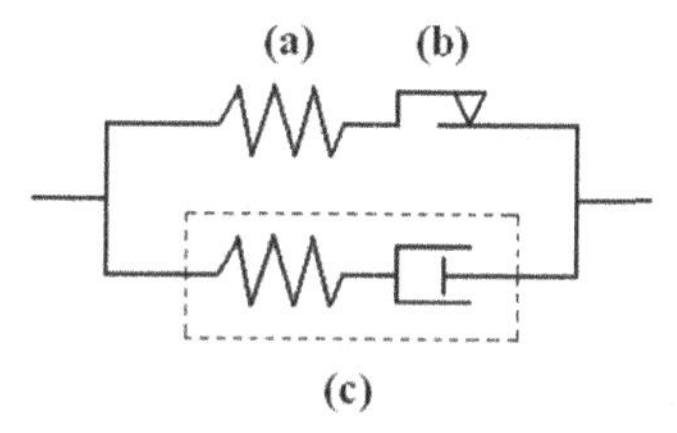

Fig. 4. A rheological damage model was applied to describe the mechanical behavior of cortical bone with microdamage accumulation. Upper part: an elastic spring (a) was series connected with a plastic pad (b); lower part: a Maxwell body (c).

parallel connected. All of these components are dependent on the instantaneous strain (ε_i), which is directly related to the damage density in the tissue.

To define Eq. (7) in a closed form, a specific function of Eq. (4) should be given. Since conventional compression tests are usually controlled by displacement, the instantaneous strain could be assumed as an exponential function of time. Hence, mathematically the derivative of the strain with respect to time still maintain as an exponential function of time (*i.e.* $(e^t)' = e^t$). Thus, the relaxation modulus of the Maxwell body can be expressed as:

$$Y(t) = E^* e^{-t/\lambda}. \tag{8}$$

where, E^* is the modulus of the spring part of the Maxwell body; λ is the viscoelastic time constant, which is usually defined by the elastic constant of the spring part and the viscosity of the dash pot in the Maxwell body (*i.e.* $\lambda = \eta/E^*$). In the previous studies by our group [6, 7] and others [11, 12], the time constant was found changing with damage accumulation. So we defined λ as a function of instantaneous strain. By inserting Eq. (8) into Eq. (4), a simple algebraic form of the convolution integral could be obtained:

$$\Delta\sigma = f(E^*, \lambda) \cdot \varepsilon_i. \tag{9}$$

where $f(E^*, \lambda)$ is a function of spring modulus and time constant in the Maxwell body, which is related to the damage accumulation. Figure 5a showed that $f(E^*, \lambda)$ had a linear relationship with the damage density ($R^2 = 0.896$). Since linear relationship is a special form of power law function when the exponent is unity, the power law function was chosen to define the function as $f(D) = A \cdot (1 - D)^a$. In fact, by curve-fitting the experimental results obtained from the progressive loading tests of bone, it was indicated that the assumption was reasonably accurate (Fig. 5b).

Finally, the specific form of Eq. (7) could be obtained for the uniaxial tests as follows:

$$\sigma = E_0 e^{-m\cdot\varepsilon}\left[\varepsilon - k\left(1 - e^{-m\cdot\varepsilon}\right)^n\right] + Ae^{-a\cdot m\cdot\varepsilon}\varepsilon. \tag{10}$$

4 Experimental Verification

In addition to the progressive tests, we also performed the monotonic tests with similar samples. We fitted the monotonic stress-strain curve to determine the parameters in

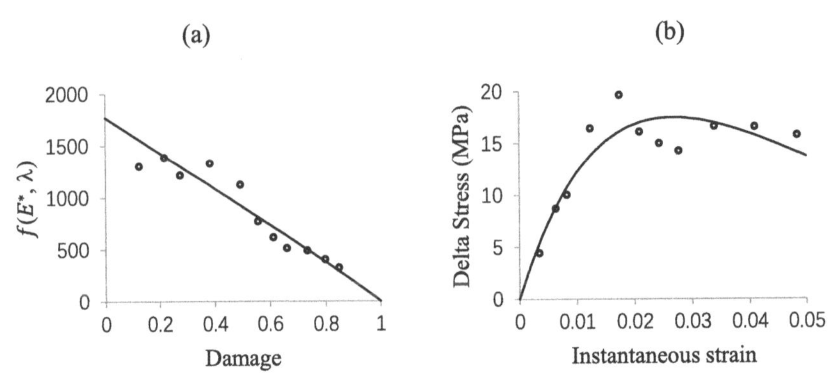

Fig. 5. For a special case in this study, the convolution integral (Eq. 4) had a simple algebraic form ($\Delta\sigma = f(E^*, \lambda) \cdot \varepsilon_i$). The relationship between f and damage density (a) was described by a power law function. The final equation ($\Delta\sigma = Ae^{-a \cdot m \cdot \varepsilon}\varepsilon$) was used to fit the experimental results (b).

Eq. (10), and compared them with the results obtained from the progressive cyclic loading tests using a Student's **t**-test. All calculations were done by a custom MATLAB script.

The nonlinear regression of the instantaneous strain versus the instantaneous modulus was done by a nonlinear least square method. The result demonstrated the decayed exponential relationship of $E_i = 7780 \cdot e^{-39.01} \cdot \varepsilon_i$ (Fig. 2). And the plastic strain indeed had a power law relationship with damage accumulation: $\varepsilon_p = 0.0287 \cdot D^{3.105}$ (Fig. 3b). Finally, the data obtained from the monotonic tests of human cortical bone were fitted by the constitutive model shown by Eq. (10). One typical result of the fitted-curve is illustrated in Fig. 6 (solid line as curve-fitted), which is consistent with the experimental results.

A two-tail Student's t-test was applied to compare the results of the monotonic tests to the results of progressive tests (Fig. 7). The significance was determined to be $P < 0.05$. We can find that there was no significant differences between the predicted results

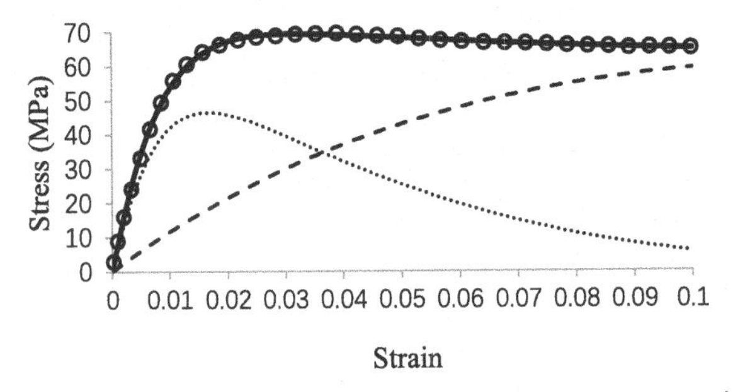

Fig. 6. The developed constitutive model (solid line) is consistent with the experimental data (open cycles). Interestingly, the pure elastic stress (dotted line) dominates the mechanical response in the early stage, while the inelastic stress (dashed line) controls the post-yield behavior.

and experimental data for all material properties. Furthermore, the results also showed the anisotropy of the bone properties which is our expectation.

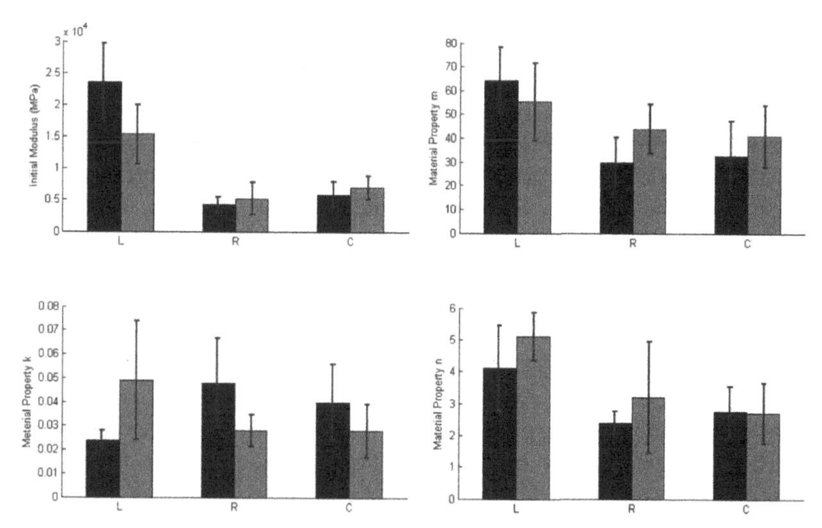

Fig. 7. To verify the constitutive model, the parameters obtained from the proposed equation were compared to the predicted material properties using progressive loading tests. There is no significant difference between the simulation results and experimental observations. Besides, the proposed constitutive model also can reveal the anisotropic elastic modulus of bone.

5 Discussion

The novel progressive loading method provides us a deep insight into the mechanism how damage accumulation influence the mechanical behavior of bone. From the strain point of view, we developed a constitutive model for bone damage, accounting for the effects of damage accumulation, plastic deformation and viscoelastic response. Damage accumulation plays a significant role in degradation of bone mechanical properties, such as toughness, strength, stiffness, loss factor [7, 8, 12–14]. The proposed model takes damage accumulation as the most important factor. It not only influences the elastic modulus, but also affects the plastic strain and viscoelastic response.

Excellent agreement for each monotonic loading test was obtained using the developed constitutive model (Fig. 6, solid line). If we calculated the averaged error square (divided the total error squares by the number of recorded data), the differences between experimental data and the simulations are less than 3 MPa of stress.

The parameters obtained from the proposed model were compared to the predicted material properties using progressive loading tests (Fig. 7). No significant differences were appeared. The initial modulus of longitudinal direction ware a little underestimated, but the results of another two directions were nearly the same. As expected, the newly constitutive model also can reveal the anisotropic elastic modulus of bone. The initial modulus of longitudinal direction were significant larger than those of another two directions. The material parameter m relates to the damage accumulation. The increasing of m implies faster accumulation of damage, and quicker loss of stiffness. There is no significant relationship between E_0 and m, indicating that damage accumulation may relate to the structural properties of bone, such as bone porosity, the interface between mineral and collagen. Both k and n show no significant differences among the three directions of bone. This may suggest that the mechanisms of plastic deformation in the three directions are the same. We had not compared the results of viscoelastic parameters, based on the reason that it is only time-dependent in progressive loading tests, while the viscoelastic response is time- and history-dependent in the monotonic loading tests.

The interesting result from our model is that when we decomposed the stress-strain curve in the monotonic tests into two components (Fig. 6): the pure elastic part (dotted line) and the viscoelastic response (dashed line), the findings may shed light on the underlying mechanism of bone's mechanical behavior. In the early stage, the pure elastic stress dominates the material response, whereas the viscoelastic effects are not remarkable. These results are consistent with those reported in other studies [4, 15]. With increasing strain, the pure elastic stress is increased to the peak value quickly, and then is decreased as the damage accumulates. This phenomenon is similar with our other studies that suggest that in the beginning, the elastic properties sustain the most of the loading on bone. After the peak of elastic stress, bone structures are destroyed gradually with damage accumulation. And the pure elastic stress plays only a second role. Besides, the pure elastic stress peak appears in the strain level of 0.8% to 1.5%, which may imply the yield strain level. On the other hand, the viscoelastic response increases continuously with damage accumulation, which has also observed by the other researches [7, 11, 15]. However, using the linear viscoelastic model to address the same problem described in this study would overestimate the viscoelastic effect [11]. Due to the perfect plastic theory we used, while in reality, the plastic deformation has contribution to the total stress, thus, the viscoelastic portion of stress in our model may include the stress produced by plastic deformation. It may be more exactly if we call the second part of stress as inelastic stress which includes viscoelastic and plastic responses. The inelastic behavior is most likely the consequence of damage formation in bone. In the present model, the inelastic stress governs the post-yield behavior of bone when damage induced plastic and viscoelastic responses dominate the mechanical behavior of cortical bone. Thus, it can be supposed that the most part of energy dissipated in post-yield stage are plastic strain energy and viscous energy rather than elastic energy.

Garcia et al. [16] also developed a 1-D constitutive law for cortical bone based on elastic plastic damage to describe the damage accumulation under tensile or compressive overloading. However, their rheological model was similar with the one proposed by Fondrk et al. which was suitable to model the behaviors of cortical bone under tension. Therefore, using the constitutive equations in [16] to depict the mechanical behaviors of cortical bone under compression was not as well as we did in this study (Fig. 8). In addition, most of the constants in the constitutive law proposed by Garcia had no physical meanings.

In this study, a generalized constitutive law for cortical bone in compression is presented. For numerical computation, we need to define the viscoelastic response of bone in the second part of Eq. (7). Equation (10) is a special case that the strain is an exponential function of time. As the linear viscoelastic model may be not suitable for the post-yield behavior, we have some alternative options, such as a linear combination of two KWW functions, or a linear combination of a Debye function and a KWW function, but it will become more complicated, and take much more time to do curve fitting.

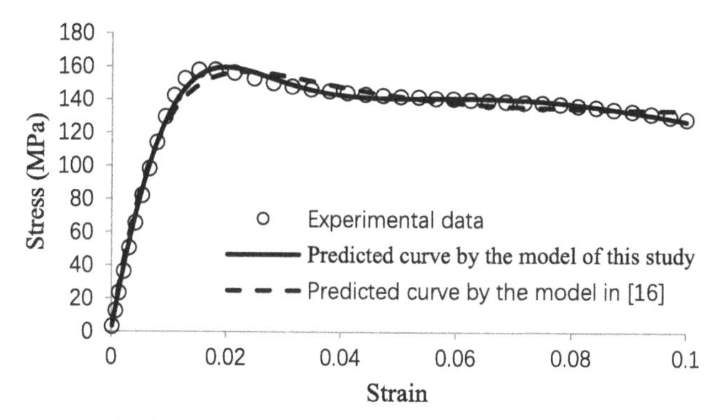

Fig. 8. Using two constitutive models to predict the experimental stress-strain history for monotonic compressive test. The model of our study (solid line) fitted the experimental data better than that proposed by Garcia (dashed line).

6 Conclusion

A nonlinear constitutive model of cortical bone was developed considering the effects of damage accumulation, plastic deformation and viscoelastic response. Based on the test data of bone specimens, the developed model can predict material properties which are in consistent with the results from the progressive loading tests. The constitutive model can be applied to study the mechanical behavior of cortical bone numerically in the future. In addition, the effects of elastic and inelastic behaviors are uncoupled in our constitutive model, so we can better understand the post-yield behavior of cortical bone.

Acknowledgements. This work was supported by National Natural Science Foundation of China (No. 11872074).

References

1. Mullins, L.P., Bruzzi, M.S., McHugh, P.E.: Calibration of a constitutive model for the post-yield behaviour of cortical bone. J. Mech. Behav. Biomed. Mater. **2**, 460–470 (2009)
2. Pawlikowski, M., Klasztorny, M., Skalski, K.: Studies on constitutive equation that models bone tissue. Acta Bioeng. Biomech. **10**(4), 39–47 (2008)
3. Natali, A.N., Carniel, E.L., Pavan, P.G.: Constitutive modelling of inelastic behaviour of cortical bone. Med. Eng. Phys. **30**, 905–912 (2008)

4. Fondrk, M.T., Bahniuk, E.H., Davy, D.T.: A damage model for nonlinear tensile behaviour of cortical bone. J. Biomech. Eng.-Trans. ASME **121**(5), 533–541 (1999)
5. Krajcinovic, D., Trafimow, J., Sumarac, D.: Simple constitutive model for a cortical bone. Biomechanics **20**(8), 779–784 (1987)
6. Nyman, J.S., Leng, H.J., Dong, X.N., Wang, X.D.: Differences in the mechanical behaviour of cortical bone between compression and tension when subjected to progressive loading. J. Mech. Behav. Biomed. Mater. **2**(6), 613–619 (2009)
7. Leng, H.J., Dong, X.N., Wang, X.D.: Progressive post-yield behaviour of human cortical bone in compression for middle-aged and elderly groups. Biomechanics **42**, 491–497 (2009)
8. Nyman, J.S., et al.: Age-Related factors affecting the postyield energy dissipation of human cortical bone. J. Orthop. Res. **25**(5), 646–655 (2007)
9. Wang, X.D., Nyman, J.S.: A novel approach to assess post-yield energy dissipation of bone in tension. Biomechanics **40**, 674–677 (2007)
10. Kim, Y.R., Allen, D.H., Seidel, G.D.: Damage-induced modeling of elastic-viscoelastic randomly oriented particulate composites. J. Eng. Mater. Technol.-Trans. ASME **128**(1), 18–27 (2006)
11. Bredbenner, T.L., Davy, D.T.: The effect of damage on the viscoelastic behaviour of human vertebral trabecular bone. J. Biomech. Eng.-Trans. ASME **128**(4), 473–480 (2006)
12. Yeni, Y.N., et al.: The effect of yield damage on the viscoelastic properties of cortical bone tissue as measured by dynamic mechanical analysis. J. Biomed. Mater. Res. **82A**(3), 530–537 (2007)
13. Burr, D.B., et al.: Does microdamage accumulation affect the mechanical properties of bone. Biomechanics **31**(4), 337–345 (1998)
14. Zioupos, P.: Accumulation of in-vivo fatigue microdamage and its relation to biomechanical properties in ageing human cortical bone. J. Microsc. **201**, 270–278 (2001)
15. Fondrk, M., Bahniuk, E., Davy, D.T., Michaels, C.: Some viscoplastic characteristics of bovine and human cortical bone. Biomechanics **21**(8), 623–630 (1988)
16. Garcia, D., Zysset, P.K., Charlebois, M., Curnier, A.: A 1D elastic plastic damage constitutive law for bone tissue. Arch. Appl. Mech. **80**, 543–555 (2010)

Pedestrian Detection Based on Improved YOLOv3 Algorithm

Ao Li, Xiuxiang Gao, and Chengming Qu(✉)

Key Laboratory of Electric Drive and Control of Anhui Higher Education Institutes, Anhui Polytechnic University, Beijing Middle Road, Wuhu, Anhui, China

Abstract. Pedestrian detection in security surveillance video is a hot spot in the field of AI (Artificial Intelligence) algorithms. Aiming at the influence of background information in different environments and the difficulty of small target detection, the HOG (Histogram of Oriented Gradient) feature extraction method of the improved algorithm is applicate as the preprocessing layer of YOLOv3 feature extraction part, which can highlight the pedestrian contour features, especially the small target pedestrian features, and reduce the interference caused by background information on the detection results. During training, the different data sets are mixed as inputs to increase the diversity of the sample data. Hybrid data is also used to validate the generalization capability of the algorithm during the final test phase.

Keywords: Pedestrian detection · Feature extraction · Yolov3 · K-means clustering

1 Introduction

Under the influence of artificial intelligence and deep learning, computer vision and machine vision technology have been widely used. Whether in the field of unmanned driving or intelligent monitoring, the application of pedestrian detection technology has become more important, and the accuracy of target detection has also become a problem studied by many scholars. Especially the accidents of unmanned driving remind us of the importance of pedestrian detection research.

Whether using machine learning or deep learning algorithms, the first task of detection is to extract the feature information [1] of a target in an image or video before making a judgment. Guolong Gan et al. A pedestrian detection algorithm combining HOG with LBP (Local Binary Pattern) algorithm [7]. It features an image extraction using a block that can be changed in size. J. Redmon et al. proposed the YOLO (You Only Once Look) network structure [10] in 2016. The idea of YOLO algorithm is to use regression method to realize detection and classification. The YOLOv3 algorithm is a combination of advanced methods to overcome short board (not good at detecting small target objects, etc.) in the YOLO series.

In this paper, the improved HOG feature extraction method is used as the pretreatment layer of YOLOv3 to highlight the features of small target pedestrians and reduce the

M. Fei et al. (Eds.): LSMS 2021/ICSEE 2021, CCIS 1467, pp. 221–231, 2021.
https://doi.org/10.1007/978-981-16-7207-1_22

interference caused by background information to the detection results. It can have a good detection effect on the pedestrians who bend over and greatly incline. Then, K-value clustering results and IOU (Intersection over Union) scores are used as evaluation criteria in combination with the given K-means data set. Finally, the experimental environment is built, and the improved YOLOv3 algorithm is trained, tested and analyzed with mixed data sets.

2 Network Structure of 2 YOLOv3 Algorithms

2.1 Overall Network Structure

The improvement of the YOLOv3 algorithm compared to the YOLOv1 algorithm and the YOLOv2 algorithm is the introduction of multi-scale predictions and better classification networks. Multi-scale prediction solves the problem of small target leaks. The classification network uses logistics loss instead of softmax loss in YOLOv2 algorithm. Good classification results can be achieved when the predicted target category is more complex (Fig. 1).

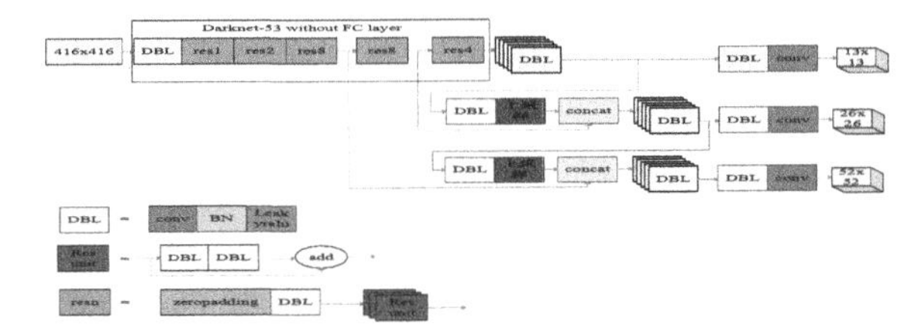

Fig. 1. Network structure of YOLOv3 algorithm.

The YOLOv3 algorithm divides an image with an input size of 416 × 416 into 13 × 13 cells. If the center coordinate of a target is located in a cell, the cell will be used to detect the target. Each cell needs to predict 3 × 3 bounding boxes. The prediction of bounding box coordinate values is consistent with the method adopted by YOLOv2 algorithm.

YOLOv3 algorithm adopts multi-scale prediction method considering different sizes of targets in the prediction stage. There is a passthrough layer in the YOLOv2 algorithm. If the scale of the finally obtained feature map is 13 * 13, this layer can connect the previously obtained 26 * 26 feature map with the 13 * 13 feature map of this layer. The YOLOv3 algorithm adopts the methods of up-sampling and scale fusion similar to those in FPN (Feature Pyramid Network). The three scales of fusion are 13 * 13, 26 * 26 and 52 * 52 respectively. Multi-scale fusion can improve the effect of small target detection.

2.2 Feature Extraction Network Layer

YOLOv3 algorithm extracts feature with a new network structure. The author uses Darknet-19 network in YOLOv2 algorithm. Later, ResNet was combined in this network. And the Darknet-53 network structure used in YOLOv3 algorithm was expanded.

The Darknet-53 network structure is mainly composed of a series of 1×1 and 3×3 convolution layers. Each convolution layer will be followed by a BN (Batch Normalization) and a LeakyReLU layer. There are 53 convolutional layers in the network, so it's called Darknet-53. Each network is trained with the same settings. It is faster in computing speed compared with Darknet-19 network and ResNet network.

2.3 ResNet

ResNet's design idea is to assume that a network layer is involved in the training process. There is an optimal layer in this network layer. We usually design deep-level networks with many redundancy layers. If these redundant layers can realize identity mapping and the input and output passing through this layer are exactly the same, the accuracy of the algorithm will not decrease. Redundancy layers are judged by network training, and these network layers can form a residual block.

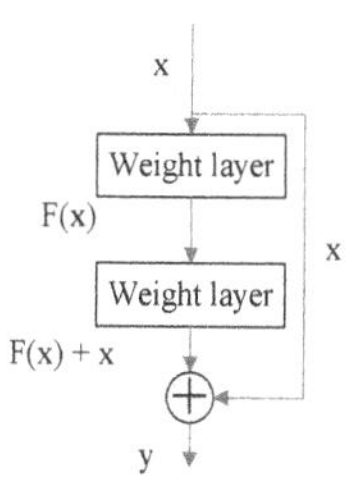

Fig. 2. Residual block structure.

x in the Fig. 2 represents the input of the residual layer, also known as the residual of $F(x)$. The first layer is linearly transformed and activated to obtain $F(x)$. $F(x)$ add x after the second level linear transformation. Curves are the connection mode of "cutting corners". The final output is $y = F(x) + x$. If the layer is redundant, an identity mapping can be realized if $F(x)$ is 0.

3 YOLOv3 Algorithm Adds Preprocessing Network Layer

3.1 Improved HOG Algorithm

The improved idea of HOG is to divide the scanning mode by circle [11]. And RGT (Radial Gradient Transform) is used to transform the gradient of pixels on the circle. Then the feature vector that keeps rotation invariant can be obtained (Figs. 3 and 4).

Each ring represents a cell, and w is the width of the ring. Starting from the innermost circle, scan from inside to outside by stepping one ring at a time. Rotation invariant eigenvectors can be obtained by RGT transform.

g represents the gradient of m points. **r** stands for radial unit vector. **t** represents a tangent unit vector. Then, by projecting **g** in the direction of **r** and **t** respectively, we can get the base vector of **r** and **t** gr, gt.

$$G_{RGT} = \sqrt{(gr)^2 + (gt)^2}. \tag{1}$$

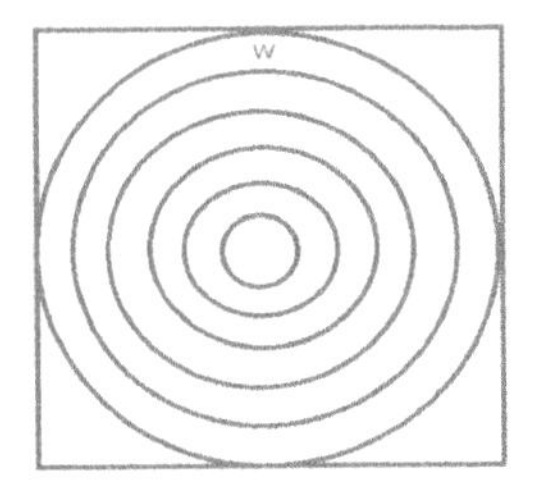

Fig. 3. Circular division.

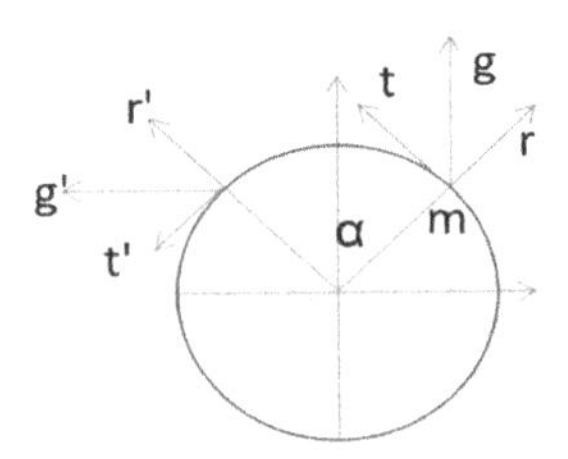

Fig. 4. RGT transform.

$$\alpha_{RGT} = \tan^{-1}\left(\frac{gt}{gr}\right). \tag{2}$$

The RGT transform is usually not sought for each point in a circular region. Divide the circular areas into sectors [12] to simplify the calculation. Each sector is 45°. Eight sector areas are represented by R [i] (i = 1… 8). Quantization of each sector pixel. The radial and tangent unit vectors of R [i] pixels are represented by r [i] and t [i]. For 416 × 416 images, 26 rings are obtained if the selected ring width (w) is 8. A circular scanning method is adopted. Then 234-dimensional feature vectors can be obtained after scanning. Obviously, the obtained feature vector can't express the information of the image well, but loses a certain amount of information. This paper proposes a method to optimize the processing for the amplitude adjustment of histogram. This method enhances the representation ability of the image by adjusting the histogram. The weight adjustment method adopted in this paper is different from that in the literature [13]. First, calculate the average value of the amplitudes of all bin intervals μ, take the maximum value of the histogram max and minimum min after making a difference, divide by μ The characteristic description parameters of histogram are obtained α. The larger a is, the better the representation ability of the image descriptor is. If α is less than 1, it indicates that the amplitude of gradient histogram is relatively concentrated and the image representation ability is not strong. At this time, take a reciprocal and bring it into the following formula for calculation.

$$\mu = \frac{\sum_{i=1}^{bin} f(x, y)}{bin}. \tag{3}$$

$$\alpha = \frac{\max - \min}{\mu}. \tag{4}$$

$$F(x, y) = \begin{cases} f(x, y) * \alpha, & (f(x, y) > \mu) \\ f(x, y), & (f(x, y) = \mu) \\ f(x, y) * \frac{1}{\alpha}, & (f(x, y) < \mu) \end{cases} \tag{5}$$

Via the above formula, the gradient histogram is adjusted for weights, $F(x, y)$ representing the amplitude of the interval after weight optimization.

3.2 Feature Extraction Base on Improved HOG and DarkNet-53

The feature extraction layer of YOLOv3 algorithm adopts Darknet-53 network structure. In the feature extraction stage, if 416 × 416 three-channel images are directly input to Darknet-53 network layer, the adjustment of parameters in the network training process will be affected. Because an image removes pedestrians, the background information will also interfere with online learning, which will eventually affect the accuracy of pedestrian detection. A network preprocessing layer is added before the network extraction layer to improve the accuracy of pedestrian detection and reduce the difficulty of sample learning. The improved HOG algorithm is used as the preprocessing layer to preprocess the input image, which can highlight the contour information of pedestrians in the image (Fig. 5).

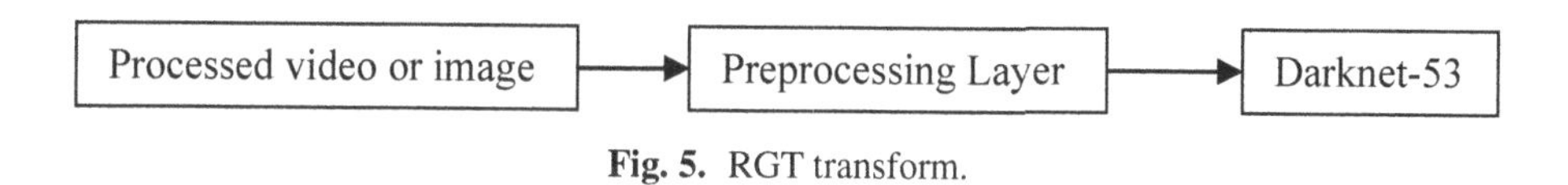

Fig. 5. RGT transform.

4 Dimension Clustering of Target Box in YOLOv3 Algorithm

Clustering algorithm, as its name implies, is to classify data [14]. In YOLOv3 algorithm, the author extends the K-means clustering algorithm in YOLOv2 algorithm.

$$E = \sum_{i=1}^{k} \sum_{x \in C_i} \|x - \mu_i\|^2. \tag{6}$$

C_i represents the category, μ_i is the mean vector of C_i, calculated as follows:

$$\mu_i = \frac{1}{|C_i|} \sum_{x \in C_i} x. \tag{7}$$

The wrong classification is not corrected by supervision algorithm in the whole process of K-means algorithm clustering. But the K value is set at the beginning to indicate how many categories to be divided. The whole classification process is completely completed by the algorithm itself, which reflects the unsupervised idea (Fig. 6).

Give K-means a data set and set K = 5 and K = 7 respectively to obtain the clustering results of the Fig. 7. In YOLOv3 algorithm, K-means algorithm is used to cluster labeled data sets. IOU score is still used as the evaluation standard. In the configuration file of YOLOv3 algorithm, we can see that the author uses 9 prior boxes, which are (10, 13), (16, 30), (33, 23), (30, 61), (62, 45), (59, 119), (116, 90), (156, 198), (373, 326). These 9 prior boxes are the results of clustering through VOC data sets.

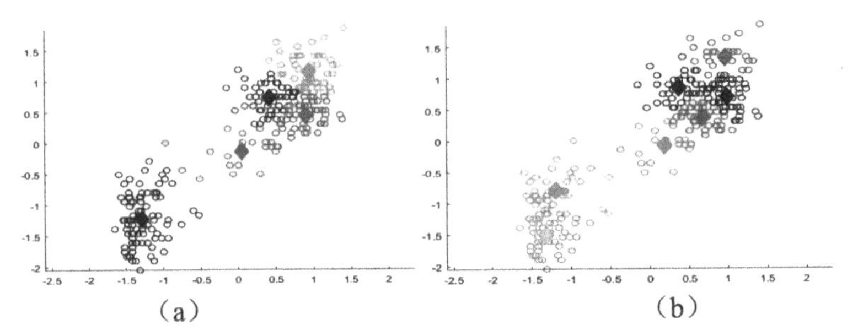

Fig. 6. (a) K = 5 Clustering, (b) K = 7 Clustering.

5 Design Experiment

The previous article highlights the target information by adding a preprocessing layer to YOLOv3 algorithm. Next, experiments are carried out to verify the real effect of the algorithm. The experimental environment of the algorithm is in Ubuntu 16.04 operating system. the CPU is Intel 8700, the graphics card is 1080Ti, and the memory is 16G.

In the training stage of this paper, the data collected by the mixed security platform and INRIA data set are applicated as training data sets. The trained algorithm is more robust by mixing data sets. Experimental comparisons will be made on INRIA data sets and private data sets respectively in the final testing phase.

The data set should be labeled first before network training. LabelImg is the annotation tool. LabelImg can be used to mark the location information of the target pedestrian on the image. And an xml format file named label.xml is generated for each picture to represent the location of the target box (x, y, w, h). 2400 pictures were collected from the security platform. 2200 pictures were labeled, of which 2000 were used as training sets, 200 were used as verification, and the remaining 200 unlabeled pictures were used as test sets. 700 pictures were selected from INRIA data set, and 600 pictures were labeled, of which 500 were used for training, 100 were used for verification, and 100 unlabeled pictures were used for testing (Fig. 8).

```
gaoxiux@gaoxiux-virtual-machine:~$ git clone https://github.com/tzutalin/labelI
mg
正克隆到 'labelImg'...
remote: Enumerating objects: 4, done.
remote: Counting objects: 100% (4/4), done.
remote: Compressing objects: 100% (4/4), done.
```

Fig. 7. Installing labelImg under Ubuntu.

When labelImg installation is completed, the data can be labeled, as shown in the Fig. 9.

It can be seen that data annotation is to find out the target area in the image and set tag information. When the data annotation is completed, an xml file is obtained, which contains each image information, as shown in the Fig. 10.

The file names need to be modified in batches according to the naming rules. Since YOLOv3 algorithm uses a configuration file in txt format, it is necessary to convert xml format into txt format via Python script. Darknet provides the script on its official

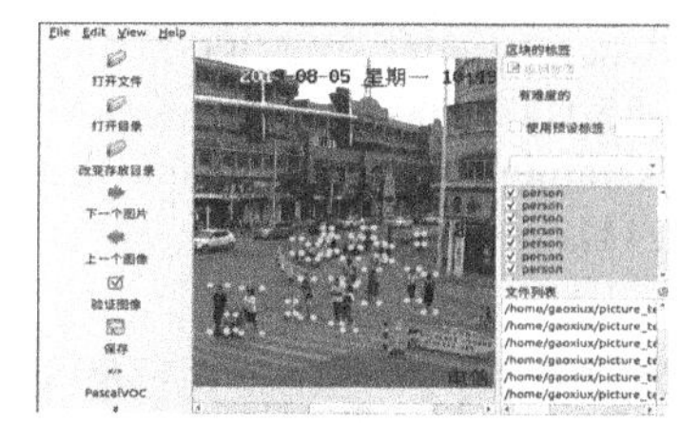

Fig. 8. Data annotation.

Fig. 9. Data annotation generation xml file.

website, which can be used directly after modifying the category information. The data preprocessing task has been completed by now, followed by network training.

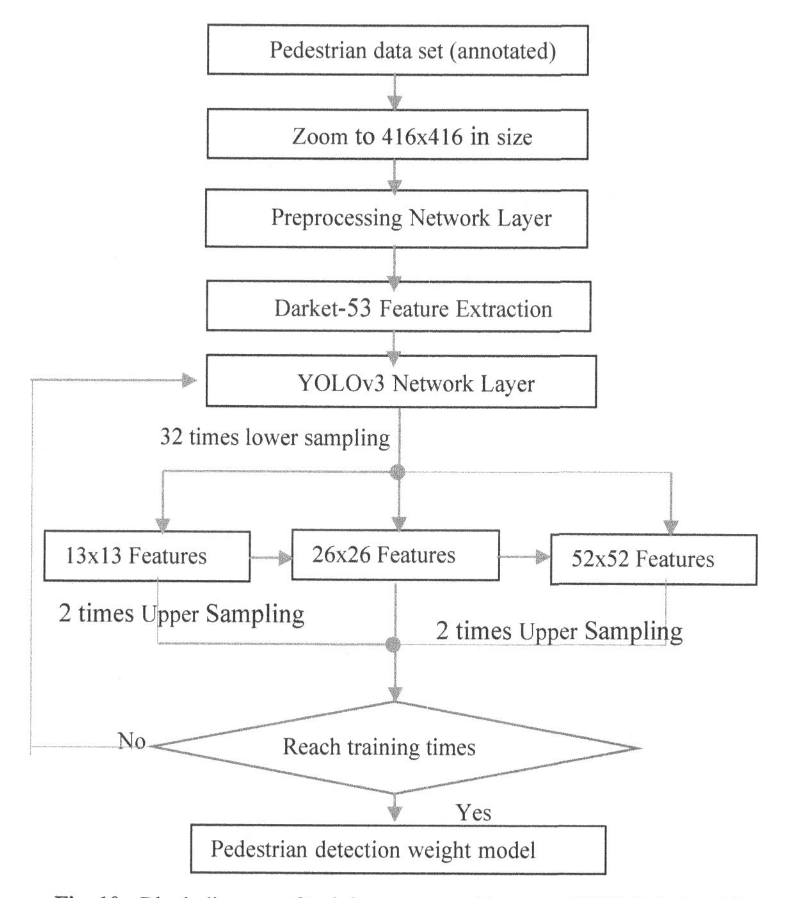

Fig. 10. Block diagram of training process of improved YOLOv3 algorithm.

Relevant parameters need to be set before training. The learning rate set for this training is 0.001, and it will be set to 0.0001 when the number of iterations reaches

20,000. The learning rate is reduced to 0.00001 when iterating over 30000 times. The loss function is further converged, with a total of 50,000 iterative trainings. After the algorithm training, the model file generated by this training can be loaded into the model file during the test to detect the sample data (Fig. 11).

```
Region 82 Avg IOU: 0.012338, Class: 0.333777, Obj: 0.897789, No Obj: 0.486502, .5R: 0.000000, .75R: 0
Region 94 Avg IOU: -nan, Class: -nan, Obj: -nan, No Obj: 0.515558, .5R: -nan, .75R: -nan,  count: 0
Region 106 Avg IOU: -nan, Class: -nan, Obj: -nan, No Obj: 0.478980, .5R: -nan, .75R: -nan,  count: 0
18: 1140.209717, 1546.560913 avg, 0.000000 rate, 0.079101 seconds, 18 images
Loaded: 0.000030 seconds
Region 82 Avg IOU: 0.001488, Class: 0.850692, Obj: 0.251911, No Obj: 0.490347, .5R: 0.000000, .75R: 0
Region 94 Avg IOU: -nan, Class: -nan, Obj: -nan, No Obj: 0.513251, .5R: -nan, .75R: -nan,  count: 0
Region 106 Avg IOU: -nan, Class: -nan, Obj: -nan, No Obj: 0.482616, .5R: -nan, .75R: -nan,  count: 0
```

Fig. 11. Partial parameter information of training process.

The first parameter Region at the beginning of each line represents the region parameter in the Fig. 12. Avg IOU represents the degree to which the predicted bounding box coincides with the annotated bounding box. If the coincidence degree is greater, the predicted bounding box is more accurate. Class represents the classification accuracy of the label target. obj is the output confidence level, the corresponding No obj is the probability that there is no target.5R represents the recall rate when IOU is 0.5. .75R represents the recall rate when IOU is 0.75. The curve of the loss function obtained after training is as shown in the Fig. 13.

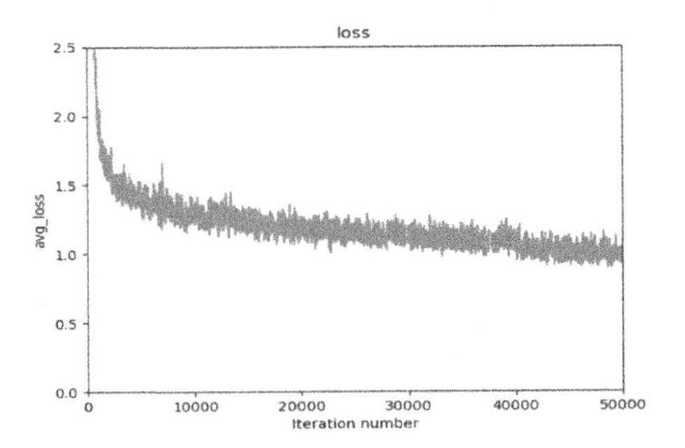

Fig. 12. Loss value function curve.

6 Analysis of Experimental Results

The network model is tested by using the test data set of security platform and INRIA test data set. The final test results are shown in the Fig. 14, 15 and 16.

From the detection results in the Fig. 14, 15 and 16, it can be directly seen that when the image to be detected contains a large amount of background information and the proportion of pedestrians relative to the whole image is small, redundant feature information is relatively large in the process of feature extraction. Pedestrian characteristics can be highlighted and the detection effect of the algorithm can be improved via adding preprocessing layer. The data sets obtained from real-time monitoring come

Fig. 13. INRIA dataset test comparison.

Fig. 14. YOLOv3 algorithm.

Fig. 15. Improved YOLOv3 algorithm.

from different equipment types and different shooting angles. Therefore, there are also differences in image clarity. These factors will bring challenges to the detection results of the algorithm. The above test uses picture data. Next, a monitoring real-time video data will be used for comparative test. The video scene is a shopping mall, with a video duration of one hour and eight minutes and a total of 375 targets. The test results are shown in the Fig. 17, 18, and the final results pass the accuracy rate (P) and recall rate (R) to compare the two evaluation indexes.

Fig. 16. YOLOv3 algorithm.

Fig. 17. Improved YOLOv3 algorithm.

It can be seen that the accuracy and recall rate of the improved algorithm have been improved via video testing. Moreover, the confidence obtained by the improved algorithm is greater than that of the original algorithm for the same detected target.

Through testing different data sets, it can be seen that this improvement has achieved certain results in detection accuracy. But the disadvantage is that the preprocessing layer increases the complexity of the algorithm, and then reduces the real-time performance of the algorithm.

Table 1. Comparison of video test results.

Algorithm	X_{TP}	X_{TN}	X_{FP}	P (%)	R (%)
YOLOv3	327	48	52	86.3	87.2
Improved YOLOv3	338	37	39	89.7	90.1

7 Conclusions

The YOLOv3 algorithm is analyzed and improved to realize real-time pedestrian detection in surveillance video. The feature extraction part of the algorithm is Darknet-53 network structure. The idea of ResNet is used for reference in Darknet-53 network structure. This makes YOLOv3 algorithm perform better in feature extraction than the previous two generations of algorithms. For the better performance of YOLOv3 algorithm in detection accuracy, the improved HOG algorithm is used as the preprocessing layer of YOLOv3 algorithm, which can highlight pedestrian features in the image.

References

1. Li, W., Wang, P.J., Song, H.Y.: Survey on pedestrian detection based on statistical classification. J. Syst. Simul. **28**(09), 2186–2194 (2016)
2. Wren, C.R., Azarbayejani, A., Darrell, T., Pentland, A.P.: Real-time tracking of the human body. Pattern Anal. Mach. Intell. **19**(7), 780–785 (1997)
3. Papageorgiou, C., Poggio, T.: A trainable system for object detection. Int. J. Comput. Vis. **38**(1), 15–33 (2000)
4. Viola, P., Jones, M.: Robust real-time object detection. Int. J. Comput. Vis. **57**(2), 137–154 (2004)
5. Dollar, P., Perona, P., Belongie, S.: Integral channel features. In: International Conference on Pervasive Computing, Manchester, UK, 4–9 May 2009, pp. 2–13 (2009)
6. Wu, J., Rehg, J.M.: CENTRIST: a visual descriptor for scene categorization. IEEE Trans. Pattern Anal. Mach. Intell. **33**(8), 1489–1501 (2011)
7. Gan, G.L., Cheng, J.: Pedestrian detection based on HOG-LBP feature. In: IEEE Seventh International Conference on Computational Intelligence and Security, pp. 1184–1187 (2011)
8. Girshick, R.: Fast R-CNN. In: 2015 IEEE International Conference on Computer Vision Santiago, Chile, 7–13 December 2015, pp. 1440–1448 (2015)
9. He, K., Zhang, X., Ren, S.: Deep residual learning for image recognition. In: Computer Vision and Pattern Recognition, pp. 770–778 (2016)
10. Redmon, J., Diwala, S., Girshick, R.: You only look once: unified, real-time object detection. In: 2016 IEEE Conference on Computer Vision and Pattern Recognition, Las Vegas, NV, 27–30 October 2016, pp. 779–788 (2016)

11. Tang, B., Zuo, Z.R.: A rotation-invariant histogram of oriented gradient for image matching [EB/OL]. Science Paper Online, Peking, 24 January 2019
12. Gabriel, T., Vijay, C., Sam, T.: Unified real-time tracking and recognition with rotation-invariant fast features. In: Computer Vision and Pattern Recognition, San Francisco, pp. 934–941. Springer (2010)
13. Bao, X.A., Zhu, X.F., Zhang, N.: Pedestrian detection algorithm based PHOG feature. Comput. Meas. Control **26**(8), 158–167 (2018)
14. Shao, J.J.: Research on clustering algorithm of high dimensional data and its distance metric. Jiang Nan University (2019)
15. Bulat, A., Tzimiropoulos, G.: Human pose estimation via convolutional part heatmap regression. In: Leibe, B., Matas, J., Sebe, N., Welling, M. (eds.) ECCV 2016. LNCS, vol. 9911, pp. 717–732. Springer, Cham (2016). https://doi.org/10.1007/978-3-319-46478-7_44

Improved One-Stage Algorithm with Attention Fusion for Human Sperm Detection Based on Deep Learning

Chuanjiang Li[1], Haozhi Han[1], Ziwei Hu[1], Chongming Zhang[1](✉), and Erlei Zhi[2]

[1] College of Information, Mechanical and Electronic Engineering, Shanghai Normal University, Shanghai 201418, China
czhang@shnu.edu.cn

[2] Department of Andrology, Center for Men's Health, Department of ART, Institute of Urology, Urologic Medical Center, Shanghai General Hospital, Shanghai Key Lab of Reproductive Medicine, Shanghai Jiao Tong University, Shanghai 200080, China

Abstract. The popular methods of human sperm detection are mostly based on machine learning. Although most deep learning methods have got state-of-art performance in object detection with the advancement of deep learning, only a few of them were used successfully in human sperm detection. With the deep learning method, the detection task is divided into different scales in feature maps of different level. The sperm target is so small under the phase-contrast microscope that the detection task is hard even with deep learning technology. Feature Pyramid Network (FPN) was designed to mitigate this problem in one-stage detector. In this paper, we proposed a new FPN based method in which the feature map was fused with attention module to get the correlation information of low-level feature maps. Eventually, we got an AP of 83.77 which is higher than YOLO v4 and some other detectors.

Keywords: Sperm detection · Computer vision · One-stage detection · Attention module · Feature fusion · Feature Pyramid Network

1 Introduction

It is reported by World Health Organization (WHO) that 15–20% of couples were diagnosed with infertility in the world and 50% of the reason was caused by the male [1]. It is quite important to detect the sperm under the microscope for medical staff. Normally, the medical staff in the hospital would observe hundreds of specimens every day. Long time observation would make medical staff feel tired and lead to inaccuracy while analyzing the specimens under the microscope. More than that, in some research, it is reported that the largest gap between the results of detection of different medical staffs is up to 50% [2]. Therefore, it is meaningful to detect sperm with computer vision technology.

With the development of deep learning in the field of computer vision, object detection becomes more accurate than before. Although the speed of a one-stage detector is quicker than a two-stage detector, the accuracy of a one-stage detector is not as well

M. Fei et al. (Eds.): LSMS 2021/ICSEE 2021, CCIS 1467, pp. 232–241, 2021.
https://doi.org/10.1007/978-981-16-7207-1_23

as a two-stage detector. The lack of RPN makes a one-stage detector cannot level up to the two-stage in average precision (AP). With the development of deep learning, the detection becomes more accurate than before. In the deep learning detection field, the method could be categorized into two aspects. The difference between them is whether the detector abstracts the proposal region. The Region Proposal Network (RPN) [3] was used to do this work. But detection rate will become slower with the RPN net. Therefore, In the real environment, the use of a one-stage detector could be more common than a two-stage detector.

In a one-stage detector, normally, we set the pre-defined anchor to detect the location and category of the target and then fun-tune the parameters by back propagation. But the pre-defined anchor was set in some scales, therefore the original scale of the anchor was fixed. The RPN of two-stage detectors can provide the interest of region of target. Therefore, the precision of the one-stage detector is less than the two-stage. But the one-stage detector has natural speed privilege without the RPN, and this makes it more popular in real scene. A lot of works was developed to improve the one-stage detector. Most of the work was concentrated on the backbone and the feature pyramid network (FPN).

The motivation of CNN was based on the idea that different level feature map to collect different level information. In this architecture, the high-level feature would have the restricted capacity of collecting the information of the whole scale. But the attention module is designed to find the correlation of each pixels which can perfectly solve this problem. When high-level feature maps were fused into low-level feature maps, the normal choice was to down sample the high-level feature maps and packed them together. Through this method, the correlation of information of high-resolution feature maps was neglected. Recently, a lot of works were done to dig out more information from the FPN. Most of the work like PAnet [4], LiBra-FPN [5], MLFPN [6], ASFF [7], BI-FPN [8] etc., was trying different connections of different feature layers. The multi-level connection makes the feature maps behave better than before. But, normally, the information of each pixels in the high-level feature maps was neglected. And the correlation in the high-level feature maps was important to enhance the capacity of feature extraction. For example, in the high-level feature maps, the correlation of each pixels might represent the information of the structure of sperms. If we neglect them and just simply add them together, it will lose some important information of the structured features.

Although the development of detector with deep learning is so fast, just a few works were used in sperm detection. For example, the mobile-net [9] and the retina-net [10] were used for sperm detection in these papers, but these works did not introduce the newest algorithm in the field of object detection into sperm detection and also did not analyze the effect of different network architectures.

In the sperm detection mission, the correlation of the feature map might be considered as, for example, the relationship of the head and tail of the sperm. And the relationship between the pixels should pay more attention to that so we can decrease the error caused by the scale of the feature map. The convolution neural network performs local perception at the bottom layer and then integrates the local information at a higher layer to obtain global information. Normally, the sperm was a small target comparing to the whole image. Therefore, the receptive field of the sperm should be small to be suitable for the scale of the sperm. In the low-level feature maps,the information is more abstract than the high-level. So, to improve the capacity of representation of feature maps, we try to use the attention module to make a feature map of the receptive field of sperm behave better.

In this paper, the detector proposed by us considered the correlation of each pixels in the high-level feature maps with the attention module and use it to enhance the feature map. In this method, we not only considered the high-level feature, and the correlation also be added. Then, we might get better performance than YOLO v4, etc. detectors.

2 Materials and Method

2.1 Dataset and Settings

The dataset was created by the use of images cut from 5 different video samples under the phase-contrast microscope. And the label was annotated by an image annotation tool called LabelImg. Finally, we got 4082 annotated images in VOC format.

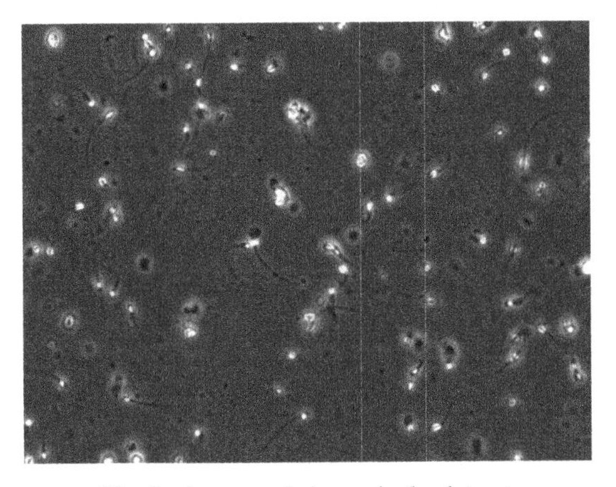

Fig. 1. An example image in the dataset.

The settings of the computer we used for both training and testing are listed in the Table 1.

Table 1. Hardware specification.

Processor	Intel(R) Xeon(R) CPU @ 2.20 GHz
RAM	12 GB
GPU	NVIDIA Tesla T4 15 GB GPU

2.2 The Structure of the Proposed Network

The sperm target should be predicted on a small scale of the receptive field, but the information of the small-scale receptive field is more fragmented. When directly condensing the high-level receptive field information, if you use downsampling to do it directly, it will ignore the information of high-level features like the correlation of feature maps. Therefore, attention module is to obtain high-latitude cross-correlation rows for stacking. One is to compensate for the loss of features when the scale is small, and the other is to bring high-level information.

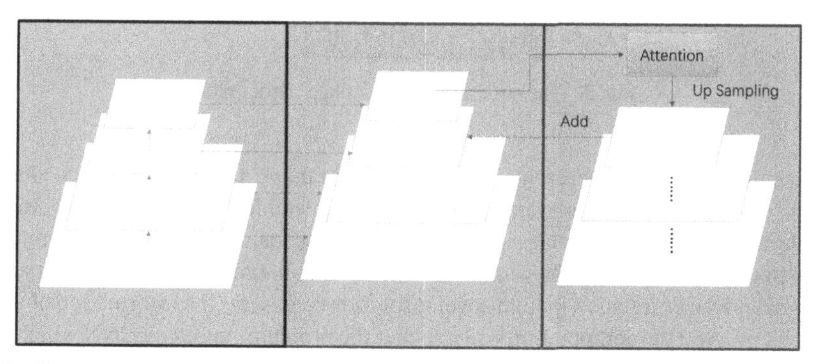

Fig. 2. From top-down, we put an attention module after the feature extraction. This attention module was used to abstract the correlation of each pixels from a higher level, then we add them together to obtain the abundant information of high-level features.

As showed in Fig. 2, the image was first put into a feature abstraction network which we use to abstract different level features. Then the multi-level feature was captured that can be used to detect a target with different scales.

Through the backbone, the feature maps of the input were abstracted. And the Cross Stage Partial darknet53 (CSPdarknet53) [11] was chosen as the feature abstraction network. The reason we use CSPdarknet53 was the shortcut structure connected two different parts which one of the parts can use multiple residual blocks. In this trick, the computation costs will be cheap, and the reason of that is the channel was split into two parts by ratio to enhance the presentation of the gradient. At the same time, the residual block makes the propagation of the gradient more easily. Through the CSPdarknet53, three different feature layers were captured to detect targets with different scales (Fig. 3).

In the feature abstraction network, the Batchnorm was replaced by the cross mini-batch Normalization (CmBatchnorm) [12]. The function of Batchnorm is to fix the problem of the Internal Covariate Shift [13] behavior of the deep network. Through scaling normalization computation, the probability distribution was narrowed in a fixed area and kept the scale and distance covariation by the scale and distance factors. CmBatchnorm was designed to use more batch information to make decision areas not too individually and the use of batch information less redundant.

After the CSPdarknet53, we obtain three different scale layers. The next step is to fuse the feature maps to be suitable for a small target. In the traditional PANet, the information went through upsampling and downsampling two computations and add them together to

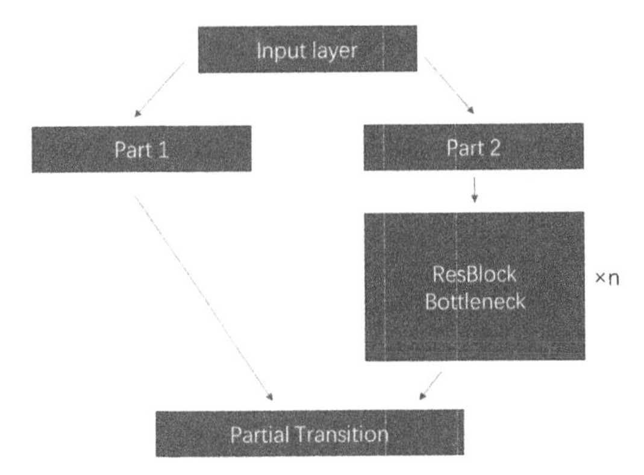

Fig. 3. The structure of cross stage partial block.

enhance the capacity of representation of the feature maps. In this module, the attention module was put into the downsampling layer. As showed in Fig. 1, we use the attention module to abstract the correlation of each high-level feature maps before connecting layers to the next layers. In this computation, the correlation of each feature pixel was captured and the correlation in high-level resolution represents the information of spatial relationship. And the computation will conquer the problem caused by CNN which lacks long-term dependency. The information of the connected layer will more abundant.

2.3 The Abstraction of Correlation of Deep Feature Module

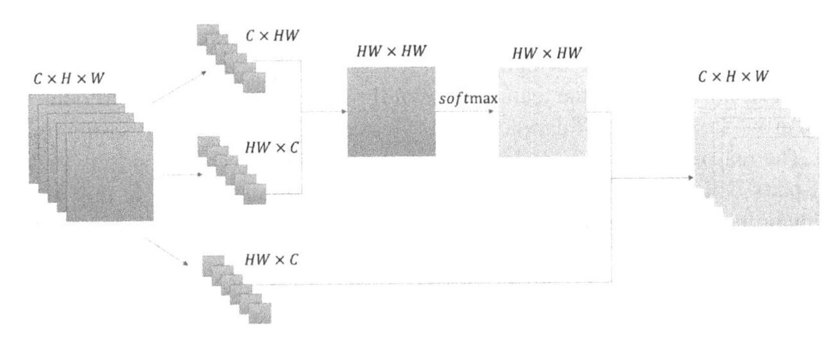

Fig. 4. The nonlocal module we used to abstract the correlation of feature map.

Because the concatenation of big and small resolution feature maps may neglect the correlation of each feature pixel. Therefore, the attention module was used here to obtain the correlation. The attention module we used here is nonlocal [13]. After the feature abstraction network, the fusion of adjacent layers was the next step. Here, the nonlocal module was used to abstract the correlation of each feature pixel. At first, the scale of a batch of a feature map was changed into $C \times HW$ and $HW \times C$ by 1×1 convolution separately. Next, the matrix multiplication was used to generate a $HW \times HW$ matrix that represent the correlation of feature map. Then the attention weight was got by passing the feature map through softmax function. Eventually, the $HW \times C$ feature map was used to multiply the attention weight to get the nonlocal block. The function of the last multiply computation was the same as the residual block which could make the propagation of gradient more convenient. The nonlocal block uses three different functions to obtain the correlation of location in the feature map and the concatenation function makes the information of the layer more abundant and the gradient more easily propagated. And the nonlocal operation abstracted the correlation through the matrix multiplication to generate the feature map which contains $HW \times HW$ pixels that represent the correlation of the feature map. Moreover, the location attention breaks the limit of CNN that only focuses on the receptive field and makes the information more robust (Fig. 4).

2.4 Target Dropout

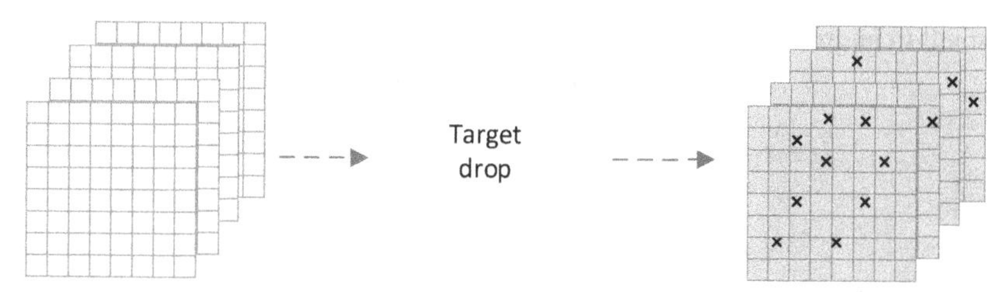

Fig. 5. The target drop module we used to decrease the overfitting effect raised by the nonlocal operation.

In the process of training, the training loss was much lower than the validation loss which means the overfitting. To solve the problem, we introduce the target drop (TDrop) [14] to balance the effect raised by the attention module (Fig. 5). The motivation for that we used the target drop was abstracting the attention and suppressing the attention correlation to decrease the overfitting. The process of TDrop was first to get the channel attention which used average mean to get the channel map. Then it used the convolution to replace the full connection layer to abstract the non-linear feature channel map. After the channel information was represented by average mean pooling, the channel feature map was passed through the relu function to make the channel attention more robust and have more high-level information. Here, instead of using softmax as the channel attention-weighted activation function, we used sigmoid to abstract the channel information to be suitable for our sperm target. And using the attention channel to locate the feature information of what we needed is the next step. Eventually, using the mask block which has certain probability to drop out the correlation of the activated feature pixels. In this way, the overfitting arises from the attention module was decreased.

3 Result

Table 2 showed the performance of Efficientdet d0,YOLO v3,YOLO v4,YOLOv4 with Attention Fusion and YOLOv4 with Attention Fusion and Target Drop. In the experiment,

Table 2. The comparison of performance of Efficientdet, YOLOv3 [15], YOLOv4 [16], and our algorithm (YOLOv4 with attention fusion module and target drop module).

	Backbone	IOU	Recall	Precision	F1	AP
Efficientdet d0	Efficientnet	0.5	50.34	86.93	0.64	74.10
YOLO v3	Darknet53	0.5	53.12	84.11	0.65	72.30
YOLO v4	CSPDarknet53	0.5	71.24	85.15	0.78	82.34
YOLOv4+AF	CSPDarknet53	0.5	**73.01**	**85.33**	**0.79**	**83.46**
YOLOv4+AF+TD	CSPDarknet53	0.5	**75.34**	**84.06**	**0.79**	**83.77**

we can tell from the AP that the proposed algorithm performs better than other algorithms in average precision. And the increase of AP also means that target drop can decrease the overfitting risen by attention module. Through the comparison, the cost of the network was small to improve the AP.

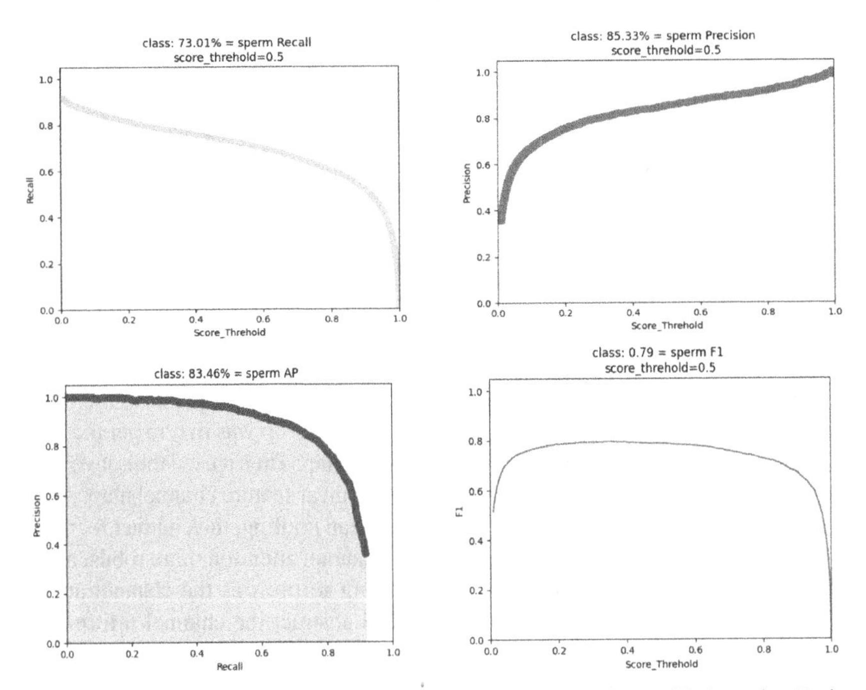

Fig. 6. The Precision, Recall, AP and F1 curve based on our algorithm with Attention Fusion in FPN.

From Fig. 6 and Fig. 7, we can see that the target drop added into the network makes the recall of the algorithm improve from 73.01% to 75.04%. In this way, we decreased the effect of overfitting lifted by the attention module, and the non-linear function in the target drop may help to improve the effectiveness of the feature map. And the result of

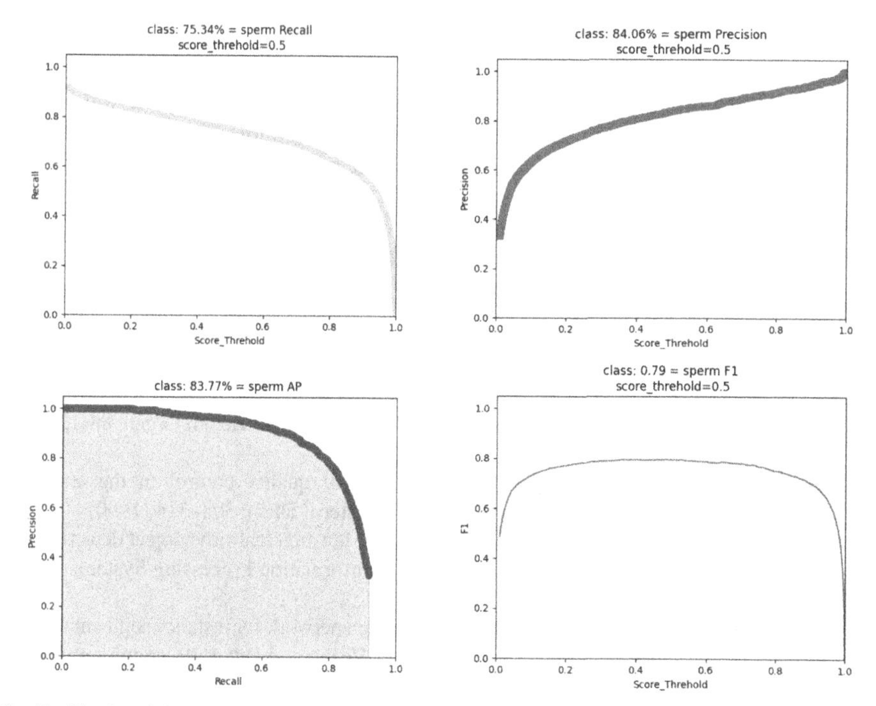

Fig. 7. The Precision, Recall, AP and F1 curve based on our algorithm with Attention Fusion in FPN and Target Drop.

the new method to fuse the attention may improve the performance. And it also turns out that the correlation of the feature map makes the information more abundant. Instead of concatenating the feature map directly, in this connection way, the problem of the scale of sperm target was decreased.

4 Conclusion

The traditional structure of FPN just connected the feature maps simply by either adding them tighter or concatenating them together to fuse the feature maps. Clearly, some information could be neglected by this traditional way. In our opinion, the correlation of high-level feature maps could affect the result of the feature fusion. Therefore, the attention module was added to enhance the performance in the FPN. And, considering the problem of overfitting, we try a new drop out algorithm named Target Dropout. Eventually, a new fusion network for sperm detection was proposed which can fuse the feature map in different scale feature layers by the attention module, and the result turned out to be more accurate than that of some state-of-art detectors. And the algorithm was tested well in the performance under the real condition.

5 Future Work

The future work will mainly focus on the architecture of the feature abstraction network for sperm detection and try to do some analysis of the morphology of the sperm. Other than that, the tracking algorithm will also be considered based on our detector.

Acknowledgments. This work was supported by the medical-engineering cross fund of shanghai jiaotong university (YG2019QNA70).

References

1. Cui, W.: Mother or nothing: the agony of infertility. Bull. World Health Organ. **88**(12), 881–882 (2010)
2. Neuwinger, J., Behre, H.M., Nieschlag, E.: External quality control in the andrology laboratory: an experimental multicenter trial. Fertil. Steril. **54**(2), 308–314 (1990)
3. Ren, S., He, K., Girshick, R., Sun, J.: Faster R-CNN: towards real-time object detection with region proposal networks. In: Advances in Neural Information Processing Systems (NIPS), pp. 91–99. IEEE (2015)
4. Liu, S., Qi, L., Qin, H., Shi, J., Jia, J.: Path aggregation network for instance segmentation. In: Proceedings of the IEEE Conference on Computer Vision and Pattern Recognition (CVPR), pp. 8759–8768. IEEE (2018)
5. Pang, J., Chen, K., Shi, J., Feng, H., Ouyang, W., Lin, D.: Libra R-CNN: towards balanced learning for object detection. In: Proceedings of the IEEE Conference on Computer Vision and Pattern Recognition (CVPR), pp. 821–830. IEEE (2019)
6. Zhao, Q., et al.: M2Det: a single-shot object detector based on multi-level feature pyramid network. In: Proceedings of the AAAI Conference on Artificial Intelligence (AAAI), vol. 33, pp. 9259–9266 (2019)
7. Wang, J., Chen, K., Yang, S., Loy, C.C., Lin, D.: Region proposal by guided anchoring. In: Proceedings of the IEEE Conference on Computer Vision and Pattern Recognition (CVPR), pp. 2965–2974. IEEE (2019)
8. Tan, M., Pang, R., Le, Q.V.: EfficientDet: scalable and efficient object detection. In: Proceedings of the IEEE Conference on Computer Vision and Pattern Recognition (CVPR). IEEE (2020)
9. Ilhan, H.O., Sigirci, I.O., Serbes, G., Aydin, N.: A fully automated hybrid human sperm detection and classification system based on mobile-net and the performance comparison with conventional methods. Med. Biol. Eng. Comput. **58**(5), 1047–1068 (2020). https://doi.org/10.1007/s11517-019-02101-y
10. Rahimzadeh, M., Attar, A.: Sperm detection and tracking in phase-contrast microscopy image sequences using deep learning and modified CSR-DCF. arXiv preprint arXiv:2002.04034 (2020)
11. Wang, C., Mark Liao, H., Wu, Y., Chen, P., Hsieh, J., Yeh, I.: CSPNet: a new backbone that can enhance learning capability of CNN. In: 2020 IEEE/CVF Conference on Computer Vision and Pattern Recognition Workshops (CVPRW), pp. 1571–1580 (2020)
12. Yao, Z., Cao, Y., Zheng, S., Huang, G., Lin, S.: Cross-iteration batch normalization. arXiv preprint arXiv:2002.05712 (2020)
13. Yan, X., et al.: PointASNL: robust point clouds processing using nonlocal neural networks with adaptive sampling. In: 2020 IEEE/CVF Conference on Computer Vision and Pattern Recognition (CVPR). IEEE (2020)

14. Zhu, H., Zhao, X.: TargetDrop: a targeted regularization method for convolutional neural networks. arXiv preprint arXiv:2010.10716 (2020)
15. Redmon, J., Farhadi, A.: Yolov3: an incremental improvement. arXiv preprint arXiv:1804.02767 (2018)
16. Bochkovskiy, A., Wang, C.-Y., Liao, H.-Y.M.: Yolov4: optimal speed and accuracy of object detection. arXiv preprint arXiv:2004.10934 (2020)

Simulation of Pedicle Screw Extraction Based on Galerkin Smooth Particle Meshless Method

Xiumei Wang, Yuxiang Sun, Hao Chen, Xiuling Huang, Shuimiao Du, and Zikai Hua(✉)

School of Mechatronic Engineering and Automation, Shanghai University, Shanghai, China
{wangxm,vincentsum,zikai_hua}@shu.edu.cn

Abstract. The purpose of this study is to use the newly developed truly mesh-free method by Ls-dyna, the smooth particle Galerkin (SPG) method, to simulate the material failure process caused by the extraction of pedicle screws from the trabecular bone. The experimental and theoretical results are compared to verify the effectiveness of the SPG method. In addition, the trabecular bone material of Johnson-Cook elastoplastic material is used to compare the convergence difference between SPG and traditional finite element method in this test. The results show that the failure curve simulated by the SPG method is more reasonable, and it is not only faster than the FEM method in terms of mesh convergence but also solves the defect that the traditional FEM cannot capture the peak force.

Keywords: Smooth particle Galerkin method (SPG) · Finite element · Screw extraction · Trabecular bone failure

1 Introduction

Pedicle screw fixation is one of the most commonly used methods in spinal surgery to stabilize the thoracolumbar injured vertebrae. It is used in cases where patients need to be fixed with internal implants due to initial fractures. The factors that affect the strength of the bone-screw interface are more complicated. At present, most of the mechanical strength of the bone screw interface is evaluated by the axial pull-out test. Although experimental studies can provide direct results, there are still exist limitations. Primarily, different densities are usually required in this test for the reason that samples are more precious. Additionally, it is not ideal to study the influence of complex factors on the pull-out stability through experiments because the experimental samples are not reproducible [1].

Simultaneously, many domestic and foreign researchers have also used traditional finite element methods to study the stability of pedicle screw extraction and the individual factors that affect the stability of bone screw extraction. However, most of these studies have many shortcomings. To exemplify, most of the bone materials used in the analysis are elastic, it is difficult to simulate the failure of the bone when the screw is pulled out [2]. Or although reasonable material parameters are used, it ignores the loss of energy and mass caused by traditional finite element erosion technology to simulate material failure,

M. Fei et al. (Eds.): LSMS 2021/ICSEE 2021, CCIS 1467, pp. 242–249, 2021.
https://doi.org/10.1007/978-981-16-7207-1_24

thus underestimating the force response value and cannot be captured in a physical form Peak failure.

The newly developed smooth particle Galerkin method (SPG) is used in this study to analyze the mechanical behavior of pedicle screws under pure pull-out load. The grid sensitivity analysis was performed on SPG and FEM models with different mesh sizes, and the difference in mesh convergence between the two methods was mainly compared [3]. Furthermore, by comparing with the published experimental data, the rationality of the SPG model in this research is verified, so as to lay a solid foundation for further research.

2 Material and Method

2.1 Modeling

The international standard size of 6.5 mm diameter pedicle screw with threaded elevation angle and simplified bone model were created in NX 12.0. Figure 1 is a cross-sectional view of a standard screw with a diameter of 6.5 mm [4]. The detailed dimensions are listed in Table 1. In order to simplify the bone-screw model, the bone is built as a cylinder with a cross-sectional diameter of 20 mm, and its height is determined by the length of the screw purchased. It is assumed that the screw and the cylinder are coaxial, and the Boolean operation assembles the two parts to ensure that the bone channel and the screw shape are consistent to achieve the purpose of perfect contact. In the process of screw extraction, the node-to-surface contact is adopted to simulate the continuous contact between the bone and the thread and the contact effect of each other at the trailing edge. The screw element is defined as the main surface, and the node set on the bone is defined as the slave node., and according to the research of Liu et al., the friction between the screw and the bone is set to 0.2 [5].

Table 1. The specific dimensions of the pedicle screw thread geometry.

	CDH6.5
R(mm)	3.25
r(mm)	2.25
a_1(°)	5
a_2(°)	25
e (mm)	0.2
p (mm)	2.7

A total of four groups of models were created, namely, single-thread-low-density foam model,single-thread-high-density foam model and the remaining two groups were three thread models using SPG and FEM methods. Among them, two foam models with different densities are used for verification, while the other two models using trabecular bone materials are used to explore the difference in the sensitivity of the two methods to the mesh.

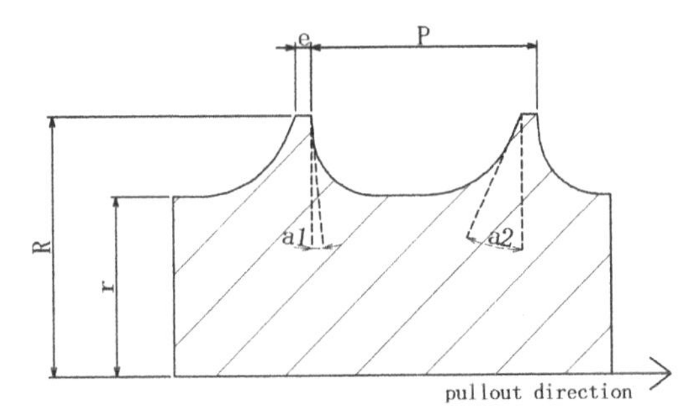

Fig. 1. Cross section of standard 6.5 mm pedicle screw.

2.2 Material Properties

In this article, it is assumed that the trabecular bone material is isotropic and homogeneous. In order to simulate the material failure of the bone, the model integrates the Johnson-Cook elastoplastic. Material constitutive model to simulate this situation [6]. The material of the screw is set to the most commonly used titanium alloy material [4], and by the reason of the elastic modulus of the screw is larger than that of the bone, which is close to the mechanical behavior of the rigid body, the rigid body attribute is used for all the nodes of the screw to reduce the solution time. The material properties of trabecular bone and screws are listed in Table 2. This material will be applied to the mesh convergence test.

Table 2. Material Properties of trabecular bone and pedicle screws.

Material Properties	Trabecular bone	Pedicle Screw
Density(Kg/mm^3)	2.0E–7	4.43E–3
Young modelus, E(Mpa)	48.75	110000
Poisson ratio, v	0.25	0.3
Yield stress, a(Mpa)	1.95	860
Hardening modelus, b(Mpa)	16.3	
Hardening exponent, n	1	
Strain Rate Constant, c	0	
Thermal SofteningExponent, m	0	
Failure plastic strain, ε_{max}	0.04	

2.3 Boundary and Loading Conditions

In the global coordinate system, the pull-out direction of the screw is defined as the positive direction of the Z axis, and the X and Y axes are the circumferential direction

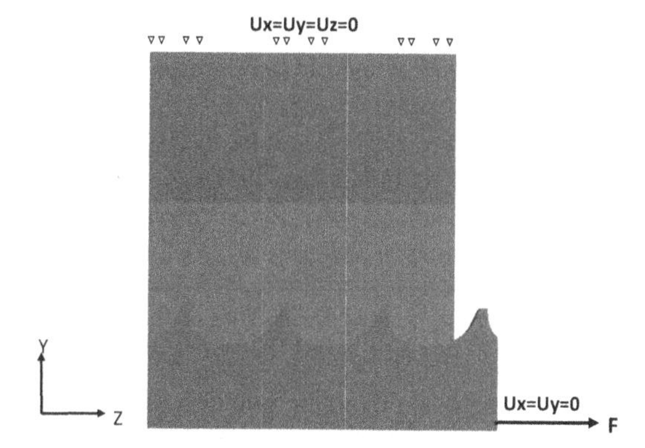

Fig. 2. Boundary conditions and loading conditions of the bone-screw model.

of the screw. All nodes and degrees of freedom on the circumferential surface of the bone are constrained, and all surface areas on the cross-section of the bone-screw model are restricted to the X and Y directions, respectively. Apply a displacement load in the Z direction to the front surface of the screw [2], and since the model only simulates a quarter part, the obtained load value should be multiplied by four. The boundary and load conditions applied in the model are shown in Fig. 2.

2.4 Mesh Convergence Test

The meshing work is carried out in ANSYS Workbench 19.2, the model processed in NX12.0 is imported into ANSYS Workbench 19.2, and the bone model is divided into two parts, namely "inner" and "outer", the two parts are connected by a common node. Since SPG takes more CPU time than traditional FEM, SPG particles are only used for the inner part of the bone that close to the screw (Fig. 3), and which are automatically generated in Ls-dyna from the ELFORM = 47 contained in the keyword *SECTION_SOLID_SPG

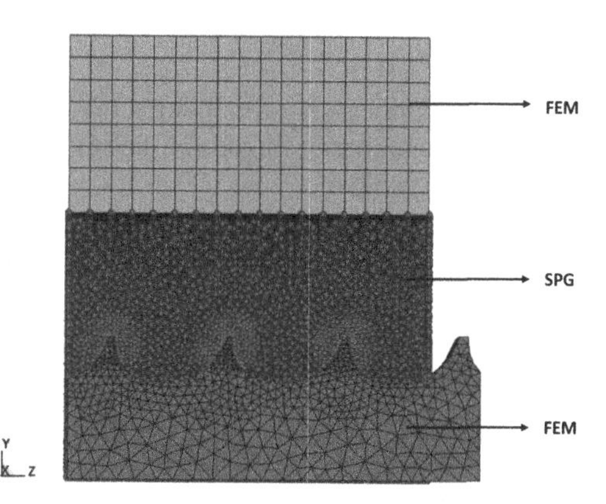

Fig. 3. Coupling of traditional finite element and SPG.

according to the nodes of the input FEM entity elements. Furthermore, the feasibility of coupling FEM and SPG has been explained in the published literature [3].

Trabecular bone was used to establish two analysis models including SPG coupled finite element model and pure finite element model to compare the convergence of the mesh. The seed sizes of the smallest thread sizes of 0.1 mm, 0.15 mm and 0.2 mm were used for this analysis. The load-displacement curve is extracted from the result, and the contact force F is used as the evaluation criterion for the convergence test [7].

2.5 Validation

The current verification is carried out by using two polyurethane foam-single threaded screw models to compare with the pullout strength evaluated by the experimental data of Hashemi et al. [8]. Low-density and high-density polyurethane foam blocks are used to replace human cancellous bone in osteoporosis and normal conditions, respectively (Table 3) [1, 2], and verify the extraction of the standard pedicle screw of CDH6.5 simulated by the SPG method based on the experimental results strength.

Table 3. Polyurethane material properties used for verification.

Foam type	Density	Tension		Shear	
	(Kg/mm^3)	Modulus (Mpa)	Strength (Mpa)	Modulus (Mpa)	Strength (Mpa)
Low density	1.6e−7	57	2.2	23	1.41
High density	3.2e−7	267	5.9	67	3.83

Since the thread number of the implanted test block of Hashemi et al. is 13. In order to facilitate comparison with its research, this research is based on the following Eq. (1) to calculate the peak screw pull-out force [9].

$$\mathrm{F} = 4*\left(\frac{F_i}{N_i}\right)*\mathrm{N}. \quad (1)$$

Where, F = total strength of complete N thread extraction, F_i = 1/4 of the pullout strength of N_i threads, N_i = the number of threads in the study, $N_i = 1$ in this study, N = total number of threads to be verified.

Furthermore, since the pull-out strength depends on the shear strength and shear area of the material, the theoretical pull-out strength of the screw can be roughly estimated by the following Eq. (2) [1].

$$F_S = \mathrm{S}*A_s = \{\mathrm{S} * \mathrm{L}*\pi*\mathrm{D}\}*\mathrm{TSF}. \quad (2)$$

Where: TSF = Thread form factor = $\left(0.5 + 0.57735*\frac{(D-d)}{2\mathrm{p}}\right)$,

F_S = Theoretical shear failure force(N), S = Ultimate shear strength (Mpa), L = thread engagement depth (mm), D = major diameter (mm), d = minor diameter(mm), p = thread pitch(mm), A_s = Thread cutting area $\left(\mathrm{mm}^2\right)$.

3 Results

3.1 Mesh Convergence Test

Figure 4 shows the convergence study of the contact force using SPG and traditional finite element. Compared with traditional finite element, SPG is easier to converge. When different mesh sizes or coarser mesh sizes are used, the peak force of SPG is very Consistent, but the results of FEM fluctuate greatly. Moreover, the SPG is easier to capture the peak force than FEM [3]. For the reason that this article mainly uses the SPG method for research, and the resultant contact force of SPG (0.1), SPG (0.15), and SPG (0.2) are 54.1 N, 55.2 N, and 53.6 N respectively, the difference between them is less than 5%. The analyses demonstrated that with a seed size of 0.15 mm convergency was reached. The final model contains 463,455 elements and 86,433 nodes.

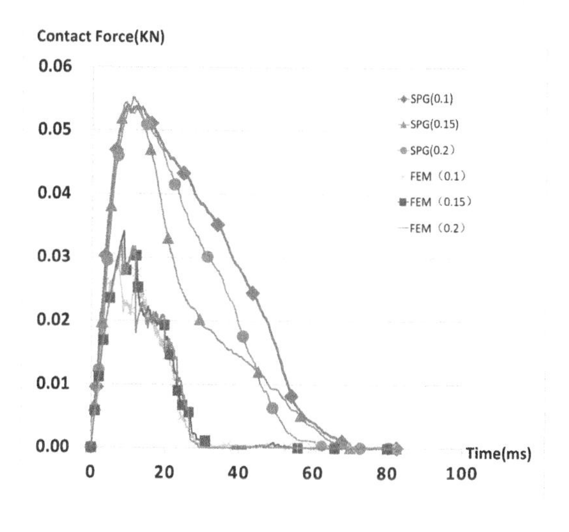

Fig. 4. Convergence study on contact force.

3.2 Validation

The theoretical results in this study are calculated by Eq. (2) in 2.5. The predicted and theoretical values of the 13-threaded high-density foam model are 2059.2 N and 1959.22 N, respectively, which are lower than the experimental value of Hashemi et al. (2132.5 ± 119.3) N, but the predicted value is within one standard deviation; In a like manner, The predicted and theoretical values of the low-density foam model are 722.8 N and 721.28 N, respectively, which are higher than the experimental value of Hashemi et al. (688.2 + 91.4) N [8], and both are within one standard deviation (Fig. 5).

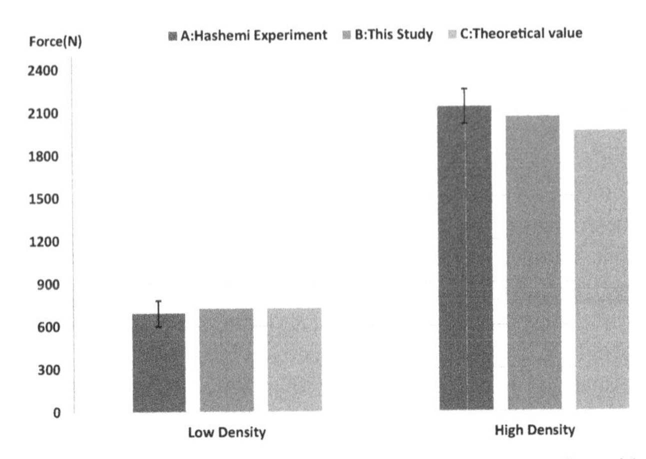

Fig. 5. Comparison of Hashemi et al.'s experimental measurement value with this research and theoretical value.

4 Discussion

In previous studies, finite element simulation of screw extraction was mainly used, ignoring the problems of non-conservation of energy and mass and inaccurate force response values caused by its failure criterion. Secondly, in many studies, linear elasticity assumptions are made on bone materials, and it is difficult to restore the simulation of bone failure during the drawing process [2]. Based on the above two deficiencies, this paper proposes the use of the SPG meshless method, and the Johnson-Cook nonlinear material constitutive model is used for the failure process of trabecular bone [6]. The previously published experimental data is used to verify the rationality and feasibility of the model in this study.

Convergence tests were performed on the SPG and FEM models respectively, and it was observed that SPG not only better captures the peak force, but also has much lower sensitivity to the grid than FEM. Meanwhile, which can be capable of obtaining more reliable results and failure trends compared with FEM. In addition to these, after the elements failed, the FEM model may cause unsatisfactory response values due to the direct contact between the screw and the bone, while SPG uses a phenomenological bonded-based failure mechanism to separate materials, and the adaptive anisotropic Lagrangian kernel is used for dealing with large deformation and restrain tension instability [10], so this problem is effectively avoided, simultaneously, the conservation of quality and energy is ensured. Afterwards, the verification of the SPG model in this study was carried out by using a single thread to obtain the predicted pull-out strength by assigning low-density and high-density polyurethane foams, and then comparing them with theoretical and experimental values. The result shows that whether high-density or low-density polyurethane foam is used, the predicted value of this study is within one standard deviation of the experimental value. This analysis confirms that SPG can be used to capture the peak screw pull-out force.

Up to now, this study has only verified the model and compared the difference in mesh convergence between SPG and traditional finite element. However, in surgery, bone material, thread embedding depth and implant angle may all be important factors affecting the extraction force, which will be further studied in the future.

References

1. Chapman, J.R., Harrington, R.M., Lee, K.M., et al.: Factors affecting the pullout strength of cancellous bone screws. J. Biomech. Eng. **118**(3), 391–398 (1996)
2. Zhang, Q.H., Tan, S.H., Chou, S.M.: Effects of bone materials on the screw pull-out strength in human spine. Med. Eng. Phys. **28**(8), 795–801 (2006)
3. Hu, W., Wu, C.T., Hayashi, S.: The immersed smoothed particle galerkin method in LS-DYNA® for material failure analysis of fiber-reinforced solid structures. In: 15th International LS-Dyna Users Conference (2018)
4. Chatzistergos, P.E., Magnissalis, E.A., Kourkoulis, S.K.: A parametric study of cylindrical pedicle screw design implications on the pullout performance using an experimentally validated finite-element model. Med. Eng. Phys. **32**(2), 145–154 (2010)
5. Liu, C.L., Chen, H., Cheng, C.K., et al.: Biomechanical evaluation of a new anterior spinal implant. Clin. Biomech. **13**(1), S40–S45 (1998)
6. Bianco, R.J., Arnoux, J.P., Wagnac, E., et al.: Minimizing pedicle screw pullout risks: a detailed biomechanical analysis of screw design and placement. Clin Spine Surg **30**(3), E226–E232 (2017)
7. Einafshar, M., Hashemi, A., van Lenthe, G.H.: Homogenized finite element models can accurately predict screw pull-out in continuum materials, but not in porous materials. Comput. Methods Prog. Biomed. **202**, 105966 (2021)
8. Hashemi, A., Bednar, D., Ziada, S.: Pullout strength of pedicle screws augmented with particulate calcium phosphate: an experimental study. The Spine J. **9**(5), 404–410 (2009)
9. Yan, Y.B., Pei, G.X., Sang, H.X., et al.: Screw-bone finite element models for screw pullout simulation. Chin. J. Traumatol. Orthop. **15**(1), 28–31 (2013)
10. Wu, C.T., et al.: Numerical and experimental validation of a particle Galerkin method for metal grinding simulation. Comput. Mech. **61**(3), 365–383 (2017). https://doi.org/10.1007/s00466-017-1456-6

Human Grasping Force Prediction Based on Surface Electromyography Signals

Yunlong Wang, Zhen Zhang(✉), Ziyi Su, and Jinwu Qian

School of Mechatronic Engineering and Automation,
Shanghai University, Shanghai 200444, China
zhangzhen_ta@shu.edu.cn

Abstract. To realize grasping force control in surface electromyography (sEMG) based prosthesis, this paper investigated the information of motion intention in sEMG signals. An eight-channel Myo armband was used to collect the sEMG signals of nine subjects. A series of grab weights were set up in the experiments. The raw signals were pre-processed by low-pass filtering using 4th Butterworth filter, data normalization, and short-time Fourier transform extracting the muscle activation region. Three classifiers, including k-nearest neighbour, decision tree, and multi-layer perceptron model were trained. The k-nearest neighbour model achieved high robustness and prediction accuracy up to 99%. It can be concluded that by decomposing the motion intention of sEMG signals, the grasping force can be accurately predicted in advance, which can significantly benefit real-time human-prosthetic interaction.

Keywords: Grasping force prediction · sEMG · Classifier · Motion intention

1 Introduction

Recently, it has become a research hotspot to use surface electromyography (sEMG) signal to recognize the hand movements and control the prosthesis for the patients with stump limbs [1]. The nervous system can still function despite the loss of limbs. Not only precise control of prosthesis can be realized by using sEMG, but also can benefit patients' gesture communication in daily life.

sEMG is generated before the body movement, which can most directly reflect the real intention of the human' motion intention, so as to achieve the best control of the prosthetic hand [2]. Pervious studies showed that the sEMG signal is used to identify the real intention of the performer accurately, and revealed the connection between the sEMG signal and different type of movement [3]. Time domain features are widely used due to their convenience [4].

Grasping force prediction firstly extracting sEMG signals from gripping and twisting actions. The preprocessed data is employed as the input of machine learning algorithm to fulfill pattern recognition tasks, such as force prediction [5]. The studies [6–8] carry out a linear regression method, gene expression programming (GEP), Linear discriminant analysis (LDA) method were used to grip strength prediction.

M. Fei et al. (Eds.): LSMS 2021/ICSEE 2021, CCIS 1467, pp. 250–255, 2021.
https://doi.org/10.1007/978-981-16-7207-1_25

Because of the positive correlation between hand grip strength and grip weight, the grasping motion can be predicted according to the mapping relation between the sEMG signal and corresponding targets' weight. The grip force could be predicted in advance through the early grip force information contained in the sEMG signal. A machine learning classification model based on the grasping motion intention prediction of sEMG signals is proposed. Three classifiers k-nearest neighbour (KNN), decision tree (DT) and multi-layer perceptron (MLP) for comparison.

2 Materials and Methods

2.1 Experimental Protocol

The subjects, including 9 healthy males whose aged from 22 to 25, participated in the study have no history of muscular or neurological disease. The experimental setup mainly consisted of 6 standard weights (0.75 kg, 1 kg, 1.25 kg, 1.75 kg, 2 kg) and a cylindrical box (the diameter of the box is 5 cm, the height of the box is 14 cm) (see Fig. 1).

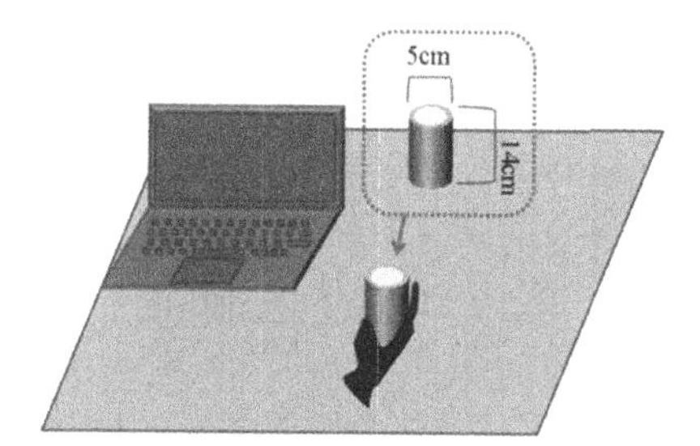

Fig. 1. Experimental setup.

We use Myo armband (Thalmic Labs, Waterloo, Canada) to acquisition of sEMG data. The Myo armband has a sampling frequency at 200 Hz, which consists of eight bipolar dry electrodes. Data from Myo armband is transmitted and stored to PC in real time through Bluetooth.

The participants were asked to all right arm wear Myo armband to grasp and each label was executed for 10 times. According to the indicator on the table, a set of grasping actions consists of 5 parts, is performed, The details of grasping protocol are carried as follows (1) Holding palms tightly and relax, (2) After 2 s the hand starts to grip and lift the target object. (3) Holding for 2 s and keeping 10 cm from the desktop, (4) Then slowly put the target back to desk within 2 s and (5) turning to relaxed posture to complete a gripping series.

2.2 Data Preprocessing

To obtain more discriminative features, we preprocess the raw data before feature extraction. The raw signal are normalized at first, with each element of each matrix T = (t1, t2…t8) being in the range [0, 1]. The 4th order Butterworth filter whose cut-off frequency is set at 5 Hz. It is used to remove the noise of sEMG signal so that the collected information can better represent the response of force. Through short-time Fourier transform [9], the signal energy spectrum is calculated to detect the muscle activity region. Apply the sliding window techniques (40 frames length with a step of 1 frame) to divide the intact analysis region into segments. The analysis region in this paper is defined as the transient state of muscle activity which is recorded as a 400 frames window from the start of the sEMG response.

2.3 Feature Extraction

In this study, we set up seven time domain features [10]. The s(k) is the kth amplitude sample, N is the sample size. Equations show in Table 1.

Table 1. Equations of time domain features.

Time domain features	Equation
Root mean square value (RMS)	$\mathrm{RMS} = \frac{1}{N}\sum_{k=1}^{N} \lvert s(k)\rvert$
Waveform length (WL)	$\mathrm{WL}=\sum_{k=2}^{N} \lvert s(k)-s(k-1)\rvert$
Number of slope sign changes (SSC)	$\mathrm{SSC} = \sum_{k=2}^{N} \lvert (s(k)-s(k-1)) \times (s(k)-s(k+1))\rvert$
Mean absolute value (MAV)	$\mathrm{MAV}=\frac{1}{N}\sum_{k=1}^{N} \lvert s(k)\rvert$
Activity parameter (AP)	$\mathrm{AP} = \mathrm{VAR}(s(k))=\frac{1}{N-1}\sum_{k-1}^{N} s(k)^2$
Mobility parameter (MP)	$\mathrm{MP} = \sqrt{\frac{\mathrm{VAR}(\frac{ds(k)}{dk})}{\mathrm{VAR}(s(k))}}$
Complexity parameter (CP)	$\mathrm{CP} = \frac{\mathrm{Mobility}(\frac{ds(k)}{dk})}{\mathrm{Mobility}(s(k))}$

3 Classifier

We use 75% samples to form the train set and the rest (25%) to form the test set.

KNN classifier decision rule commonly used by the KNN classifier is to take the category with the most occurrence times in KNN training data as the category of the input new instance. The k value of the classifier is 7, and the default distance between the number of neighbours and the data points is Euclidean distance.

The decision process of DT is from the root node of the decision tree to the leaf node, and the test data is compared with the nodes in the characteristic decision tree. In order to prevent over fitting and stop tree growth early, use the pre-pruning strategy and set the depth of tree full expansion to 4.

MLP also is known as feed forward neural network, can be regarded as a generalized linear model. The number of iterations is set at 500. The default number of other parameters is 100 hidden nodes, and the activation function is set as ReLU.

4 Results

4.1 Recognition Accuracy

The confusion matrix of the KNN model is illustrated in Fig. 2. Among them, the label with the weight of 2 kg has the highest prediction accuracy (99%). The prediction accuracy of 1.75 kg weight label ranks secondly, and the accuracy is 98%. The accuracy of the rest categories are similar, 1 kg and 1.5 kg have the same predicted progress of 95%.

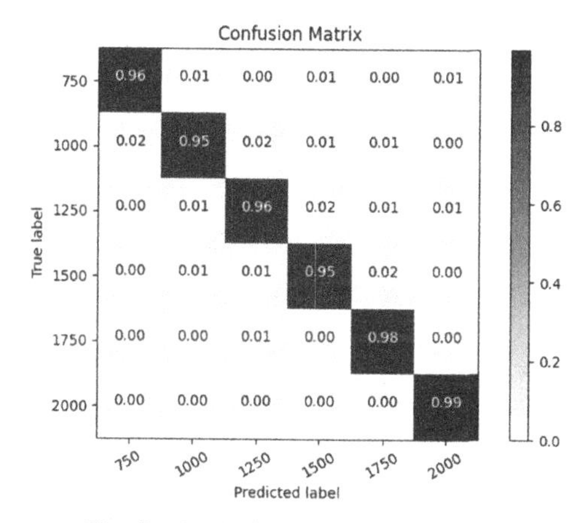

Fig. 2. Confusion matrix of all labels.

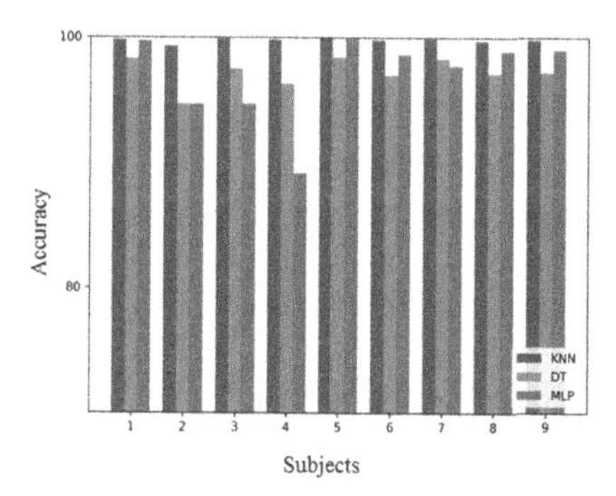

Fig. 3. Accuracy comparison of different models.

4.2 Classification Models

The average accuracy of the three models for each subject is compared, as shown in the Fig. 3. Among them, the KNN model has the best performance on the test set. The performance of DT model followed behind, the performance of MLP model is the worst in subject 4, and the rest of the performance is unstable as well. The accuracy of the MLP model was higher than that of the DT model in 5 subjects.

4.3 Sliding Windows Size

We use different sliding window sizes in the grasping motion feature extraction process for subject 1. We test the value of the window size from 50 frames to 400 frames (step size is 50 frames). When the window size increases from 150 frames to 400 frames, the recognition rate of the KNN model hardly increases. At an earlier time (50 frames to 100 frames), KNN get a higher recognition rate, and then it goes a little bit down from 100 frames to 150 frames. Sliding window size and KNN model accuracy are shown in the Fig. 4.

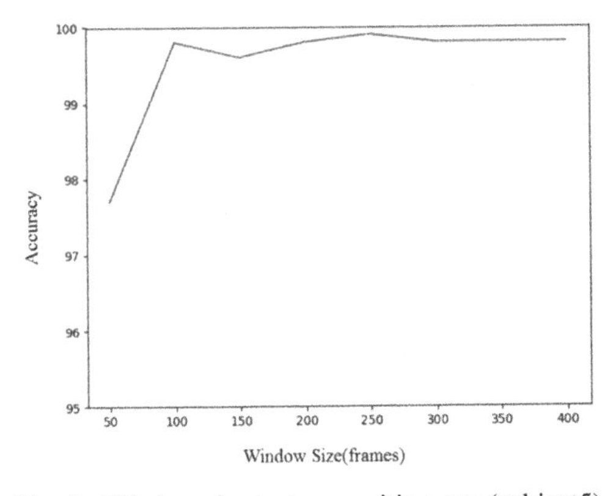

Fig. 4. Window size test recognition rate (subject5).

5 Discussion

When grasping a small weight, the sEMG is relatively weak, which has a great impact on the recognition progress and leads to a low accuracy. The larger target weight to grasp, the stronger sEMG signal amplitude response and the higher classification accuracy can be obtained.

The KNN model has the best performance and achieved the highest prediction accuracy. The performance of DT model is next, and the accuracy is slightly lower than that of KNN model. After 500 iterations of the MLP model, the performance of subject 2 to subject 4 is comparatively poor. The poor performance of MLP in subject 4 was believed to be caused by the weak sEMG signal strength of the subject. And the accuracy of the rest subjects is relatively high.

In this study, 400 frames (2 s) of motion extraction interval is set for evaluation, and the sliding window size is adjusted to test the influence on the accuracy of KNN model. The information contained in the sEMG signal can be predicted as soon as possible by reducing the size of the sliding window. The sliding window size is 50 frames, the prediction accuracy of KNN model decreases. The prediction accuracy of the sliding window size fluctuates slightly between 100 and 300 frames, and the model accuracy is stable when the size is larger than 300 frames.

6 Conclusion

In this paper, machine learning models KNN, DT and MLP classifiers are used to predict and recognize the grasping motion intention. The prediction and recognition accuracy of the KNN model reached 99%. The method proposed in this study has high robust. Reducing the size of the sliding window can get high prediction accuracy. In the future, the prediction of gripping motion intention can be applied to the control of soft prosthetic hands to achieve better human-computer interaction.

References

1. Potvin, J.R., Norman, R.W., McGill, S.M.: Mechanically corrected EMG for the continuous estimation of erector spinae muscle loading during repetitive lifting. Eur. J. Appl. Physiol. **74**, 119–132 (1996)
2. Jordanić, M., Rojas-Martínez, Mónica., Mañanas, M., Alonso, J.: Prediction of isometric motor tasks and effort levels based on high-density EMG in patients with incomplete spinal cord injury. J. Neural Eng. **13**(4), 046002 (2016)
3. Li, C.J., Ren, J., Huang, H.Q., et al.: PCA and deep learning based myoelectric grasping control of a prosthetic hand. Biomed. Eng. Online **17**, 1–18 (2018)
4. Gijsberts, A., Atzori, M., Castellini, C., Muller, H., Caputo, B.: Movement error rate for evaluation of machine learning methods for sEMG-based hand movement classification. IEEE Trans. Neural Syst. Rehabil. Eng. **22**(4), 735–744 (2014)
5. Yusuke, Y., Soichiro, M., Ryu, K., Hiroshi, Y.: Development of myoelectric hand that determines hand posture and estimates grip force simultaneously. Biomed. Signal Process. Control. **38**, 31–321 (2017)
6. Martinez, I., Mannini, A., Clemente, F., Sabatini, A., Cipriani, C.: Grasp force estimation from the transient EMG using high-density surface recordings. J. Neural Eng. **17**(1), 016052 (2020)
7. Liang, Y.Z., Miao, Y.C., Chuan, T.Z., Ping, J.W.: Surface EMG based handgrip force predictions using gene expression programming. Neurocomputing **207**, 56–579 (2016)
8. Feng, W.N., Yi, L.K., Hao, Z.X., Fan, L.J., Min, Z.X.: The recognition of grasping force using LDA. Biomed. Signal Process. Control **47**, 393–400 (2019)
9. Zhen, Z., Kuo, Y., Jinwu, Q., Lunwei, Z.: Real-Time surface EMG pattern recognition for hand gestures based on an artificial neural network. Sensors **19**, 3170 (2019)
10. Zhiyuan, L., Chen, X., Zhang, X., Tong., K.-Y., Zhou, P.: Real-time control of an exoskeleton hand robot with myoelectric pattern recognition. Int. J. Neural Syst. **27**(05), 1750009 (2017)

Stroke Identification Based on EEG Convolutional Neural Network

Jun Ma[1], Banghua Yang[1(✉)], Wenzheng Qiu[1], Xuelin Gu[1], Yan Zhu[2], Xia Meng[3], and Wen Wang[4(✉)]

[1] School of Mechatronic Engineering and Automation, School of Medicine, Research Center of Brain Computer Engineering, Shanghai University, Shanghai 200444, China
yangbanghua@shu.edu.cn

[2] Shanghai Second Rehabilitation Hospital, Shanghai 200441, China

[3] China National Clinical Research Center for Neurological Diseases, Beijing 100070, China

[4] Department of Radiology and Functional and Molecular Imaging Key Lab of Shaanxi Province, Tangdu Hospital, Fourth Military Medical University, Xi'an 710038, Shaanxi, China
wangwen@fmmu.edu.cn

Abstract. Stroke is a disease with high incidence and high disability rate. Stroke rehabilitation training technology based on motor imagery-brain computer interface (MI-BCI) has been widely studied. However, brain computer interface (BCI) technology can not realize auxiliary diagnosis. In this paper, the auxiliary diagnosis method of stroke disease is studied, and a stroke disease diagnosis experiment based on electroencephalogram (EEG) is designed. The EEG of 7 healthy subjects and 7 stroke patients in resting state and motor imagery (MI) task were collected and studied as follows: Step 1: the resting state and task state data are analyzed by power spectral density (PSD) according to different rhythms, and the frequency domain characteristics of all EEG signals are observed. Step 2: common spatial pattern (CSP) is performed on resting state and MI task data according to different rhythms, and the spatial features of all EEG signals are observed. Step 3: The EEG-CNN deep learning network architecture is designed, and the labels are designed according to whether they are stroke or not. Six healthy subjects and six patients are randomly selected as the training set, and the rest as the test set for 7-fold cross-validation training and testing. After the results of the above experiments and analysis, it is found that the EEG of the α rhythm of the MI task has the highest degree of discrimination, and the average correct rate on the EEG-CNN network model is 79.0%. Compared with other feature extraction and deep learning algorithms, EEG-CNN prediction accuracy has obvious advantages. The auxiliary diagnosis method and analytical modeling algorithm proposed in this paper provide a reliable research basis for stroke disease diagnosis and severity diagnosis.

Keywords: Brain computer interface (BCI) · Deep learning · Stroke identification · Electroencephalogram (EEG) · Convolutional neural network (CNN)

M. Fei et al. (Eds.): LSMS 2021/ICSEE 2021, CCIS 1467, pp. 256–266, 2021.
https://doi.org/10.1007/978-981-16-7207-1_26

1 Introduction

Stroke is a high incidence and high disability rate disease. Stroke rehabilitation training based on MI-BCI has been widely studied [1]. At present, MI-BCI can only be used for rehabilitation, not for diagnosis. Doctors can only make diagnosis through patients' examination reports in the hospital, and lack of automatic machine diagnosis method. EEG reflect the potential changes of the cerebral cortex and are widely used in MI and visual evoked potentials. Stroke diagnosis based on EEG has also been widely studied. Rafay M. F. analysis of the EEG features of neonates with seizures was used to distinguish stroke from hypoxic ischemic encephalopathy [2]. Bentes C conducted a prospective longitudinal study of potential patients with transient ischemic attack (TIA) based on EEG, which proved the diagnostic value of EEG features in TIA [3]. Shreve L used resting state EEG to compare the electrical power of both sides of the patient's brain. The correlation between larger infarct volume and higher brain power was shown to predict the risk of large acute ischemic stroke in emergency [4]. Gottlibe M quantified the symmetry of the power spectrum between the two hemispheres by modifying the brain symmetry index to distinguish between healthy subjects and stroke patients, and observed statistical differences [5]. The study [3–5] takes EEG as the basis for diagnosis but there is no modeling to assist diagnosis.

Although EEG had high practical value in stroke diagnosis, the modeling and classification method of EEG remains to be further studied. The mainstream EEG modeling methods are deep learning, which used a large training set and depth neural network to improve the prediction results. Studies related to deep learning include the evaluation of the severity of ischemic stroke and the classification of ischemic and hemorrhagic stroke. Omar used artificial neural network to classify the severity of ischemic stroke, and the accuracy was 85%. But the author only studied the ischemic stroke and ignored hemorrhagic stroke [6]. Li F used fuzzy entropy and hierarchical theory to get 96.72% accuracy in the classification of hemorrhagic stroke and ischemic stroke [7]. The author also improved the accuracy classification of hemorrhagic stroke and ischemic stroke to 97.66% by combining wavelet packet energy, fuzzy entropy and hierarchical theory [8]. The content above was from the research point of view to consider that different samples of the same subjects participated in the training set and the test set respectively. The algorithm designed in this paper was completely in accordance with the actual application of all the data that were included in the test set do not participate in the training set. EEG modeling methods also include traditional CSP [9] feature extraction and typical deep learning methods such as FBCNet [10], DeepConvNet [11] and EEGNet [12]. These methods can not only solve the classification problem of motor imagery but also other EEG feature classification problems. They are more mature EEG modeling classification methods in the BCI.

In this paper, a stroke diagnosis experiment based on EEG was designed, which included eyes-opening and eyes-closing resting states and motor imagery task. The EEG data of the experimental group of stroke patients and the control group of healthy subjects were recorded in the experiment. In the stage of data analysis, firstly, the feature difference in frequency domain was observed by power spectral density. Then the feature differences in the spatial domain were observed based on the feature matrix brain topographic map generated by CSP. Finally, the EEG-CNN model was used to train the original data and predict it across subjects to find out the best model for classifying healthy subjects and stroke patients.

2 Materials

2.1 Subjects

This study was conducted in research center of brain computer engineering of Shanghai University and Shanghai second Rehabilitation Hospital. Seven healthy subjects (age 24.57 ± 2.23, 5 females) and seven stroke patients (age 58.85 ± 11.51, 1 female) were recruited for the study. Ethical approval was given by Shanghai second Rehabilitation Hospital Ethics Committee (Approval number: ECSHSRH 2018–0101). Written informed consent was signed by each subject before the experiment.

2.2 Experimental Paradigm

The experiment included rest state and motor imagery task. The rest state included one minute eyes-opening and one minute eyes-closing. The motor imagery task (Fig. 1) asked the subjects to imagine the continuous grasping of the right hand. At the beginning of the experiment, the subjects sat comfortably in front of the screen for 1 m and were asked to focus on the course of the experiment. The experiment took 5 min, including two-minute resting state (eyes-opening and eyes-closing, Fig. 2a) and three-minute motor imagery task. The motor imagery task consists of 20 trails, and each trail time showed in Fig. 2b contains a 3-s prompt, a 4-s motor imagery and a 2-s break. The experimental paradigm was designed by E-Prime3.0.

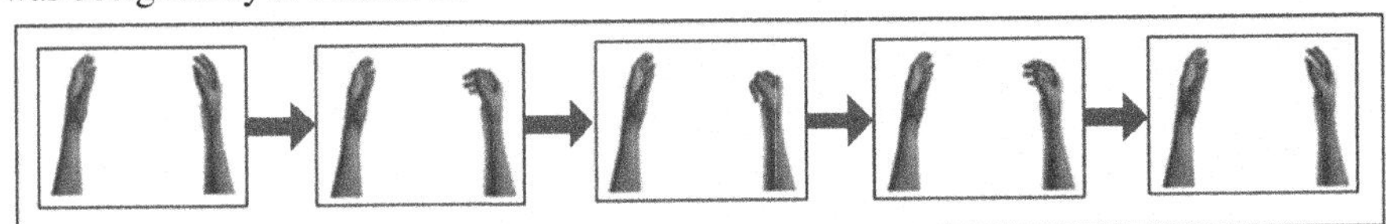

Fig. 1. Hand grasping motor imagery task

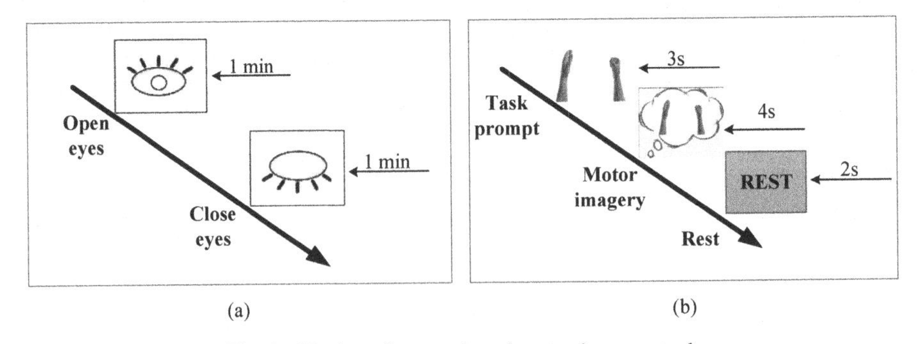

Fig. 2. Timing of rest task and motor imagery task

2.3 Data Collection and Preprocessing

The 30-channel wireless dry electrode EEG headsets produced by American CGX company was adopted. The advantages of the dry electrode experiment were easy to wear and no need to clean the conductive paste after the experiment was completed. The purpose of the experiment with dry electrode was to make the patients more comfortable and easier to complete the EEG collection experiment. The device contains 30 electrodes positions as shown in Table 1 and acquisition frequency is 1000 Hz. Data preprocessing includes 250 Hz frequency reduction, de-baselining and 1–40 Hz bandpass filtering. The 60 s resting state data was divided into 15 trails according to every 4 s. The task state takes 4 s MI data as the training set. All healthy subjects were classified as class 1, and all stroke patients as class 2.

Table 1. All channels of the EEG headset

Number	1	2	3	4	5	6	7	8	9	10
Channel	Fp1	Fp2	AF3	AF4	F7	F8	F3	Fz	F4	FC5
Number	11	12	13	14	15	16	17	18	19	20
Channel	FC6	T7	T8	C3	Cz	C4	CP5	CP6	P7	P8
Number	21	22	23	24	25	26	27	28	29	30
Channel	P3	Pz	P4	PO7	PO8	PO3	PO4	O1	O2	A2

3 Method

3.1 Analysis of PSD

In order to study the features of each task data in frequency domain, we performed power spectral density analysis on the EEG signals between healthy subjects and stroke patients. Referring to the study of [6–8], the PSD of Fp1 and Fp2 in the prefrontal lobe of stroke patients was different from that of healthy subjects. The Fp1and Fp2 electrodes were selected to calculate and observe the PSD features of healthy subjects and stroke patients under different rhythms. In order to intuitively observe the PSD features of EEG signals with different rhythms, θ (4–7 Hz), α (8–13 Hz) and β (14–30 Hz) and all frequency (1–35 Hz) were selected respectively. Welch method [13] was used to calculate the PSD shown in Fig. 3. These PSD maps showed the differences in brain power between healthy subjects and stroke patients during eyes-opening, eye-closing and motor imagery.

Figure 3 showed the brain power of healthy subjects and stroke patients with different tasks and different rhythms, where rows represent the same rhythm, and each column represents the same task. The θ rhythm power of healthy subjects was higher than that of patients in all tasks and there were significant differences. The α rhythm can only be seen in the eye-opening task that the power of the healthy person was higher than

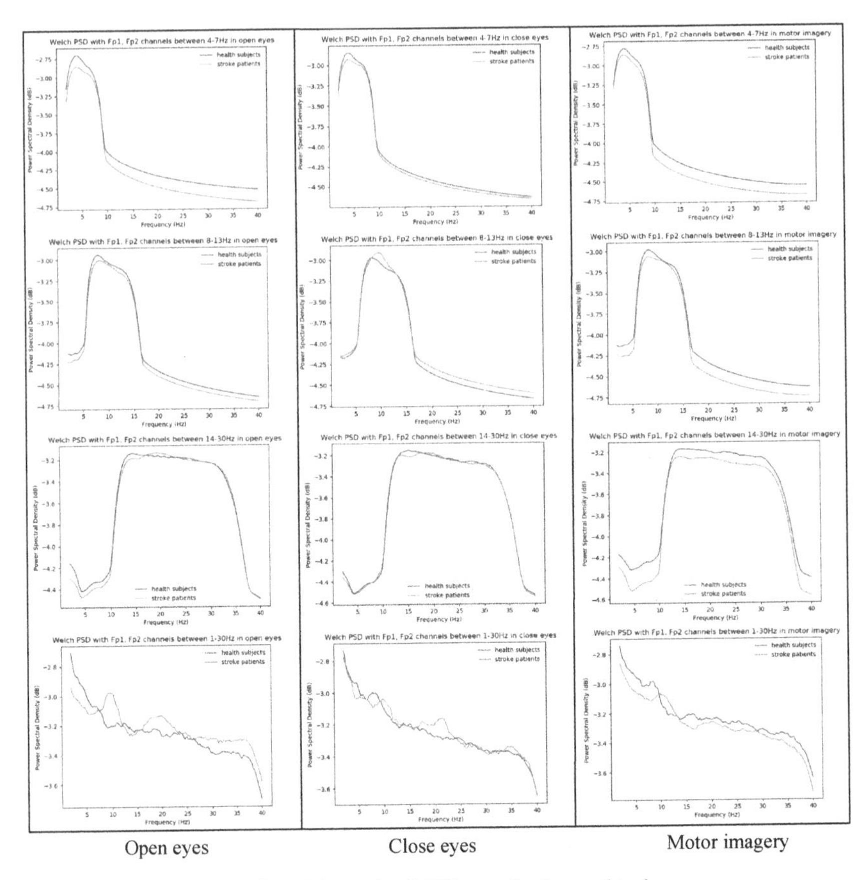

Fig. 3. PSD result of different rhythm and tasks

that of the patient. The MI task of β rhythm showed that the power of healthy subjects were significantly higher than that of patients. In the all frequency diagram of all tasks, there was no significant difference in the alternation of power between healthy subjects and stroke patients. The results of PSD show that the gap between healthy subjects and stroke patients was not significant, although data from different tasks were used, it is difficult to draw a clear distinction from the perspective of PSD.

3.2 Spatial Filtering Brain Topographic Map

CSP feature extraction and linear discriminant analysis (LDA) classification were used as benchmark modeling methods. Resting state data and task state data were used as inputs, and healthy subjects and stroke patients were trained as two labels. The data of 7 healthy subjects and 7 patients were used as the training set, and the remaining 1 healthy subjects and 1 patients were used as the test set. CSP algorithm used supervised method to create an optimal public space filter to maximize one class of variance while minimizing another class of variance. The covariance matrix of orthonormal whitening of two kinds of feature matrices [10] is projected into the CSP eigenmatrix $Z = WE$ by diagonalizing the eigenmatrix at the same time. Z is the spatially filtered signal, W is the feature matrix, and E is the EEG data of one trail. The first column and the last column of the feature matrix of CSP represent the maximum differentiation of the two kinds of data, respectively. According to the location of EEG channel, the corresponding eigenvalues of each point are mapped to the brain electrode plane model to obtain the brain topographic map based on CSP features.

Figure 4 shows the CSP brain topographic map of healthy subject and stroke patients with different tasks and rhythms. The discrimination degree of the θ and α rhythm of the eyes-opening and eyes-closing rest state was larger than that of the MI task state, while the discrimination degree of the brain topographic map of the θ and α rhythms of the eyes-opening and eyes-closing was larger. The difference of α rhythm under the same task was also the most obvious. There was no significant difference between healthy subjects and stroke patients in the CSP brain topographic map, so CSP can only be used as a benchmark algorithm.

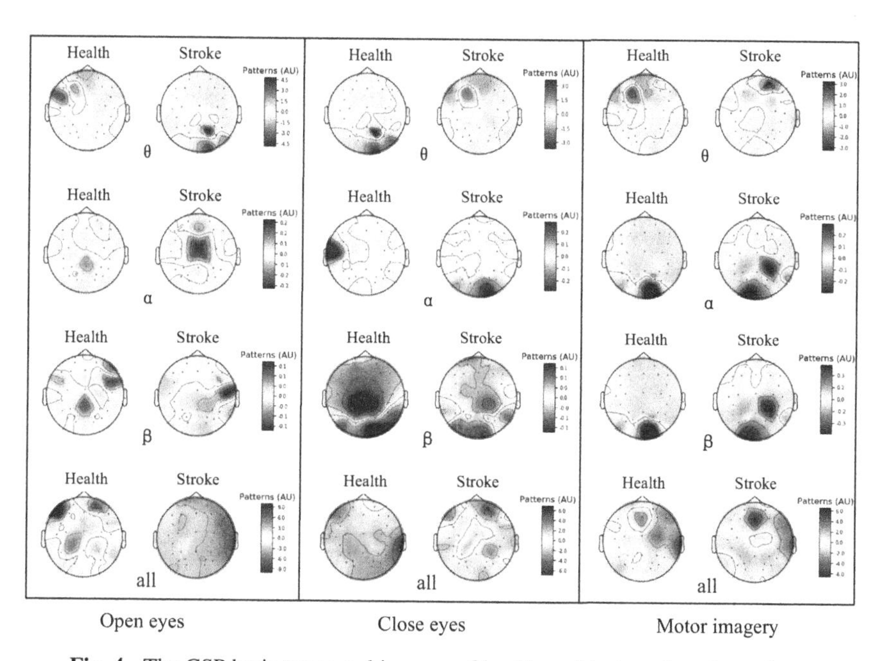

Fig. 4. The CSP brain topographic maps of healthy subjects and stroke patients

3.3 EEG Convolutional Neural Network

Convolution neural network has been paid more and more attention by researchers in the modeling and classification of EEG signals. At present, the more mature EEG classification method is FBCNet, DeepConvNet, EEGNet and so on. The processing method of FBCNet is similar to that of CSP, which classifies and models EEG which has obvious spatial features of EEG. The architectures of DeepConvNet and EEGNet networks are similar. Both of them construct convolution-pooling model based on the original EEG signal. Because the feature discrimination of the EEG of heath and stroke data was very small, we improved the network architecture of EEGNet [12] to make it more suitable to distinguish. The network architecture named EEG-CNN and its structure and parameters were shown in Fig. 5 and Table 2.

The EEG-CNN contained a total of four layers, the first layer contained two-dimensional convolution (Conv2D), normalization (BatchNorm2d), exponential liner unit (ELU), average pooling (AvgPool2d) and dropout. The second layer contained Conv2D and AvgPool2d. The third layer contained Conv2D and the last layer was soft-max regression. The value of stride was set to 4 in AvgPool2d in layer1, and set to 8 in layer2 to compress the feature s in the time domain. Conv2D was used on layer3 to fully compress the time domain feature, and finally all inputs were normalized to predicted values in the softmax of layer 4. We trained and predicted all the subjects with different rhythms and different tasks. The accuracy of α rhythm of eye-opening motor imagery task was the highest, and the correct rate and loss rate of its training set and test set are shown in Fig. 6.

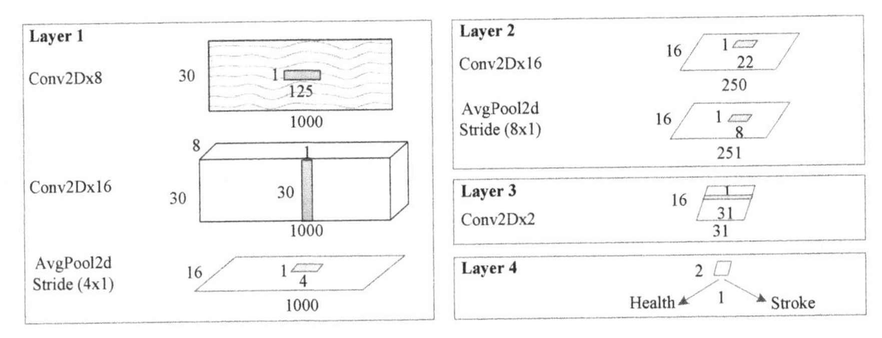

Fig. 5. EEG convolutional neural network architecture

Table 2. EEG convolutional neural network parameters

Layer	Input	Operation	Output	Number of parameters
1	[1,30,1000]	Conv2D (1×125)	[8, 30, 1000]	1000
	[8, 30, 1000]	BatchNorm2d	[8, 30, 1000]	16
	[8, 30, 1000]	Conv2D (30×1)	[16, 1, 1000]	3840
	[16, 1, 1000]	BatchNorm2d	[16, 1, 1000]	32
	[16, 1, 1000]	ELU	[16, 1, 1000]	
	[16, 1, 1000]	AvgPool2d (1×4),stride = (4×1)	[16, 1, 250]	
	[16, 1, 250]	Dropout (0.5)	[16, 1, 250]	

(*continued*)

Table 2. (*continued*)

Layer	Input	Operation	Output	Number of parameters
2	[16, 1, 250]	Conv2D (1,22)	[16, 1, 251]	352
	[16, 1, 251]	BatchNorm2d	[16, 1, 251]	32
	[16, 1, 251]	ELU	[16, 1, 251]	
	[16, 1, 251]	AvgPool2d (1×8),stride = (8×1)	[16, 1, 31]	
	[16, 1, 31]	Dropout (0.5)	[16, 1, 31]	
3	[16, 1, 31]	Conv2D (1×31)	[1, 2]	994
4	[1,2]	Softmax Regression	[1]	
Total parameters				6266

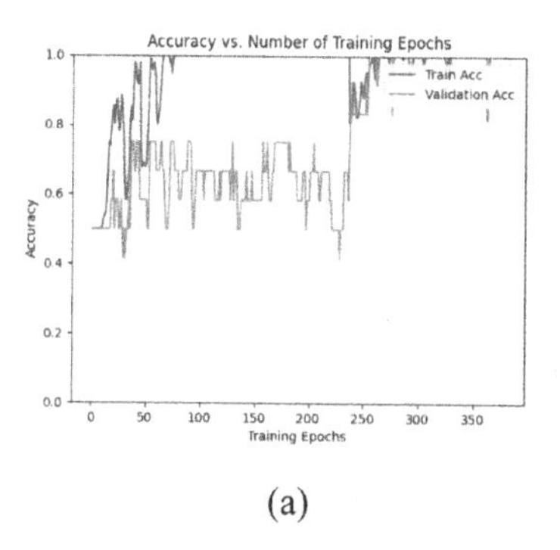

(a)

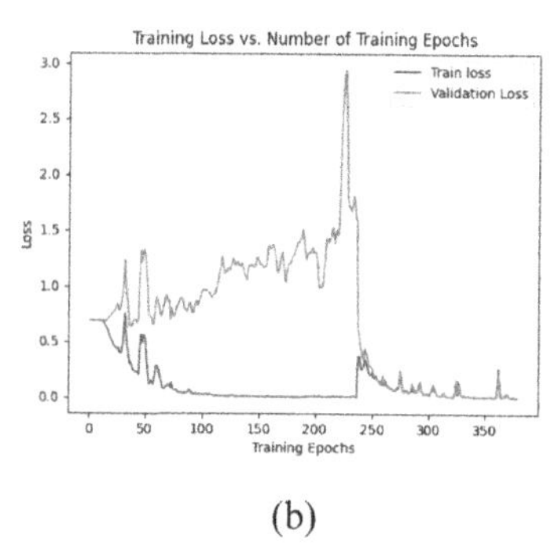

(b)

Fig. 6. The accuracy and loss of training and validation with training epochs

4 Results and Discussion

Five healthy subjects and five stroke patients were randomly selected as the training group, 1 healthy subject and 1 stroke patient as the verification group, and 1 healthy person and 1 patient as the test group. All the training data were verified by 7-fold cross-validation. We construct the data set completely according to the actual auxiliary diagnosis mode, and the data of the test set was completely separated from the training set. The subjects to which the test set data belongs do not appear in the training set and verification set, which can ensure that there was no data leakage problem. FBCNet and DeepConvNet were selected to compare the EEG-CNN designed in this article. CSP feature extraction and LDA classifier were selected as benchmark algorithms and compared with deep learning algorithms. The accuracy of the algorithm for all the rhythms of all state was shown in Table 3.

The results of all algorithms showed that CSP and FBCNET based on spatial filtering have no effect on classification. It can also be seen in Fig. 4 that there was a large gap between CSP brain topographic maps with different rhythms. Although there were differences in different health categories, the model performance was bad in classification.

Table 3. Accuracy results of different models

Task	Rhythm	CSP [9]	FBCNet-cv [10]	Deep Conv Net-cv [11]	EEG-CNN-cv
Eyes-opening	θ	0.525	0.480	0.505	0.510
	α	0.5	0.519	0.5	0.567
	β	0.5	0.519	0.5	0.752
	all	0.508	0.5	0.5	0.581
Eyes-closing	θ	0.5	0.529	0.5	0.562
	α	0.5	0.481	0.5	0.648
	β	0.5	0.5	0.505	0.700
	all	0.5	0.514	0.5	0.548
Motor imagery	θ	0.49	0.495	0.486	0.552
	α	0.49	0.543	0.495	0.790
	β	0.49	0.486	0.50	0.643
	all	0.49	0.486	0.490	0.638

Both DeepConvNet and EEG-CNN network architecture designed in this paper were based on convolution-pooling network architecture. The difference was that DeepConvNet connects the maximum pooling layer behind the convolution layer, while EEG-CNN was normalization behind the convolution layer. The function of Normalization was to normalize the convoluted data so that the model had a good generalization effect on the test set, which was important for cross-subject modeling. Because DeepConvNet lacks the generalization effect of normalization, it showed bad prediction results when using the new test model. The EEG-CNN network architecture designed in this paper has the highest accuracy of 79.0% in 7-fold cross-validation, which had higher recognition accuracy than other deep learning algorithms and benchmark algorithms.

Compared with the recognition accuracy of different tasks, the α rhythm of MI task had the highest classification accuracy. What can be determined was that better results can be achieved by selecting β rhythm for resting state data, and the highest classification accuracy can reach 75.2%. MI task showed better prediction results than resting state in distinguishing healthy subjects and stroke patients, which indicated that stroke patients showed obvious inconsistency with healthy subjects in MI.

In Fig. 3, the MI task β rhythm PSD result showed significant differences between the two types of subjects, but did not show good classification results from the point of view of modeling and prediction. We also try to use only Fp1 and Fp2 electrodes to model MI data, and the prediction accuracy is not high. Figure 3 the α rhythm PSD of eye-opening state did not show much difference between healthy subjects and stroke patients, but it had the highest correct rate of all resting state models. The above content showed that when using deep learning to classify EEG, we can not rely on simple time domain and frequency domain analysis, but more retain the original data, and fully learn the data characteristics through the deep network.

5 Conclusion

In this paper, we designed the resting state and MI task experiments for the auxiliary diagnosis of stroke group and the control group of health. In the experiment, all the subjects' EEG were recorded and analyzed by PSD and CSP, and the power map and brain topographic map were drawn. It can be seen that there were some differences in the characteristics of the two types of subjects in the frequency domain and spatial domain, but it was difficult to model and classify depending on these features. EEG-CNN was designed to model two types of subjects and other deep learning methods were compared. The experimental results showed that the EEG-CNN network architecture designed in this paper was more consistent with the cross-subject EEG data classification. In the data analysis phase, three task states (eye-opening resting state, eye-closing resting state and MI task) and four rhythms (θ, α, β and all rhythm) data were trained and predicted for each algorithm. The results showed that the prediction accuracy of α rhythm of MI task in EEG-CNN algorithm was 79.0%. Through the above experiments, it is proved that the auxiliary diagnosis of stroke based on EEG and deep learning was expected to be realized. The MI task of α rhythm can better distinguish the EEG features of healthy subjects and stroke patients, which provided an important research basis for auxiliary diagnosis and stratified diagnosis of stroke.

Acknowledgements. This project is supported by National Key R&D Program of China (No.2018YFC1312903, No.2018YFC0807405), National Natural Science Foundation of China (No. 61976133), Major scientific and technological innovation projects of Shan Dong Province (2019JZZY021010).

References

1. Mane, R., Chouhan, T., Guan, C.: BCI for stroke rehabilitation: motor and beyond. J. Neural Eng. **17**(4), 041001 (2020)
2. Rafay, M.F., Cortez, M.A., deVeber, G.A., et al.: Predictive value of clinical and EEG features in the diagnosis of stroke and hypoxic ischemic encephalopathy in neonates with seizures. Stroke **40**(7), 2402–2407 (2009)
3. Bentes, C., Canhão, P., Peralta, A.R., et al.: Usefulness of EEG for the differential diagnosis of possible transient ischemic attack. Clin. Neurophysiol. Pract. **3**, 11–19 (2018)
4. Shreve, L., Kaur, A., Vo, C., et al.: Electroencephalography measures are useful for identifying large acute ischemic stroke in the emergency department. J. Stroke Cerebrovasc. Dis. **28**(8), 2280–2286 (2019)
5. Gottlibe, M., Rosen, O., Weller, B., et al.: Stroke identification using a portable EEG device–a pilot study. Neurophysiol. Clin. **50**(1), 21–25 (2020)
6. Omar, W. R. W., Mohamad, Z., et al.: ANN classification of ischemic stroke severity using EEG sub band relative power ration. In: 2014 IEEE Conference on Systems, pp. 157--161. Process and Control (2014)
7. Li, F., Wang, C., Zhang, X., et al.: Features of hierarchical fuzzy entropy of stroke based on EEG signal and its application in stroke classification. In: 2019 IEEE Fifth International Conference on Big Data Computing Service and Applications,. pp. 284--289 (2019)
8. Li, F., Fan, Y., Zhang, X., et al.: Multi-feature fusion method based on EEG signal and its application in stroke classification. J. Med. Syst. **44**(2), 1–11 (2020)

9. Ramoser, H., Muller-Gerking, J., Pfurtscheller, G.: Optimal spatial filtering of single trial EEG during imagined hand movement. IEEE Trans. Rehabil. Eng. **8**(4), 441–446 (2000)
10. Mane, R., Robinson, N., Vinod, A.P., et al.: A multi-view CNN with novel variance layer for motor imagery brain computer interface. In: 2020 42nd Annual International Conference of the IEEE Engineering in Medicine & Biology Society,. pp. 2950–2953 (2020)
11. Schirrmeister, R.T., Springenberg, J.T., Fiederer, L.D.J., et al.: Deep learning with convolutional neural networks for EEG decoding and visualization. Human Brain Map. **38**(11), 5391–5420 (2017)
12. Lawhern, V.J., Solon, A.J., Waytowich, N.R., et al.: EEGNet: a compact convolutional neural network for EEG-based brain–computer interfaces. J. Neural Eng. **15**(5), 05601 (2018)
13. Welch, P.: The use of fast Fourier transform for the estimation of power spectra: a method based on time averaging over short, modified periodograms. IEEE Trans. Audio Electroacoust. **15**(2), 70–73 (1967)

Classification of LFPs Signals in Autistic and Normal Mice Based on Convolutional Neural Network

Guofu Zhang[1], Banghua Yang[1](✉), Fuxue Chen[2](✉), Yu Zhou[2], Shouwei Gao[1], Peng Zan[1], Wen Wang[3], and Linfeng Yan[3]

[1] School of Mechatronic Engineering and Automation, Research Center of Brain Computer Engineering, Shanghai University, Shanghai 200444, China
yangbanghua@shu.edu.cn

[2] School of Life Sciences, Shanghai University, Shanghai 200444, China
chenfuxue@staff.shu.edu.cn

[3] Department of Radiology and Functional and Molecular Imaging Key Lab of Shaanxi Province, Tangdu Hospital, Fourth Military Medical University, Xi'an 710038, China

Abstract. In view of the time-consuming and expensive problems that traditional methods is used to diagnose Autism Spectrum Disorder (ASD) through scales or magnetic resonance images, this paper proposes a classification method for ASD based on intrusive Electroencephalogram signals, which assists in determining whether there is ASD. Using artificially induced ASD mice' local field potentials (LFPs) data set and normal mice' LFPs data set, this thesis takes advantage of statistical analysis methods to test the significance between the two data sets, designs an architecture based on one-dimensional convolutional neural network, and trains two classification model of LFPs through Adam algorithm. The experimental result shows there is a strong significant difference between the two comparison groups ($P < 0.001$). After multiple tests on the test set, the average classification accuracy is 99.05%, indicating that the analysis method based on LFPs is an effective auxiliary method for judging whether there is ASD.

Keywords: Convolutional neural network · Autism spectrum disorder · Local field potentials · Mice

1 Introduction

Autism Spectrum Disorder (ASD), also known as autism, is a developmental disorder with widespread neurological disorders, which is more common in boys, beginning in infants, and whose incidence is increasing year by year. The clinical manifestations are mainly social communication disorders, speech function disorders, intellectual disability, and abnormalities in interest and behavior, such as narrow range of interest, stereotyped repetitive movements, and large differences in the severity of symptoms between different individuals [1]. The American Autism and Developmental Disorders Surveillance Network's monitoring of 8-year-old children in 11 states in the United

M. Fei et al. (Eds.): LSMS 2021/ICSEE 2021, CCIS 1467, pp. 267–276, 2021.
https://doi.org/10.1007/978-981-16-7207-1_27

States showed that from 2000 to 2002, the incidence of childhood autism was 1/150 (one in every 150 children has autism); from 2010 to 2012, the incidence of autism rose to 1/68, an increase of over 122%; in 2014, the prevalence of ASD in 8-year-old children was 1/59 [2]. In addition, the incidence of childhood trauma, self-harm, suicidal behavior and ideation in autistic patients was significantly increased. About two-thirds of children with autism were unable to live independently after adulthood and need lifelong care and nursing care, which caused a heavy financial burden for the family and society. According to statistics, in the United Kingdom, the lifetime cost of ASD patients was 1.36 million US dollars [3–5]. Although the incidence of autism is increasing, the corresponding resource for early diagnosis is scarce. For this reason, people are eager to find scientific and efficient means to accurately diagnose autism in order to achieve the purpose of early detection and early treatment. The traditional diagnosis of ASD generally makes use of a scale method, such as Autism Diagnostic Interview-Revised using guardians to provide children's language, behavior, basic living abilities, interests and many other information for diagnosis [6]; and Autism Diagnostic Observation Schedule Second Edition, to assess children's ability in social interaction communication and imagination [7]. Many scholars utilize infant brain imaging data or multi-modal data, for example language, movement, and facial expressions, to study artificial intelligence-based autism diagnosis models. Hazlett et al. used deep learning algorithms to analyze the magnetic resonance images (MRI) information of high-risk autistic infants aged 6 to 12 months to predict their diagnosis at 24 months of age, whose accuracy rate is 81% [8]. Heinsfeld et al. made use of Autism Brain Imaging Data Exchange of brain activation patterns in autistic patients to identify patients based on deep learning algorithms, and the accuracy rate reached 70% [9]. Wingfield et al. took advantage of clinical image autism assessment schedule data, trained autism classification model, and evaluated its predictive performance, based on machine learning algorithms [10]. Abbas et al. used parent report questionnaires, key behavior lists in children's family videos, and clinical evaluation clinicians' evaluation questionnaires to establish a machine learning-based autism evaluation system, through a multi-site blinding clinical study, whose evaluation was proved Reliability [11].

Traditional autism diagnosis models based on deep learning or machine learning mostly use MRI images or historical scale information, while scale diagnosis methods require long-term observation of infants' words and behaviors, and diagnosis through the subjective judgment of doctors. The entire diagnosis process takes a long time and affects the treatment of autism, whose results lack objectivity. There are also some problems in MRI, for instance, high cost, relatively more artifacts, many contraindications. Electroencephalogram has the advantages of low cost, high time accuracy, and can detect potential changes in milliseconds. It has become an important biological indicator for the diagnosis and treatment of children with brain developmental disorders [12, 13].

Based on artificially induced invasive EEG signals of autistic mice and normal mice, this paper utilizes statistical analysis methods and convolutional neural network (CNN) to analyze them, to study autistic mice and normal mice LFPs signals' difference, and designs CNN to classify the LFPs of autistic mice and normal mice.

2 Materials and Methods

2.1 Subjects

The mice data in this paper is obtained from [14]. Select healthy adult C57 mice, cage the mice at 18:30 according to a 1:2 male to female ratio, and allow them to mate. They were kept in an SPF animal room, and the vaginal plug was tested at 8:00 in the morning of the next day. If a vaginal plug was detected, the pregnant mice was marked as the 0.5 th day of gestation, and the male mice that was caged with pregnant mice was removed, and the pregnant mice was raised in a single cage. Female mice were gestated to 12.5 days and received a single intraperitoneal injection of Valproic Acid (VPA) [15, 16], (500 mg/kg dissolved in 0.85% saline, taking VPA during pregnancy can significantly increase the risk of autism in offspring), but the pregnant mice in the control group received the same dose of normal saline under the same conditions. After the injection was completed, the pregnant mouse, returned to the original cage, was raised until it was produced. The pups were fed by female mice and weaned at 14 days. Male offspring were selected as autism model mice. All animal experiments performed in this study were approved by the Animal Protection and Utilization Committee (Department of Experimental Animal Science, Shanghai University) and were strictly performed in accordance with the Animal Care and Used Program of the Experimental Animal Institute of Shanghai University.

2.2 Data Acquisition

Using a brain stereotaxic instrument, the five-channel electrode array was implanted into the CA3 of the mice hippocampus. The Plexon multi-channel recording system was used to collect the LFPs signal at CA3 in real time, the LFPs was amplified to a few volts, and the signal was transmitted to the computer for storage after digital-to-analog conversion. The LFPs of autistic mice and normal mice were recorded for 50–60 min respectively, as shown in Fig. 1.

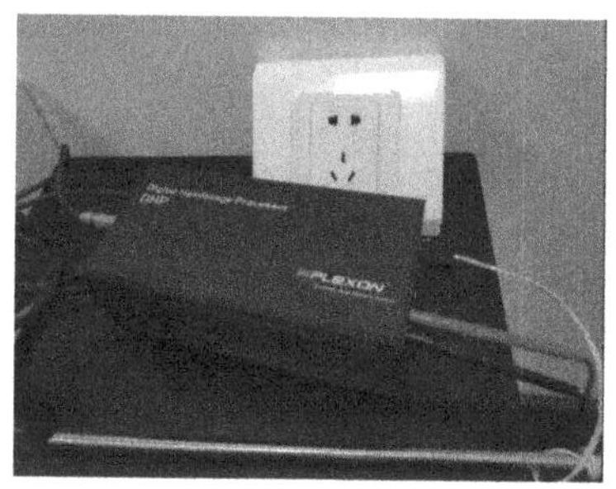

Fig. 1. Collecting LPFs of mice.

2.3 Statistical Analysis

All statistical analyses are performed using SPSS windows version 26.0. Paired sample t-test is used to evaluate the effect of LFPs of mice brain and ASD relationships. When $P < 0.05$, it indicates that the result is significant and statistically significant; and the smaller the P value, the more significant the result.

2.4 Model and Training

CNN Model. CNN has not only achieved good results in the field of image classification and recognition, but also has good classification results in the field of biosignal analysis [17, 18]. This paper proposes an ASD classification model based on CNN, which adopts an end-to-end training method and directly uses time domain data as model input, eliminating the need for manual extraction of data features. The focus of this article is to design a reasonable and effective CNN structure to better extract mice LFPs features and improve classification accuracy. The classification model of ASD mice based on CNN adopts one-dimensional convolutional neural network, which consists of input layer, hidden layer and output layer. The hidden layer consists of three layers of convolution layer, activation layer, pooling layer and two layers of full connection layer (FC). The model is shown in Fig. 2, and the detailed parameters are shown in Table 1.

Table 1. Network parameters.

Network layers		Input	Output	Number of kernels	Size of kernel	Stride	Padding
Input layer		(1, 1000)	(1, 1000)				
Hidden layers	Convolution1	(1, 1000)	(8, 1000)	8	1 × 25	1	12
	Activation1	(8, 1000)	(8, 1000)				
	Pooling1	(8, 1000)	(8, 500)		1 × 2	2	
	Convolution2	(8, 500)	(16, 500)	16	1 × 25	1	12
	Activation2	(16, 500)	(16, 500)				
	Pooling2	(16, 500)	(16, 250)		1 × 2	2	
	Convolution3	(16, 250)	(32, 250)	32	1 × 25	12	12
	Activation3	(32, 250)	(32, 250)				
	Pooling3	(32, 250)	(32, 125)		1 × 2		
	FC4	(1, 4000)	(1, 100)				
	FC5	(1, 100)	(1, 2)				
Output layer	Softmax	(1, 2)	1				

CNN Model Training. This experiment is based on the Pytorch deep learning framework, preprocessing the mice data set and building a convolutional neural network learning framework. Experimental software environment: Pytorch1.7.0 + cuda10.1,

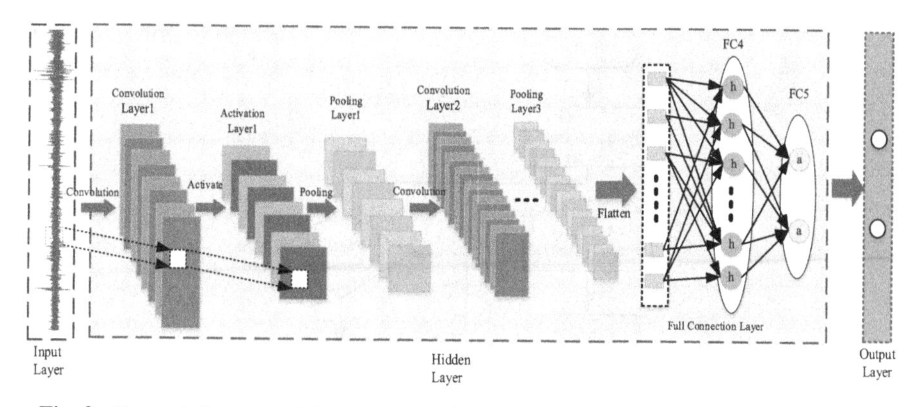

Fig. 2. Network Structure. It is composed of Input Layer, Hidden Layer and Output Layer.

Python3.8.6, Ubuntu18.04, Hardware environment: i5–4210 CPU, 8G RAM. The one-dimensional CNN is used to train the LFPs classification model of autistic mice. The flow chart is shown in Fig. 3. The Adam algorithm is used to train the CNN model, the learning rate is set as 0.001, and the iteration is 2000 times. Proceed as follows:

(1) Generate training sets and test sets. The collected mice LFPs is split into training sets and test sets, and in the light of the sampling frequency (1000 Hz), each sample contains 1000 data points (features). There are 5284 samples respectively in the control group and the autistic group, a total of 10568 samples. The training sets and the test sets are randomly divided according to the ratio of 7:3, therefore the training sets contain 7678 samples, and the test sets contain 3290 samples;
(2) 64 training samples randomly selected and input into the network. Set mini batch sizes to 64, which means that 64 training samples are randomly selected from the training sets and input to the one-dimensional CNN model. Only 64 samples are used in each iteration to improve the training speed;
(3) Calculate the loss and update the weights. The forward propagation algorithm calculates the loss of the model, the back propagation algorithm updates the weights and bias of the model, and prints the loss function curve of the training sets;
(4) Determine whether the training is complete. When the number of iterations reaches the upper limit—2000 iterations, the model training is completed, the model is saved, and the test is performed on the test sets.

3 Results and Discussion

3.1 Statistical Analysis Results

The paired sample T is used: $t = 16.995$, $P < 0.001$, as shown in Table 2, and the box diagram is shown in Fig. 4. The samples are from the same distribution, the mean of the control group and the autistic group is statistically significant, and there is a significant difference, indicating that the mice LFPs can be used to determine whether the mice have autism.

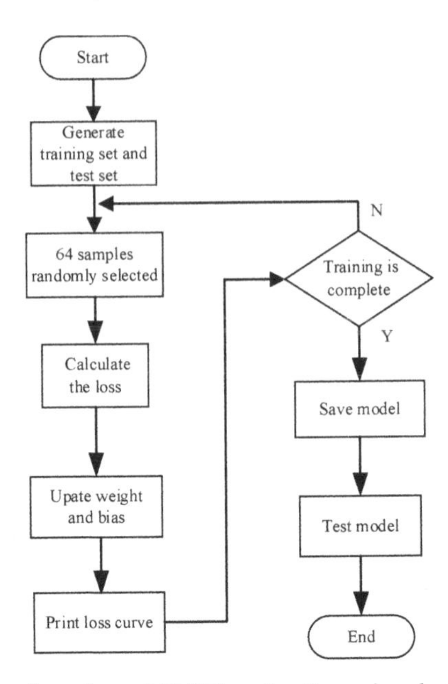

Fig. 3. Training flow chart of CNN based autism mice classification model.

Table 2. Statistical results.

	Paired differences					t	df	Sig. (2-tailed)
	Mean	Std. deviation	Std. error mean	99% confidence interval of the difference				
				Lower	Upper			
Control - autism	0.396	0.737	0.023	0.336	0.456	16.995	999	0.000

3.2 Classification Results of the Model

The training samples are input into the model, and as the number of iterations increased, the change process of the loss function value is shown in Fig. 5. In the first 50 iterations, the loss function value decreases rapidly; in 1800 iterations, the loss function value tends to be stable and fluctuated within the range of 0.02 to 0.05. After the iteration is completed, the model is saved; then input the test set into the model and output the prediction results.

In order to verify the performance of the model, the model is trained for ten times, whose classification results on the test sets are shown in Table 3. The test sets consist of 3,290 samples, including 1,645 autistic mice (labeled 1) and 1,645 control mice (labeled 0). Ten times of testing, the highest classification accuracy is 99.54%, the worst classification accuracy is 97.60%, and the average classification accuracy is 99.05%. For

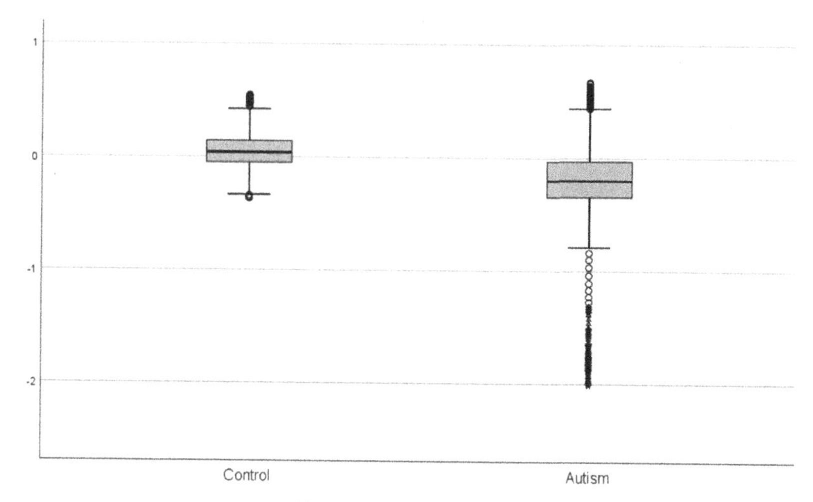

Fig. 4. The box diagram.

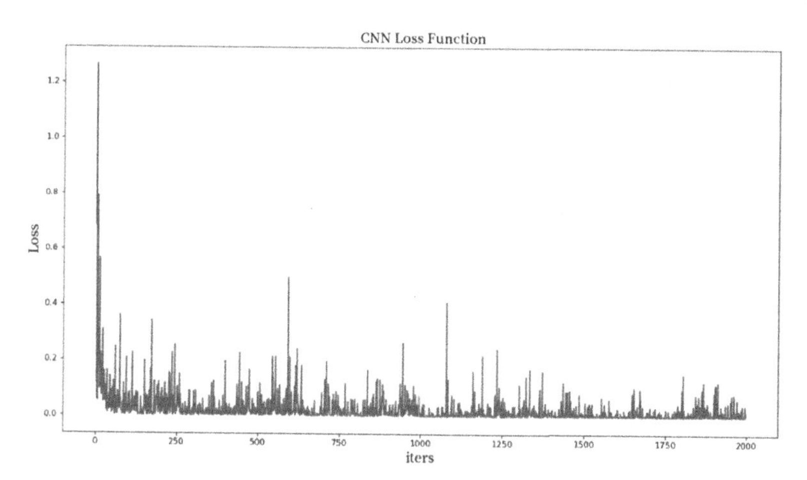

Fig. 5. Change curve of loss function value.

normal mice and autistic mice, the average accuracy rate of the ten tests is 99.07% and 99.03%, respectively.

Table. 3. Model classification accuracy (%).

Model Number	Normal Mice	ASD Mice	Average
1	99.21	97.75	98.48
2	99.15	99.45	99.30
3	98.72	99.64	99.18

(continued)

Table. 3. *(continued)*

Model Number	Normal Mice	ASD Mice	Average
4	98.66	99.88	99.27
5	99.09	99.82	99.46
6	99.03	96.17	97.60
7	99.03	99.33	99.18
8	99.15	99.64	99.40
9	99.21	99.76	**99.54**
10	99.09	99.15	99.12
Average	99.07	99.03	99.05

Taking the highest classification accuracy of 99.54% as an example, the model output results are reduced to 2 dimensions through t-SNE technology, and visualized, as shown in Fig. 6. 0 (yellow) means the classification is correct, which represents the LFPs of control group (normal mice) test sets is predicted to be the LFPs of control group mice; 1 (purple) means the classification is correct, which represents the LFPs of the autistic group is predicted to be the LFPs of autistic mouse; 2 (blue) indicates misclassification, predicting the LFPs of control group mice as the LFPs of autistic mice; 3 (red) indicates the misclassification, predicting the LFPs of autism group mice as the LFPs of control group mice. The confusion matrix of classification results is shown in Table 4. Precision rate, recall rate and F1-Socre were obtained respectively:

$$Precision = \frac{1637}{1637 + 7} = 0.9957.$$

$$Recall = \frac{1637}{1637 + 8} = 0.9951.$$

$$F1 = \frac{2 \times 0.9957 \times 0.9951}{0.9957 + 0.9951} = 0.9954.$$

Table. 4. Confusion matrix of classification results.

Predict true	Autism	Control
Autism	1637	8
Control	7	1638

Using the SVM algorithm for two classification, the classification accuracy is 94.16%, whose classification effect is not as good as the classification method based on CNN proposed in this paper.

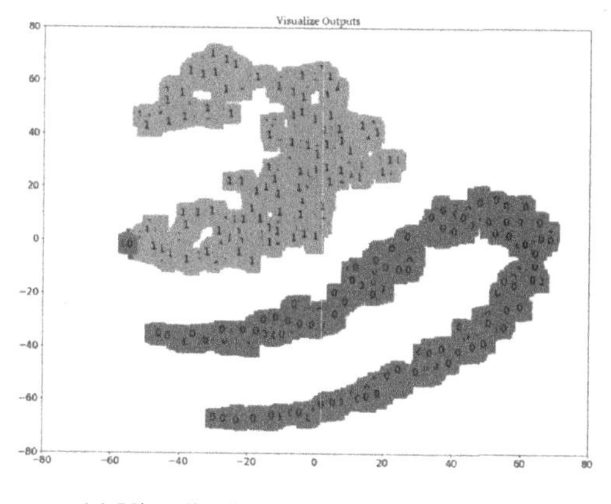

(a) Visualization of classification results

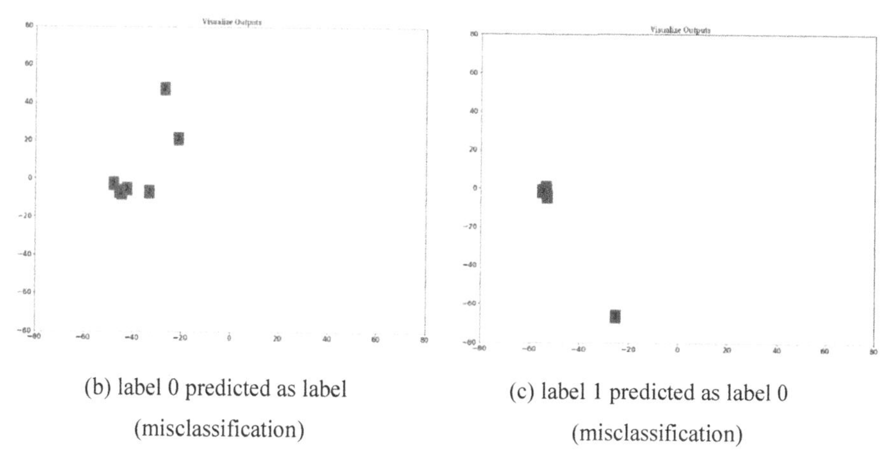

(b) label 0 predicted as label (misclassification)

(c) label 1 predicted as label 0 (misclassification)

Fig. 6. Visualization of output. (Color figure online)

4 Conclusion

The LFPs classification method of autistic and normal mice based on CNN that was proposed in this paper can better classify mice LFPS signals, laying a foundation for later research on the diagnosis of children's autism based on children's LFPs, which has great application prospects and value. The LFPs used in this article is an invasive EEG signal (the electrode is inserted into the brain of a mice), whose signal quality is good, but it causes trauma to the subjects. Later, non-invasive EEG signals will be used for analysis. In addition, the subjects are mice in this article, and we will collect and analyze the non-invasive EEG signals of children with autism and normal children to study the difference of EEG signals between autistic children and normal children in the future.

Acknowledgments. This project is supported by National Science Foundation of China (No. 61976133).

References

1. Lord, C., Elsabbagh, M., Baird, G., et al.: Autism spectrum disorder. The Lancet **392**(10146), 508–520 (2018)
2. Baio, J., Wiggins, L., Christensen, D.L., et al.: Prevalence of autism spectrum disorder among children aged 8 years—autism and developmental disabilities monitoring network, 11 sites, United States, 2014. MMWR Surveill. Summ. **67**(6), 1–23 (2018)
3. Warrier, V., Baron-Cohen, S.: Childhood trauma, life-time self-harm, and suicidal behaviour and ideation are associated with polygenic scores for autism. Mol. Psychiatry **26**, 1–15 (2019)
4. Masi, A., DeMayo, M.M., Glozier, N., et al.: An overview of autism spectrum disorder, heterogeneity and treatment options. Neurosci. Bull. **33**(2), 183–193 (2017)
5. Buescher, A.V.S., Cidav, Z., Knapp, M., et al.: Costs of autism spectrum disorders in the United Kingdom and the United States. JAMA Pediatr. **168**(8), 721–728 (2014)
6. Rutter, M., Le, C.A., Lord, C.: Autism diagnostic interview-revised. West. Psychol. Serv. **15**(19), 1–5 (2003)
7. Pruette, J.R.: Autism diagnostic observation schedule-2 (ADOS-2). In: Google Scholar, pp. 1–3 (2013)
8. Hazlett, H.C., Gu, H., Munsell, B.C., et al.: Early brain development in infants at high risk for autism spectrum disorder. Nature **542**(7641), 348–351 (2017)
9. Heinsfeld, A.S., Franco, A.R., Craddock, R.C., et al.: Identification of autism spectrum disorder using deep learning and the ABIDE dataset. NeuroImage. **17**, 16–23 (2018)
10. Wingfield, B., Miller, S., Yogarajah, P., et al.: A predictive model for paediatric autism screening. Health Inf. J. **26**(4), 2538–2553 (2020)
11. Abbas, H., Garberson, F., Liu, M.S., et al.: Multi-modular AI approach to streamline autism diagnosis in young children. Sci. Rep. **10**(1), 1–8 (2020)
12. Tang, Y., Chen, D., Wang, L., et al.: Bayesian tensor factorization for multi-way analysis of multi-dimensional EEG. Neurocomputing **318**, 162–174 (2018)
13. Ke, H., Chen, D., Shah, T., et al.: Cloud-aided online EEG classification system for brain healthcare: a case study of depression evaluation with a lightweight CNN. Softw. Pract. Exp. **50**(5), 596–610 (2020)
14. Tang, Y., Liu, Y., Tong, L., et al.: Identification of a β-arrestin 2 mutation related to autism by whole-exome sequencing. BioMed Res. Int. **2020**, 1–9 (2020)
15. Roullet, F.I., Wollaston, L., Decatanzaro, D., et al.: Behavioral and molecular changes in the mouse in response to prenatal exposure to the anti-epileptic drug valproic acid. Neuroscience **170**(2), 514–522 (2010)
16. Schneider, T., Przewocki, R.: Behavioral alterations in rats prenatally exposed to valproic acid: animal model of autism. Neuropsychopharmacology **30**(1), 80–89 (2005)
17. Schirrmeister, R.T., Springenberg, J.T., Fiederer, L.D.J., et al.: Deep learning with convolutional neural networks for EEG decoding and visualization. Human Brain Mapp. **38**(11), 5391–5420 (2017)
18. Kwon, O.Y., Lee, M.H., Guan, C., et al.: Subject-independent brain–computer interfaces based on deep convolutional neural networks. IEEE Trans. Neural Netw. Learn. Syst. **31**(10), 3839–3852 (2019)

Based on Advanced Connected Domain and Contour Filter for CASA

Tianfang Zhou[1], Yang Zhou[2](✉), Xiaofei Han[2], Yixuan Qiu[1], and Bo Li[1]

[1] The Faculty of Engineering, Architecture and Information Technology, The University of Queensland, Brisbane 4072, Australia
{tianfang.zhou,yixuan.qiu,bo.li3}@uqconnect.edu.au
[2] School of Mechatronic Engineering and Automation, Shanghai University, Shanghai 200444, China
hanxiaofei@shu.edu.cn

Abstract. Computer-assisted sperm analysis (CASA) is a method that can help the doctor detect sperm efficiently. It can reduce patients' waiting time and pain while the patient is surgering for obtaining sperm. In this paper, a CASA system is developed for finding sperms on the microscopic image, which can raise the detection speed and increase the success rate of finding sperms. A method is proposed which combines advanced connected domain algorithm and contour filter to improve CASA performance. The advanced connected domain algorithm uses 8-connected domains to improve the ability which can detect small targets from the background. Sperms can be distinguished and positioned from lots of small objects by the contour filter. The experiment result shows that our method outperforms traditional human eye recognition and can produce promising results for detecting sperm mixed in the blood.

Keywords: Sperm positioning · Image segmentation · Three-frame difference · Target recognition · CASA

1 Introduction

In recent years, more and more research pay attention to computer-assisted sperm analysis (CASA) [1]. This is because the incidence of infertility is increasing year by year. According to the statistics of the relevant investigation, the incidence of infertility in China has reached more than 12% [2]. When the concentration of the patient's sperm is very low and cannot be collected by non-surgical methods, doctors inevitably use surgery to collect sperm [3]. During the operation, the patient needs to lie on the operating table until the doctor confirms that the appropriate sperm has been collected [4]. This is time-consuming and laborious work due to the complexity of multiple cells and tiny size of the sperm. However, the results of [5] suggest that traditional CASA requires further improvement for a wider application in clinical practice.

CASA uses image segmentation technology to position sperm from a complex background. The quality of segmentation will directly affect the accuracy and recall of sperm

M. Fei et al. (Eds.): LSMS 2021/ICSEE 2021, CCIS 1467, pp. 277–285, 2021.
https://doi.org/10.1007/978-981-16-7207-1_28

recognition. Al-amri et al. undertook the study of segmentation image techniques by using five threshold methods as Mean method, P-tile method, Histogram Dependent Technique (HDT), Edge Maximization Technique (EMT) and visual Technique [6]. To detect faces, Fan et al. proposed a segmentation method combining an improved isotropic edge detector and a fast entropic thresholding technique [7]. Li et al. proposed a region-based active contour model that draws upon intensity information in local regions at a controllable scale to overcome the difficulties caused by intensity inhomogeneities [8]. In [9], Kirillov et al. presented a PointRend (Point-based Rendering) neural network module which performs point-based segmentation predictions at adaptively selected locations based on an iterative subdivision algorithm. Minaee et al. [10] believed the broad success of Deep Learning (DL) has prompted the development of new image segmentation approaches leveraging DL models. Hesamian et al. presented a critical appraisal of popular methods that have employed deep-learning techniques for medical image segmentation [11]. The contour filter was designed to filter sperm in many kinds of objects, such as blood cells and other impurities. Canny edge detector that can improve its detection accuracy was proved by Ding et al. [12].

In this paper, a CASA system is created for finding sperms on the microscopic image, which can raise the detection speed and increase the success rate of finding sperms. Base on traditional connected domain methods, an advanced connected domain method that can remarkably segment objects was claimed. The filter method was developed by combining features of sperm and Canny edge detector. In this system, the sperm picture mainly through the high-definition microscope and cameras to collect. Then the captured image will be processed and sperm will be positioned.

2 Advanced connected domain algorithm

Due to the influence of light and other external environments, it is difficult to segmente objects directly from a noisy background.

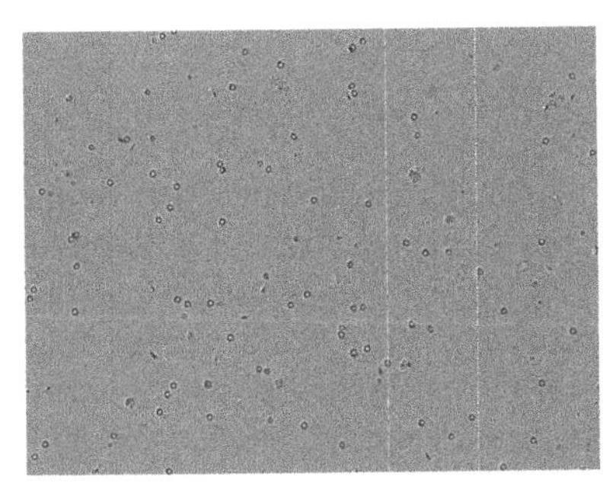

Fig. 1. A picture of sperm extract under a high-time microscope.

To suppress noise, multi-channel histogram equalization has been used in this paper. Then as Fig. 2. shows, the target stands out in the overall environment. This paper proposes an improved threshold segmentation algorithm based on connected domain, which can divide sperm and background accurately.

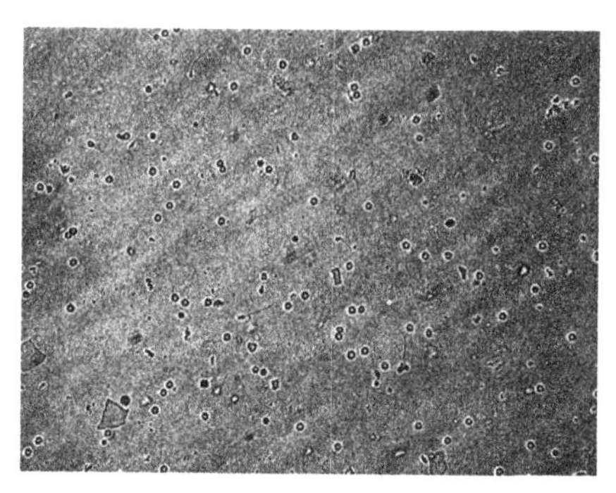

Fig. 2. Sperm picture before image processing and picture after multi-channel histogram equalization

Connected domain algorithm is generally to find out and mark areas with the same properties. It base on the two-value connected domain which refers to the image pixel value adjacent to the area of the same "0" or "1" composed of the area marked out. There are two connected domain marking methods. One is a 4-connected domain and the other is an 8-connected domain. The 4-connected domain refers to an area consisting of the pixel and its top, bottom, left, and right pixels adjacent to it. The 8-connected domain not only refers to the area like 4-connected domain, but also refers to four extra pixels which are the top left, bottom left, top right, and bottom right. These two connected domain methods are shown in Fig. 3.

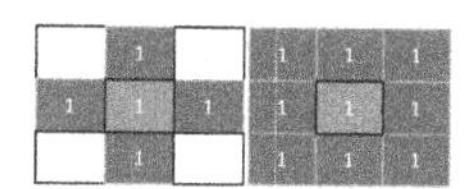

Fig. 3. 4-connected domain and 8-connected domain.

For the same region, marked separately with 4-connected domain method, it will be divided into many separate zones. But 8-connected domain method divides the zoon into relatively few separate regions. The 8-connected domain method is used here to filter out the small area noise in the image and preserve as many target areas as possible.

To find and mark the connected area in the image, this process we call connected region analysis (connect component analysis), connected domain algorithm mainly has Two-Pass method and Seed-Filling method 2 kinds. Inspired by the idea of Seed-Filling method, this paper puts forward an improved threshold segmentation method based on the connected domain, which can better divide the connected target objects. The process is as follows:

Firstly, traversing the entire enhanced image until finding the pixel which values of R, G, B three-channel are less than the thresholds we set. Secondly, seed the current pixel and assign the area to 255. If the pixel value difference between the surrounding neighborhood points and the point we find before is less than the threshold, add the eligible locations in its surrounding realm to the stack. Then pops out the pixel at the top of the stack and assigns the area to 255. Repeat these operations until the stack is emptied. Finally, continue to traverse the image and do as above until the picture traversal is complete (Fig. 4).

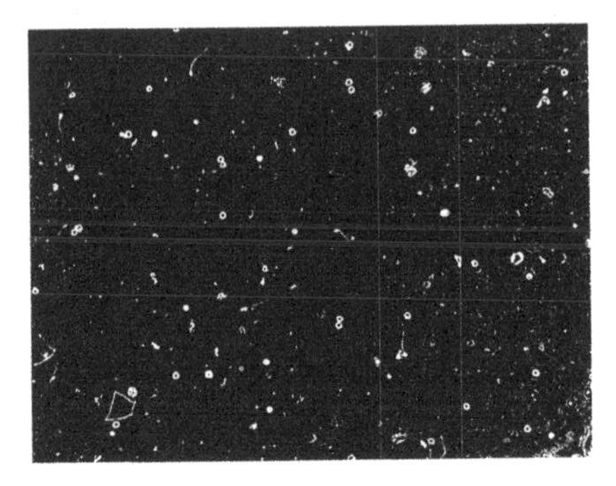

Fig. 4. Sample image after improved threshold segmentation.

The black area in the figure is the split background area and the white area is the split foreground area. It can be seen from the figure that promising targets such as sperm and blood cells can be more accurately separated. This helps to discover the characteristics of various objects to detect sperm.

3 Contour Filter for Sperm

Based on the results obtained from threshold segmentation, it is easy to find that there are several other objects left, including sperm, blood cells, and other impurities. For the above characteristics, some of the impurities that do not meet the target size characteristics of sperm can be removed firstly and blood cells are removed secondly (Fig. 5).

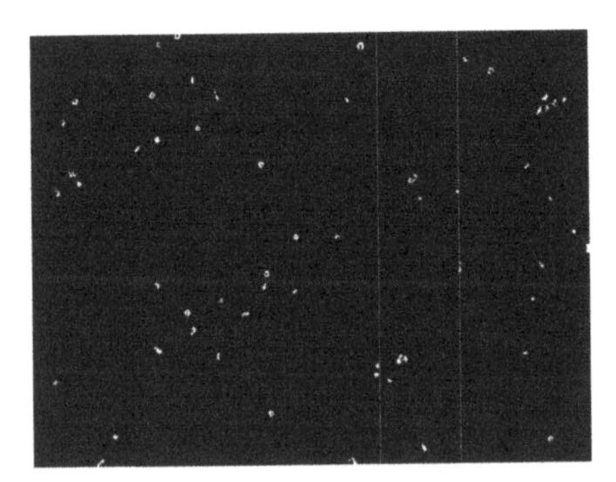

Fig. 5. Picture use the preliminary treatment method.

The black area in the figure is the split background area and the white area is the split foreground area. After removing blood cells and some impurities that are too large or too small, the remaining alternative targets are significantly reduced. This reduces interference in the identification of sperm and improves the accuracy of the target.

After the preliminary treatment above, only a small amount of impurities and sperm targets remain in the remaining segmentation area. At this time, sperm characteristic recognition has become a key step in sperm positioning. In this paper, the characteristics of sperm shape are selected to analyze the geometric laws. The sperm is often made up of two parts of the head and tail, with a slender posture as a whole.

The feature can be expressed by formula:

$$E_x = \frac{\sum_n (x_i - \bar{x})^2}{n}. \tag{1}$$

$$E_y = \frac{\sum_n (y_i - \bar{y})^2}{n}. \tag{2}$$

Horizontal direction variance E_y refers to the variance of horizontal coordinates in the same horizontal direction of the sperm image. Vertical direction variance E_y refers to the variance of horizontal coordinates in the same horizontal direction of the sperm image.

Then calculating averages for horizontal and vertical variances respectively:

$$\bar{E_x} = \frac{\sum_n E_{x_i}}{n}. \tag{3}$$

$$\bar{E_y} = \frac{\sum_n E_{y_i}}{n}. \tag{4}$$

Where $\bar{E_x}$ is the average variance of the horizontal direction, $\bar{E_y}$ is the horizontal variance of the vertical direction. The variance of the calculated segmented region is:

$$E = \frac{\sum_n (x_i - \bar{x})^2 + (y_i - \bar{y})^2}{n}. \tag{5}$$

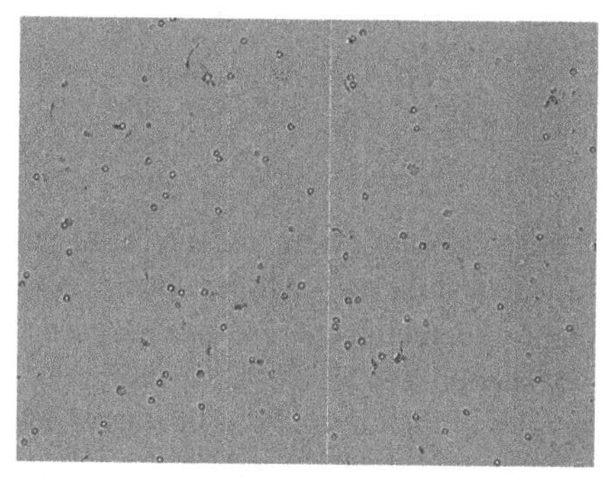

Fig. 6. The final result figure after detecting sperm in Fig. 1. (Color figure online)

E is the variance of a single segmented region. a is a larger threshold and b is a smaller threshold. The thresholds are selected to meet $a >> b$. If $E > a$ and $\bar{E_y} < b$, or $E < a$ and $\bar{E_x} > b$ is satisfied, the target region can be considered as sperm (Fig. 6).

The red marker is sperm. It is easy to find that there are a small number of missed detection targets. However, for hospital staff, the workload has been greatly reduced. Therefore, it can be proved that the method proposed in this paper is true and effective.

4 Experiment and Result

4.1 Data Preparing and Experiment Setting

The image of sperm collected by this system is the image of intercepting sperm at a rate of 25 frames per second under an electron microscope. The hardware equipment used in this paper is a PC and microscope. The main hardware of the computer includes: CPU using Intel Core i7-8700K, host memory is 32G. The microscope uses single-layer microscopy which is named Olympus IX73 made in Japan.

4.2 Experiment Result

The contour filter is necessary for this CASA system. So the proposed CASA system is compared with the system which uses the traditional methods, such as 2-channel Histogram equalization and 4-connected domain segmentation method. Five different sperm pictures were chosen and be detected in the system which uses the traditional methods and the system proposed in this paper.

In Fig. 7, column (a) includes three different origin pictures of sperm, column (b) includes results of CASA system with traditional methods, column (c) includes results of CASA system proposed in this paper. In evach picture, image in blue frame are symbol part of complete picture. The targets which are recognized as sperm by systems are marked by red frames.

According to group 1 pictures, when the number of sperm and noise in the pictures is very small, the difference between the two is not obvious. However, if the contour of the sperm is not very clear, CASA system with proposed methods can find more sperm in group 2. When the noise in the picture occupies the majority, the traditional system usually has a very high false detection rate and the proposed system performs well as group 3.

4.3 Data Analysis

To show the advantage of the proposed system, we selected 5 patients and took 20 sperm pictures of each patient as the test dataset. All the data below are for these 100 sperm pictures. In this experiment, two detection indicators were used to verify the performance of the system, which are precision and recall. TP is the target detected by the system, and it is the number of samples of the target; FP is the target detected by the system, but it is not the number of samples detected by the algorithm; TN is the number of samples that have not been detected. All pictures are detected by systems 5 times (Table 1).

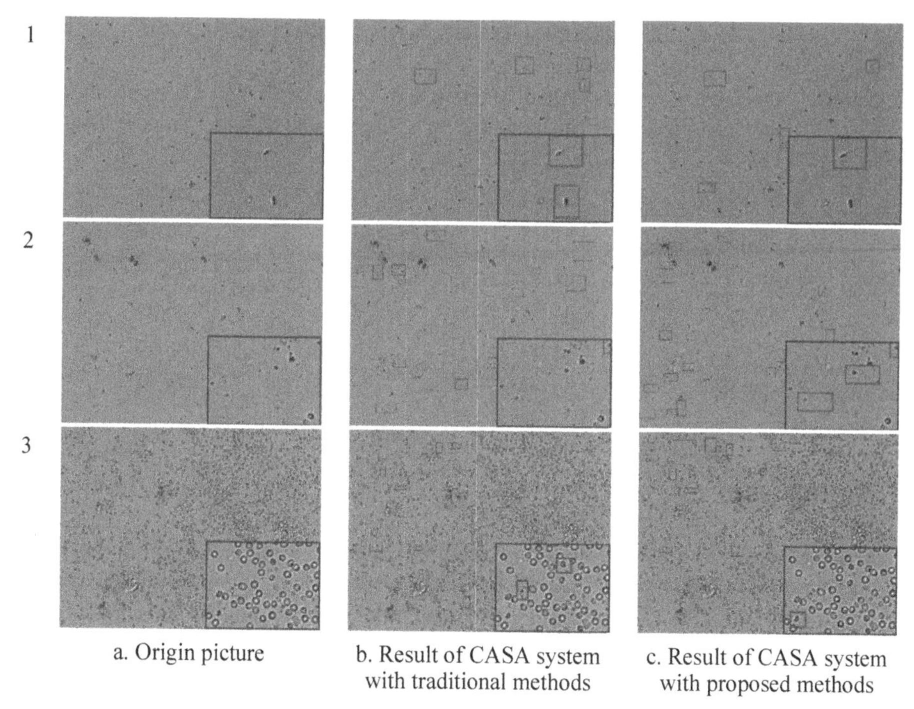

Fig. 7. Sample pictures of sperm and results of CASA system. (a) Origin picture (b) Result of CASA system with traditional methods (c) Result of CASA system with proposed methods. (Color figure online)

The precision is defined as:

$$Precision = \frac{TP}{TP + FP} \times 100\%. \tag{6}$$

The recall is defined as:

$$Recall = \frac{TP}{TP + TN} \times 100\%. \tag{7}$$

Through the above indicator statistical test results as shown in the table, it is easy to find that not only precision but also recall is strikingly improved by the proposed system. Because the sperm can be artificially positioned by doctors after CASA, recall is more important than precision. The recall of the proposed system can reach 85%, which means the proposed system can extremely improve the practicality of CASA.

Table 1. Result of traditional methods system and the proposed system.

Video No.	Precision of traditional methods system	Recall of traditional methods system	Precision of the proposed system	Recall of the proposed system
1	58.1%	52.3%	79.2%	85.1%
2	58.3%	52.9%	79.6%	85.4%
3	58.6%	52.7%	79.0%	85.3%
4	57.4%	51.5%	80.6%	86.0%
5	58.7%	51.9%	79.5%	85.0%
Average	58.2%	52.3%	79.6%	85.3%

5 Conclusion

This paper mainly introduces the living small target positioning algorithm under a single image. Given the shortcomings of the traditional method, this paper improves the histogram equalization method to better enhance the original image and puts forward an improved threshold segmentation algorithm based on the connected domain, which can better split the object, which provides a good basis for the subsequent identification work.

References

1. Amann, R.P., Waberski, D.: Computer-assisted sperm analysis (CASA): capabilities and potential developments. Theriogenology **81**(1), 5–17 (2014)
2. Zhou, Z., et al.: Epidemiology of infertility in China: a population-based study. BJOG Int. J. Obstet. Gynaecol. **125**(4), 432–441 (2018)
3. Niederberger, C., et al.: Forty years of IVF. Fertil. Steril. **110**(2), 185–324 (2018)
4. Roque, M., Haahr, T., Geber, S., Esteves, S.C., Humaidan, P.: Fresh versus elective frozen embryo transfer in IVF/ICSI cycles: a systematic review and meta-analysis of reproductive outcomes. Human Reprod. Update **25**(1), 2–14 (2019)
5. Talarczyk-Desole, J., Berger, A., Taszarek-Hauke, G., Hauke, J., Pawelczyk, L., Jedrzejczak, P.: Manual vs. computer-assisted sperm analysis: can CASA replace manual assessment of human semen in clinical practice? Ginekologia Polska **88**(2), 56–60 (2017)
6. Al-Amri, S.S., Kalyankar, N.V.: Image segmentation by using threshold techniques, 1005--4020 (2010)
7. Fan, J., Yau, D.K., Elmagarmid, A.K., Aref, W.G.: Automatic image segmentation by integrating color-edge extraction and seeded region growing. IEEE Trans. Image Process. **10**(10), 1454–1466 (2001)
8. Li, C., Kao, C.Y., Gore, J.C., Ding, Z.: Minimization of region-scalable fitting energy for image segmentation. IEEE Trans. Image Process. **17**(10), 1940–1949 (2008)
9. Kirillov, A., Wu, Y., He, K., Girshick, R.P.: Image segmentation as rendering. In Proceedings of the IEEE/CVF Conference on Computer Vision and Pattern Recognition (2020)

10. Minaee, S., Boykov, Y.Y., Porikli, F., Plaza, A.J., Kehtarnavaz, N., Terzopoulos, D.: Image segmentation using deep learning: a survey. IEEE Trans. Pattern Anal. Mach. Intell. (2021)
11. Hesamian, M.H., Jia, W., He, X., Kennedy, P.: Deep learning techniques for medical image segmentation: achievements and challenges. J. Dig. Imaging **32**(4), 582–596 (2019)
12. Ding, L., Goshtasby, A.: On the Canny edge detector. Pattern Recogn. **34**(3), 721–725 (2001)

Investigation of Wear Amounts on Artificial Hip Joints with Different Femoral Head Diameter

Zhouyao Weng[1](✉), Xiuling Huang[1], Zikai Hua[1], Qinye Wang[2], and Leiming Gao[3]

[1] School of Mechatronics Engineering and Automation, Shanghai University, Shanghai, People's Republic of China
{wzy19722033,xiulh,eddie_hua}@shu.edu.cn
[2] Department of Orthopedic, Nanxiang Hospital, Jiading, Shanghai, People's Republic of China
[3] Engineering Department, School of Science and Technology, Nottingham Trent University, 50 Shakespare Street, Nottingham NG1 4FG, UK
leiming.gao@ntu.ac.uk

Abstract. A semi-empirical model was applied to investigate the wear behavior presented in the artificial hip joints in this paper, based on the concept of an effective frictional work. Different wear conditions were considered and compared according to the testing standard ISO14242 by varying the femoral head diameters. Results reveal that the amount of wear will increase as the femoral head diameter increases.

Keywords: Artificial hip joint · Wear amount · Femoral head diameter · Standard ISO14242

1 Introduction

The primary factor limiting the longevity of hip replacements with a polyethylene component is wearparticle-induced osteolysis; wear debris released from the bearing surface accumulates in the surrounding tissues causing a cellular response and eventually loosening of the prostheses [1]. Recent advances in UHMWPE technology have revived the alternative of hip resurfacing with a polyethylene liner. State-of-the-art, highly crosslinked polyethylene (XLPE) materials provide substantially improved wear resistance [2–6]. However, reports of crosslinked liner fracture in retrieved components have increased concerns about the adverse effects of crosslinking on mechanical properties [7].

Hip simulator study is an efficient tool for basic research as well as for preclinical testing to minimize patients' risk when receiving new implant types, but the costs in equipment and time interval required on hip simulators are the main drawbacks in application on new prosthesis designs [8]. Therefore, researchers developed a variety of numerical simulation methods to estimate the force and wear rate of the prosthesis [9–11].

The sizes of the femoral head have significant influence on the wear performance of artificial hip joints [12, 13]. In this work, the influence of femoral head diameters

M. Fei et al. (Eds.): LSMS 2021/ICSEE 2021, CCIS 1467, pp. 286–293, 2021.
https://doi.org/10.1007/978-981-16-7207-1_29

on wear amounts is investigated within the standard ISO 14242–3 by using a semi-empirical model [14]. The effective frictional work concept was used to evaluate the wear condition in different femoral head diameters. Obtained results can not only reflect the force distribution at the head-liner top interface, but also give important instructions on prosthesis design.

2 Methods and Results

2.1 Mathematical Model for Hip Prosthesis Wear Assessment

This mathematical model was developed with a 28mm diameter femoral head as an example, several parameters were specified at the beginning of the code:

1. Diameter of the femoral head, r = 0.028, [m];
2. Friction coefficient at head-liner interface, f = 0.08;
3. Cross-link density [15], Xc = 0.00025, [mole/g];
4. Gait frequency, freq = 1, [Hz];
5. Gait period, T = 1, [s];
6. Proportionality constant, k' = 1;
7. C–C bond energy, $\gamma_c = 1$, [J];
8. Average cross sectional width of UHMWPE fibrils, d = 1, [mm].

The quantities of interest to be determined are as follows:

1. Cross-shear effect coefficient, Co;
2. Wear factor, K, [$mm^3/N{\cdot}m$];
3. Discretised sliding displacements in a gait cycle, pythagoras, [m].

2.1.1 Head Trajectories

Motion law and loading law of femoral component were specified in the standard ISO 14242–3. The method of fitting the known key points with harmonic function was adopted to simulate the motion laws, meanwhile, the way of broken line connecting the known key point was adopted to simulate the loading law.

A marker was set at the inner tip of the liner, which lies along ier-axis (Fig. 1). The rotation of femoral head is performed through the rotation matrix, and the rotation sequence is: AA → FE → IER.

Head trajectory in ISO 14242–3, consisting of 100 trajectory points, can be obtained on the surface of the femoral head. Minimal displacement in the direction of ier-axis can be ignored because of its tiny influence to the wear, therefore, the trajectory projected on the aa-fe axis plane was used for subsequent calculations, as shown in Fig. 2.

2.1.2 Frictional Work Vector and FMO Direction

Direction vectors were obtained through rotation matrix by "rotating back" every coordinate present in trajectories with the opposite rotation applied to find the head trajectories.

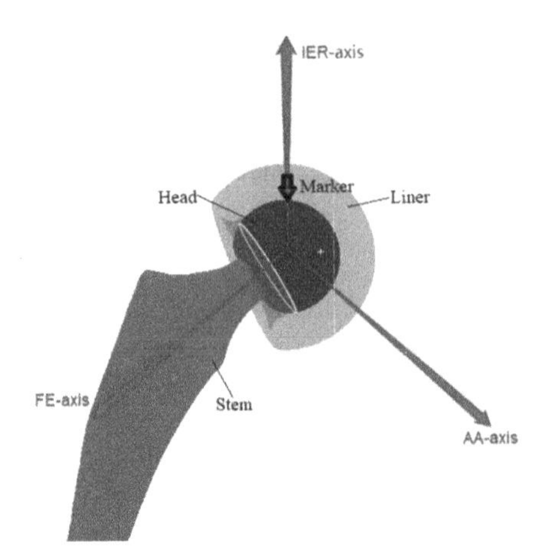

Fig. 1. Marker location on liner

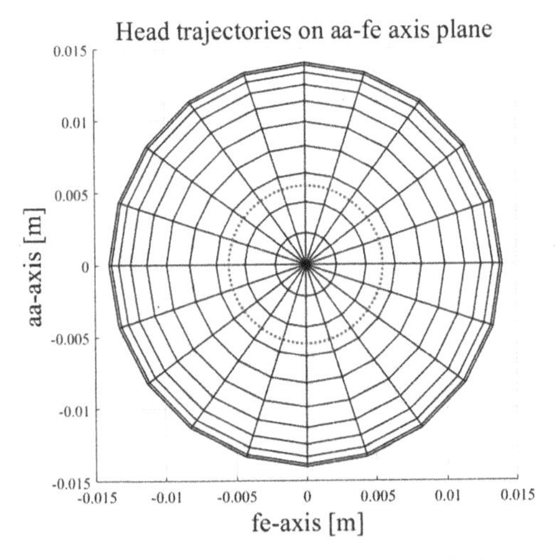

Fig. 2. Head trajectory in ISO 14242–3

The modulus of direction vector is the distance between the contiguous points that generate the path, and the corresponding direction is the instantaneous direction of each track point, which can be approximated regarded as a direction connecting the front and rear track point.

After projecting the direction vector on the fe-aa axis plane, and calculating the length of every single vector by using Pythagorean Theorem, the discretised sliding displacements in one gait cycle can be obtained by adding up all vector length.

The friction force at any moment in the gait cycle can be obtained by multiplying the loading force and the friction coefficient f. Through multiplying the friction force and the direction vector can get the friction work distribution. The sum of the modulus lengths of all frictional work vectors is the total frictional work in one gait cycle (W_{cycle}).

The fibril main orientation (FMO) is the theoretical direction that the UHMWPE fibrils assume when stretched repeatedly in a specific direction [15], so the fibrils are stretched along the direction of maximum frictional work.

The friction work distribution not only shows how the friction work is distributed in a gait cycle, but also provides information for judging the direction of the FMO. Firstly, the friction work vector of the third and fourth quadrants is rotated 180° and superimposed with the vectors of the first and second quadrants, and then the first and second quadrants are equally divided into eight fan-shaped areas, and they are numbered counterclockwise. After this, the sum of the frictional work dissipated (fkd) in each sector was calculated. The angle corresponding to the largest sector of fkd is the theoretical FMO direction.

2.1.3 Parameter Calculation

The total frictional work dissipation in the FMO direction (W_f) can be obtained step by step. Firstly, project the direction vector to the theoretical FMO direction obtained above, and then multiply the projection distance by the friction force. On this basis, the total frictional work dissipation in the orthogonal direction of the FMO Wo can be obtained by the formula W_{cycle}-W_f. Now, the value of parameter Co can be obtained.

$$Co = \frac{W_{cycle} - W_f}{W_{cycle}} = \frac{W_o}{W_{cycle}}. \quad (1)$$

Wear factor K can be obtained as follow [15]:

$$K = K' \frac{\Delta W_o d}{2X_c \gamma_c PL}. \quad (2)$$

Where k' = 1, γ_c [J] = 1, d [mm] = 1, Xc [mole/g] = 0.00025, ΔWo [J] is the frictional work released in the orthogonal direction with respect to the fibrils, P [kN] is the normal load, L [mm] is the sliding distance travelled per motion cycle.

Calculation results are shown in Table 1.

Table 1. Calculation results

Parameter	Co	K [mm^3/N·m]	W_{cycle} [J]	Pythagoras [m]
ISO 14242–3	0.4703	75.2485	0.0037	0.0343

2.2 Wear of Different Femoral Head Diameters

The same method was used to investigate the movement trajectories of the femoral head with diameters of 32 mm, 36 mm, 40 mm, 44 mm, 48 mm and 52 mm, as shown in Fig. 3. According to the running results of the model, the frictional work and movement displacement for the femoral head with different diameters in the gait cycle are shown in Fig. 4 and Fig. 5.

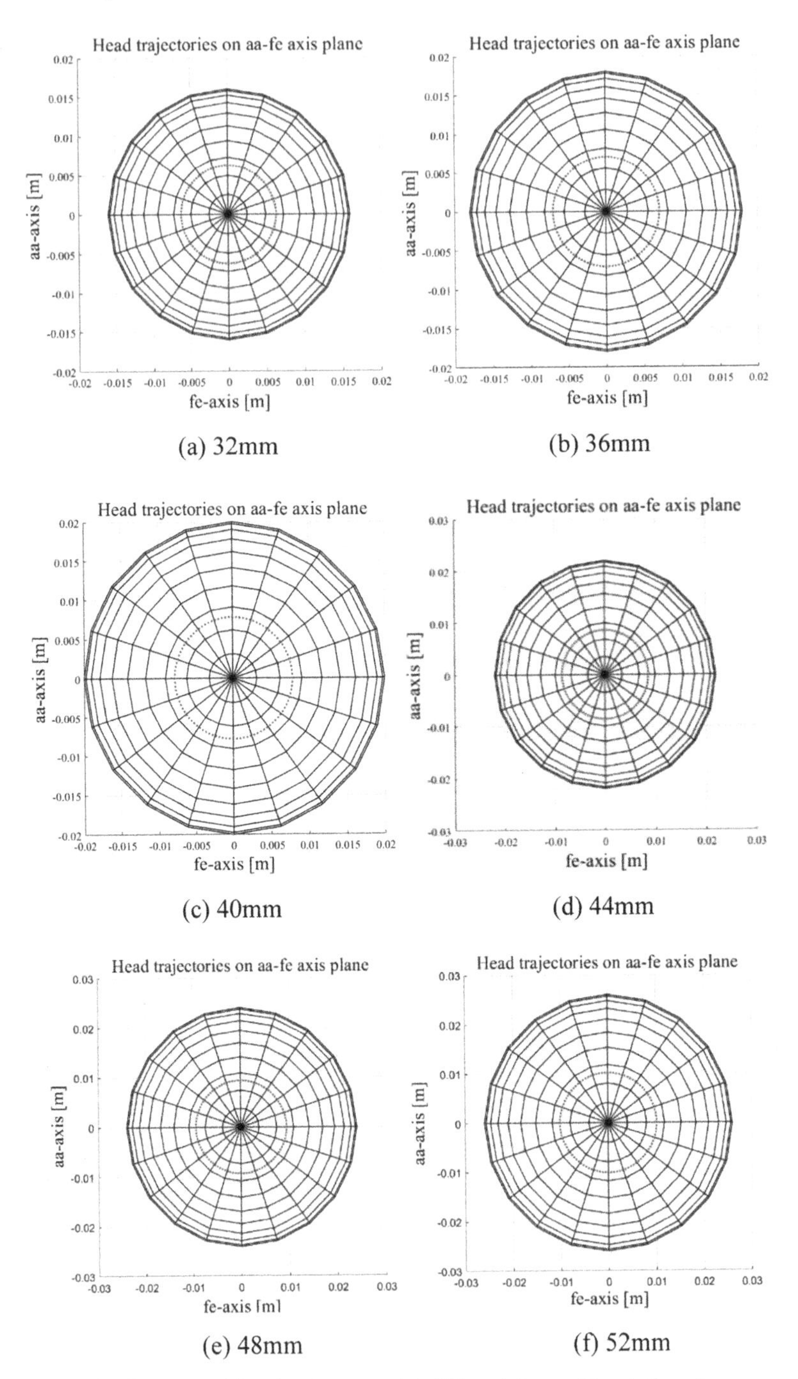

Fig. 3. Femoral head trajectory with different femoral head diameters

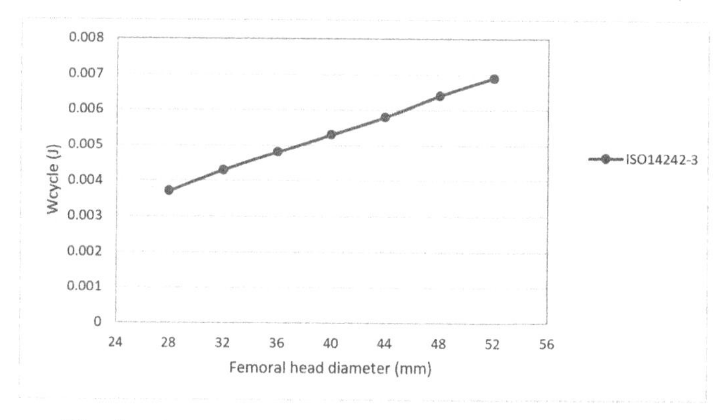

Fig. 4. Frictional work with different femoral head diameters

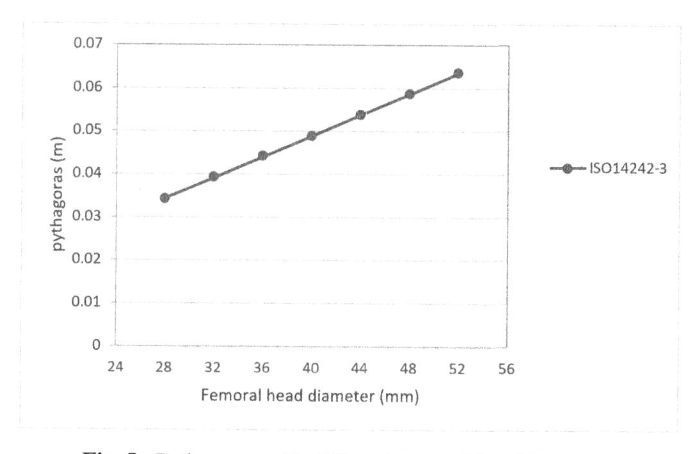

Fig. 5. Pythagoras with different femoral head diameters

2.3 Results and Disscussion

The calculation results of several important parameters were obtained through step-by-step calculation.

First of all, for the coefficient Co, which reflects the distribution of discrete sliding directions in a gait cycle. In the case of the same loading law and friction coefficient, the larger Co indicates the more work done perpendicular to the FMO direction, and the larger the displacement projected in the vertical direction, the easier it is to produce wear debris, and vice versa. The value of coefficient Co is 0.4703 in ISO 14242–3, attributes to the friction work of ISO 14242–3 is evenly distributed. In addition, the value of Co will not change with femoral head diameter because the motion law is not change.

For the wear factor K, which indicates the amount of volume wear per unit sliding distance and unit load, in the case of ISO 14242–3, its value is 75.2485. Due to the default settings of some parameters (k', γ_c, d), the wear factor K itself has no practical meaning and only makes sense when compared with each other. Same as coefficient Co, its value will not change because the value of K is only related to the law of motion and loading.

The results of frictional work dissipated per gait cycle (W_{cycle}) and cycling displacement (pythagoras) were calculated in this model. The value of W_{cycle} and pythagoras increase linearly with the diameter of the femoral head, which illustrate that the larger the diameter of the femoral head, the stronger the friction work, and the easier it is to produce wear debris.

In addition, there are many other wear cases that can be compared with each other by modifying individual parameters, such as cross-link density Xc and friction coefficient f. Even different motion law and loading law of femoral head can be considered in this mathematical model. Several unknown parameters (k', γ_c, d) are left in the code to allow a possible further development in the future.

3 Conclusion

In the present paper a mathematical model of wear assessment was performed, wear condition in different head diameters were compared. It can be concluded that, the amount of wear will increase as the diameter of the femoral head increase.

Foundation. Health Committee of Jiading (2019-ky-07).

References

1. Ingham, E., Fisher, J.: The role of macrophages in osteolysis of total joint replacement. Biomaterials. **26**, 1271–1286 (2005)
2. de Steiger, R., Lorimer, M., Graves, S.E.: Cross-linked polyethylene for total hip arthroplasty markedly reduces revision surgery at 16 years. J. Bone Joint Surg. **100**, 1281–1288 (2018)
3. Amstutz, H.C., Takamura, K.M., Ebramzadeh, E., Le, D.M.J.: Highly cross-linked polyethylene in hip resurfacing arthroplasty: long-term follow-up. Hip Int. J. Clin. Exp. Res. Hip Pathol. Therapy **25**, 39–43 (2015)
4. James, W.P.: Hip resurfacing using highly cross-linked polyethylene: prospective study results at 8.5 years. J. Arthroplasty **31**, 2203–2208 (2016)
5. Ebru, O., Steven, D.C., Arnaz, S.M., Keith, K.W., Orhun, K.M.: Wear resistance and mechanical properties of highly cross-linked, ultrahigh-molecular weight polyethylene doped with vitamin E. J. Arthroplasty **21**, 580–591 (2005)
6. Oludele, O.P., Diego, A.O.V., Craig, F.J., Kimberly, M., Alicia, R.: High cycle in vitro hip wear of and in vivo biological response to vitamin E blended highly crosslinked polyethylene. Biotribology **16**, 10–16 (2018)
7. David, T.S., Natalie, H.K., Timothy, M.W., Michael, L.P.: Retrieved highly crosslinked UHMWPE acetabular liners have similar wear damage as conventional UHMWPE. Clin. Ortho. Related Res.® **469**, 387–394 (2011)
8. Jui-pin, H., James, S.W.: A comparative study on wear behavior of hip prothesis by finite element simulation. Biomed. Eng. Appl. Basis Commun. **14**, 139–148 (2002)
9. Tina, A.M., Thomas, D.B., Douglas, R.P., John, J.C.: A sliding-distance-coupled finite element formulation for polyethylene wear in total hip arthroplasty. J. Biomech. **29**, 687–692 (1996)
10. Teoh, S.H., Chan, W.H., Thampuran, R.: An elasto-plastic finite element model for polyethylene wear in total hip arthroplasty. J. Biomech. **35**, 323–330 (2002)

11. Scott, L.B., Grant, R.B., Janaki, R.P., Anthony, J.P., Paul, J.R.: Finite element simulation of early creep and wear in total hip arthroplasty. J. Biomech. **38**, 2365–2374 (2004)
12. Hammerberg, E.M., Dastane, M., Dorr, L.D.: Wear and range of motion of different femoral head sizes. J. Arthroplasty **25**, 839–843 (2010)
13. Carmen, Z., Christian, F., Lars, M., Gerhard, F., Wolfram, M., Rainer, B.: Wear testing and particle characterisation of sequentially crosslinked polyethylene acetabular liners using different femoral head sizes. J. Mater. Sci. Mater. Med. **24**, 2057–2065 (2013)
14. Luca, F.: Mathematical Models of Wear Assessment in Hip Replacement Testing Methods According to ISO 14242, pp. 65--95. Universita Degli Studi Di Padova, Italy (2016)
15. Wang, A.: A unified theory of wear for ultra-high molecular weight polyethylene in multi-directional sliding. Wear **248**, 38–47 (2001)

Author Index

Printed in the United States
by Baker & Taylor Publisher Services